ANNALS OF THE NEW YORK ACADEMY OF SCIENCES

Volume 478

The New York Academy of Sciences
2 East 63rd Street
New York, New York 10021

METABOLIC REGULATION: APPLICATION OF RECOMBINANT DNA TECHNIQUES

ANNALS OF THE NEW YORK ACADEMY OF SCIENCES
Volume 478

METABOLIC REGULATION: APPLICATION OF RECOMBINANT DNA TECHNIQUES

Edited by Alan G. Goodridge and Richard W. Hanson

The New York Academy of Sciences
New York, New York
1986

The photograph on the cover of the paperback volume shows a computer graphics rendition of an alpha carbon model of the muscle phosphorylase dimer. Glucose and a glycogen fragment are shown bound to the dimer. The glucose is found in the cleft separating the NH_2- and COOH-terminal domains (near the lower center of the bottom subunit). The glycogen fragment is bound to the NH_2-terminal (left side of the bottom subunit). Part of the glycogen fragment is seen to be jutting away from the protein. The positions where side chains are variant between human liver and rabbit muscle are shown to be, in general, at surface locations.

See the paper by Fletterick et al. *(pp. 220-232), which includes a black-and-white version of this photograph, for more details.*

Library of Congress Cataloging-in-Publication Data

Metabolic regulation.

(Annals of the New York Academy of Sciences; v. 478)
"Result of a conference . . . held in New York, New York on November 6-8, 1985 by the New York Academy of Sciences"—Footnote, p.
Bibliography: p.
Includes index.
1. Metabolism—Regulation—Congresses. 2. Genetic regulation—Congresses. 3. Recombinant DNA—Congresses. 4. Gene expression—Congresses. I. Goodridge, Alan G. II. Hanson, Richard W. III. New York Academy of Sciences. IV. Series.
Q11.N5 vol. 478 500 s 86-28514
[QP171] [574.1'33]

PCP
Printed in the United States of America
ISBN 0-89766-351-9 (cloth)
ISBN 0-89766-352-7 (paper)
ISSN 0077-8923

ANNALS OF THE NEW YORK ACADEMY OF SCIENCES

Volume 478
October 31, 1986

METABOLIC REGULATION: APPLICATION OF RECOMBINANT DNA TECHNIQUES[a]

Editors and Conference Organizers
ALAN G. GOODRIDGE and RICHARD W. HANSON

CONTENTS

[a]This volume is the result of a conference entitled Metabolic Regulation: Application of Recombinant DNA Techniques, which was held in New York, New York on November 6-8, 1985 by the New York Academy of Sciences.

Part II. Differentiation and the Regulation of Metabolism

Part III. Role of cAMP in the Regulation of Metabolism

Part IV. Structure-Function Relationships in Metabolically Important Proteins

Part V. Expression and Function of Foreign Cells in Culture and in Intact Animals

Poster Papers

Financial assistance was received from:

- BAYER AG/MILES BIOTECHNOLOGY GROUP
- McNEIL PHARMACEUTICAL
- MERCK SHARP & DOHME RESEARCH LABORATORIES
- MERRELL DOW RESEARCH INSTITUTE
- MONSANTO
- NATIONAL INSTITUTE OF ARTHRITIS, DIABETES, AND DIGESTIVE AND KIDNEY DISEASES/NATIONAL INSTITUTES OF HEALTH
- REVLON HEALTH CARE GROUP/RESEARCH AND DEVELOPMENT DIVISION
- A. H. ROBINS COMPANY
- SANDOZ, INC.
- SCHERING-PLOUGH CORPORATION
- SMITH KLINE & FRENCH LABORATORIES
- SQUIBB CORPORATION
- STUART PHARMACEUTICALS/DIVISION OF ICI AMERICAS, INC.
- U. S. OFFICE OF NAVAL RESEARCH/DEPARTMENT OF NAVY GRANT N00014-85-G-0137

Preface

The term "metabolic regulation" is generally used to describe phenomena involved in the regulation of the flux of carbon through metabolic pathways. The physiological aspect of metabolic regulation involves identifying environmental and extracellular agents that regulate flux through specific pathways. The more molecular aspects involve 1) identifying rate-controlling enzymes, 2) analyzing the kinetic properties of regulatory enzymes or the mechanisms that control the concentrations of such enzymes, 3) postulating a hypothesis based on the kinetic properties or mechanisms that regulate concentration, and 4) testing that hypothesis in intact cells and organisms. Traditional approaches to the identification of regulatory enzymes include the measurement of flux rates with isotopically labeled substrates and intermediates, the measurement of the concentrations of the substrates and products of putative regulatory enzymes in intact systems, the use of specific enzyme inhibitors, and the analysis of the kinetic properties of purified enzymes. These approaches have led to a significant degree of understanding of many regulatory phenomena.

Despite the major advances in our knowledge of the control of metabolism, there are still technical barriers to a clearer understanding of regulatory processes. One of the principal reasons for the current diminution in interest in metabolism is a perceived lack of methodology for studying more definitively the control of metabolic pathways. Until recently, it was not possible to alter the properties of an enzyme in a metabolic pathway in a systematic manner or to modify a specific step in a series of enzymatic reactions and then test the effect of these changes on the function of the entire pathway in an intact cell. The same constraints have applied to intact animals, making it difficult to verify the mechanisms responsible for coordinating the metabolic responses between tissues. Recombinant DNA technology, however, has, since its advent, found increasingly wide application to genes of metabolic interest, and has created new research opportunities in this field. It is now possible to introduce the gene for an enzyme into a cell and to study the effects of the products of that gene on cellular processes. Although this research is still in its formative stages, it represents a different and powerful strategy for the study of metabolic regulation. This *Annal* will review some of the major advances in several different aspects of metabolic regulation.

The area in which this new technology has had its greatest impact on metabolism research is the analysis of gene expression and the regulation of this process by hormones. The availability of cDNA probes for a broad variety of genes of metabolic interest has permitted the quantitation of the levels of specific mRNAs. An increasing number of these genes are being isolated and sequenced and having their promoter-regulator regions identified and characterized. Hormone regulatory elements have been delineated for several genes, including several described in detail in this volume. Regulation of the expression of the genes for phosphoenolpyruvate carboxykinase (Hod *et al.,* Granner *et al.,* Chin and Fournier, and Gluecksohn-Waelsch), malic enzyme (Goodridge *et al.* and Bagchi *et al.*), ATP-citrate lyase (Bagchi *et al.*), fatty acid synthase (Goodridge *et al.* and Wakil), tyrosine aminotransferase (Schütz *et al.* and Gluecksohn-Waelsch), tryptophan oxygenase (Schütz *et al.*), liver protein S14 (Tao and Towle), casein (Rosen *et al.*), and lactate dehydrogenase (Jungmann *et al.*) are described in this volume. In addition to summarizing the state of knowledge with respect to gene regulation, these papers identify several cloned cDNAs and genes that are already available for the kinds of studies outlined below.

Although analysis of the regulation of gene expression is a common use of cloned DNAs, it is not the only way the recombinant DNA technology can be applied to the study of metabolic regulation. Flux through metabolic pathways is regulated by allosteric or covalent modification of key regulatory enzymes. Several examples of this type of regulation have been described for enzymes in the pathways for carbohydrate and lipid metabolism. How is a pace-setting enzyme identified? If an enzyme is subject to allosteric or covalent modifications, how can we identify the physiologically relevant mechanism and analyze the structural and kinetic bases for the observed regulatory phenomena? New genetic engineering techniques will permit definitive analyses of these and related problems.

Central to the analysis of such problems is the knowledge of the amino acid sequence of the native protein and the ability to create site-specific mutants in that protein. DNA sequencing and site-specific mutagenesis (single-base changes, deletions, or insertions) are established techniques, developed as part of the recombinant DNA "revolution." Several papers in this volume use these techniques in a variety of analyses, primarily directed at identifying regulatory sequences in DNA and RNA. In the future, however, application of this technology to structure/function problems may become equally important.

Knowing the primary structure and being able to generate specific mutations provides the raw material for analyzing the relationship of structure to function. At the molecular level, however, there is an additional requirement: access to sufficient protein to carry out complex physical and kinetic analyses. Several laboratories have developed host/vector systems and methods for the efficient expression of cloned DNAs. The properties and uses of some of these systems are described in the paper by Shatzman and Rosenberg.

The bacterial systems described by Shatzman and Rosenberg allow one to synthesize and isolate large amounts of virtually any protein, irrespective of its abundance under natural conditions. At a more physiological level there is still another requirement: the ability to express normal and mutant proteins from cloned DNA transfected into cells or animals. The papers by Hod *et al.*, Chin and Fournier, Gottesman *et al.*, Schütz *et al.*, Luskey, and Botteri *et al.* demonstrate the power of gene transfer to analyze problems of gene expression. The next step of relevance to studies of metabolic regulation will involve expression of normal and mutant proteins. Special vectors will be engineered to contain powerful promoters, which will increase the amount of protein produced by the transfected cells. The use of regulated promoters will allow the regulation of the concentration of the expressed protein. Finally, co-transfection of antisense genes (J. G. Izant and H. Weintraub, *Science* **229:** 345-352, 1985) may make it possible to inhibit expression of the natural gene while expressing a mutant version of that same gene. In sum, these techniques will make it possible to relate the primary structures of proteins to their functions. Furthermore, it will be possible to determine in intact cells the physiological significance of structures and regulatory mechanisms established in experiments with purified proteins.

Cloned genes also can be transferred into the germ lines of intact animals, producing genetically stable mutant animals. Just getting expression of a transferred gene in an intact animal is a major achievement. In addition, however, transferred genes can be constructed so that they are expressed in a tissue-specific fashion. The paper by MacDonald *et al.* analyzes the sequence requirements for tissue-specific expression of the elastase gene in transgenic mice. The same sets of experimental manipulations discussed above for genes transfected into cells in culture can be carried out *in vivo.* This technology should permit determination of the relevance of many regulatory mechanisms heretofore analyzed only in cells in culture. For the most part, constructing transgenic animals involves injection of cloned DNA into the pronucleus of fertilized egg. This procedure is technically difficult and relatively inefficient. The use of retroviral vectors, discussed by Botteri *et al.,* may make the transgenic animal technology more accessible.

We have described some of the new techniques that will allow the application of recombinant DNA technology to important problems in metabolism. The combination of these procedures with existing techniques of animal genetics and somatic cell genetics will result in even more powerful analytic approaches. Papers by Chin and Fournier and by Gluecksohn-Waelsh demonstrate this latter point. In addition to technology, however, one needs interesting and important problems to solve. The papers by Fletterick *et al.* Pilkis *et al.,* and Wakil discuss the relationship between structure and function and the regulation of the activity of some enzymes that play key roles in the metabolism of carbohydrates and lipids. Application of the recombinant DNA technology to these important problems is just beginning.

ALAN G. GOODRIDGE
RICHARD W. HANSON

Hormonal Modulation of Key Hepatic Regulatory Enzymes in the Gluconeogenic/Glycolytic Pathway[a]

S. J. PILKIS, E. FOX, L. WOLFE, L. ROTHBARTH,
A. COLOSIA, H. B. STEWART, AND
M. R. EL-MAGHRABI

Department of Molecular Physiology and Biophysics
Vanderbilt University School of Medicine
Vanderbilt University
Nashville, Tennessee 37232

INTRODUCTION

The mechanisms whereby hormones regulate metabolic processes in cells have been a focus of intensive research for decades. The study of hormonal regulation of hepatic gluconeogenesis and glycolysis has been a particularly interesting area in metabolic regulation because these two opposing processes must be regulated in a concerted fashion in order to permit unidirectional flux to occur. Regulation of these two processes can be divided for convenience into three categories. The first involves regulation of the supply of substrate. All gluconeogenic substrates as well as glucose reach the liver in subsaturating concentrations. Thus regulation of substrate release into the blood from extrahepatic tissues or provision of substrate from the diet will directly affect hepatic glucose formation and utilization. The second category deals with the very important but relatively slow (hours to days) adaptive changes in enzyme activity due to regulation of gene expression and/or enzyme degradation. The third category is concerned with the minute-to-minute regulation of pathway flux by hormones such as insulin, glucagon, and catecholamines. This report will deal primarily with the third category, with special emphasis on the role of fructose 2,6-bisphosphate (Fru 2,6-P_2) in this regulation. In addition, adaptive changes in the Fru 2,6-P_2-metabolizing enzyme activities will also be discussed. Although the emphasis will be on the minute-to-minute regulation of glycolysis and gluconeogenesis, it should be clear that regulation *in vivo* occurs in all three categories simultaneously and in an integrated fashion.

GENERAL MECHANISMS OF HORMONAL REGULATION OF HEPATIC METABOLISM

Studies from many laboratories over the last two decades have shown that hormonal regulation of metabolic processes in liver occurs by two general mechanisms (FIG. 1).

[a]This work was supported by Grant PCM-82 18661 from the National Science Foundation and Grant AM 18270 from the National Institutes of Health.

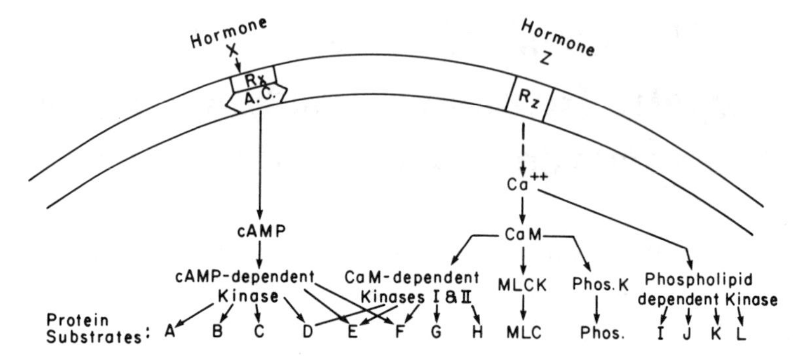

FIGURE 1. Two general mechanisms for the hormonal regulation of metabolic processes. One class of hormones acts via cAMP and cAMP-dependent protein kinase-catalyzed phosphorylations. Another class acts via changes in intracellular Ca^{2+} levels or related messengers and Ca^{2+}-linked protein kinase-catalyzed phosphorylations. Some protein substrates are specifically phosphorylated by a single protein kinase whereas others may be substrates for both cAMP-dependent kinases and Ca^{2+}-linked protein kinase(s).

The first mechanism includes those hormones (glucagon, β-adrenergic agonists) that interact with a plasma membrane receptor that is coupled to adenylate cyclase resulting in an elevation of the level of intracellular cAMP. Elevation of cAMP levels leads to activation of cAMP-dependent protein kinases that then catalyze the phosphorylation of several protein substrates resulting in a physiologic response such as the stimulation of gluconeogenesis and inhibition of glycolysis.[1-4] The second mechanism involves those hormones that act via changes in intracellular Ca^{2+} levels. These hormones (α-adrenergic agonists, vasopressin, angiotensin) also interact with plasma membrane receptors resulting in elevated intracellular Ca^{2+} levels, by a mechanism involving the generation of two intracellular messengers: *myo*-inositol 1,4,5-trisphosphate and 1,2-diacylglycerol.[5] The change in intracellular Ca^{2+} level, in combination with calmodulin and/or other effectors, leads to activation of several Ca^{2+}-linked protein kinases including the Ca^{2+}- and calmodulin-dependent protein kinase, phosphorylase kinase, and protein kinase C. These protein kinases also catalyze phosphorylation of several protein substrates leading to physiological responses that may or may not be the same as those seen with cAMP-dependent phosphorylation. In general, with regard to gluconeogenesis and glycolysis, both cAMP- and Ca^{2+}-dependent phosphorylations lead to similar changes in pathway flux.

Insulin, on the other hand, opposes the action of these hormones, probably by its ability to activate cAMP phosphodiesterase resulting in lower cAMP levels.[6] Precisely how this effect on phosphodiesterase is brought about is not known. The mechanisms of insulin's action to oppose Ca^{2+}-linked hormones is also unknown at present. Insulin's effects may thus represent an additional general mechanism for regulation of metabolic processes in liver.

SITES OF HORMONE ACTION IN THE GLUCONEOGENIC/ GLYCOLYTIC PATHWAY

During the last ten years, work from many laboratories, including those of Lardy, Katz, Blair, and Hers, and work from our own laboratory have established that the most important sites of hormone action in the pathway in mammalian liver occur at two so-called substrate cycles (FIG. 2).[1-4] The first cycle is that between pyruvate (PYR) and phosphoenolpyruvate (PEP) whereas the second involves the interconversions between fructose 6-phosphate (Fru 6-P) and fructose 1,6-bisphosphate (Fru 1,6-P_2). The maximal catalytic activity of pyruvate kinase (PK) is 50 U/g liver. This activity is opposed by pyruvate carboxylase and phosphoenolpyruvate carboxykinase (PEPCK), both at 7 U/g liver, so one might expect that PK would need to be downregulated for net gluconeogenesis to occur. At the Fru 6-P/Fru 1,6-P_2 level the maximal catalytic activity of fructose 1,6-bisphosphatase (Fru 1,6-P_2ase) is 15 U/g, and this reaction is opposed by 6-phosphofructo-1-kinase (6PF-1K) at 3 U/g. One would expect some downregulation of Fru 1,6-P_2ase for glycolysis to occur. In line with these expectations, all the above enzymes have been shown to be substrates for protein kinases that have been implicated in regulation of flux at these sites.[6]

The effect of phosphorylation by two different protein kinases on PK activity is shown in FIGURE 3. Both the cAMP-dependent protein kinase and the Ca^{2+}- and calmodulin-dependent protein kinase catalyze the phosphorylation of rat liver PK, and both phosphorylations result in inhibition of activity.[7] This inhibition is characterized by an increase in the K_m for the substrate PEP and a shift in its dependence curve to the right. Inhibition is seen when the enzyme is assayed with low PEP concentrations but is overcome by high substrate concentrations or by the allosteric activator Fru 1,6-P_2. The Ca^{2+}- and calmodulin-dependent protein kinase catalyzes phosphorylation of two sites: the first on the same seryl residue as that phosphorylated by the cAMP-dependent protein kinase and the second on a unique threonyl residue.[7] PK is an example of a protein substrate that is phosphorylated by both the cAMP and a Ca^{2+}-linked protein kinase leading to an identical response. It is also noteworthy that cAMP-dependent phosphorylation of PK is itself subject to modulation by allosteric effectors of the enzyme. Fru 1,6-P_2, in particular, is a potent inhibitor of the process that amplifies its role in favoring the active form of the enzyme.[1-3]

FIGURE 4 shows how such phosphorylation is thought to alter pathway flux, using glucagon action as an example. Glucagon acts to raise cAMP levels leading to phosphorylation and inhibition of PK and decreased recycling of PEP to PYR and enhanced flux toward glucose. In addition to glucagon the phosphorylation state of PK in intact cells has also been shown to be regulated by insulin, catecholamines, angiotensin, and vasopressin.[2-4,8] The rate-limiting portion of the gluconeogenic pathway is located between PYR and PEP,[1,2] and PK, the only enzyme in that portion of the pathway

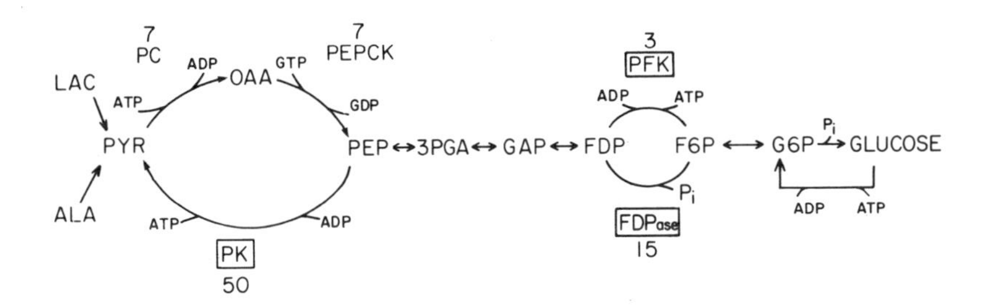

FIGURE 2. Substrate cycles in the gluconeogenic/glycolytic pathway. PK: L-type pyruvate kinase; PEPCK: phosphoenolpyruvate carboxykinase; PC: pyruvate carboxylase; PFK: 6-phosphofructo-1-kinase; FDPase: fructose 1,6-bisphosphatase. The numbers associated with each enzyme represent that enzyme's maximal activity in μmol/min/g liver. The enzymes enclosed in boxes have been shown to be substrates *in vitro* for cAMP-dependent protein kinase.

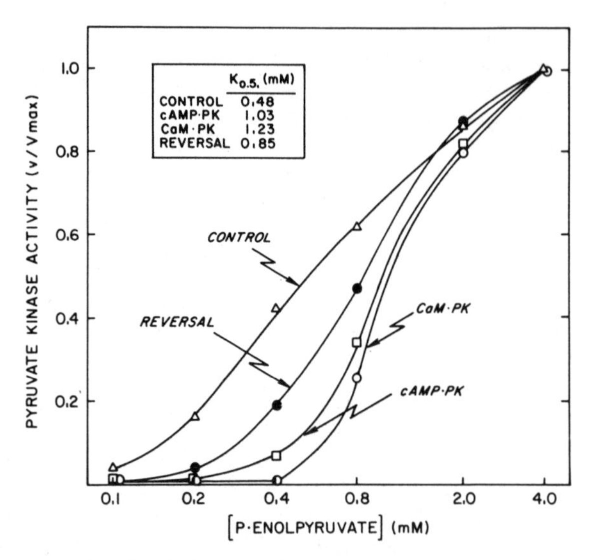

FIGURE 3. Effects of phosphorylation on PK activity. PK activity was assayed in the presence of increasing concentrations of PEP after incubation in the presence of ATP under the following conditions: Control: no added kinase; cAMP-PK: PK plus cAMP-dependent protein kinase; CaM-PK: PK plus CaM-dependent protein kinase; and reversal: PK plus CaM-dependent protein kinase for 60 min followed by a 30-min additional incubation under reversal conditions. The reversal conditions remove phosphate only from the cAMP-dependent site.

regulated by phosphorylation, is now thought to be the major site of acute hormone action for the entire pathway.

Current notions of the regulation of the Fru 6-P/Fru 1,6-P_2 substrate cycle are shown in FIGURE 5. Hormonal and substrate regulation at this cycle are mediated by changes in the level of Fru 2,6-P_2. This unique sugar diphosphate is a potent allosteric activator of 6PF-1K and acts synergistically with AMP to oppose the inhibitory action of ATP and citrate.[3,4,9-11] Fru 2,6-P_2 is also a potent inhibitor of Fru 1,6-P_2ase and acts synergistically with AMP to inhibit that enzyme.[3,4,9-11] Thus in states where Fru

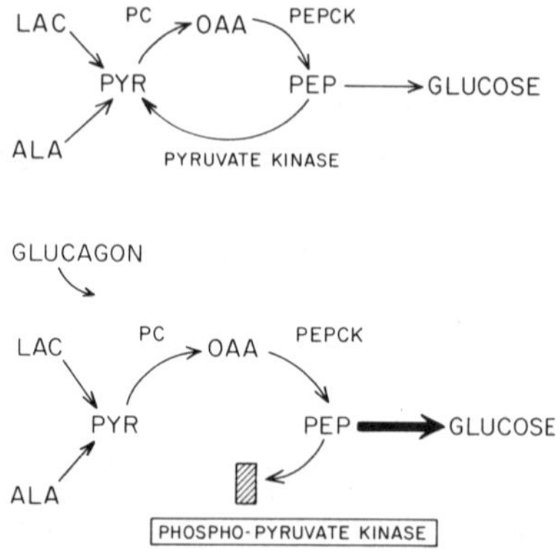

FIGURE 4. Glucagon action at the PK step in liver.

2,6-P_2 levels are high the activity of and flux through 6PF-1K is high and Fru 1,6-P_2ase is inhibited. When Fru 2,6-P_2 levels are low Fru 1,6-P_2ase activity and flux are enhanced, 6PF-1K is inhibited, and gluconeogenic flux predominates. The ensuing changes in the level of Fru 1,6-P_2, because of its effects on PK, serve to coordinate the control of both substrate cycles. Fru 2,6-P_2 can thus be thought of as a master switching signal between gluconeogenesis and glycolysis.

In line with this notion, the level of Fru 2,6-P_2 has been shown to be under acute hormonal control.[3,4,9–11] For example, addition of glucagon or cAMP to hepatocytes results in a dramatic fall in the level of the compound.[1–4,12] Insulin, which acts by counteracting glucagon's effect to elevate cAMP levels, opposes the action of glucagon to lower Fru 2,6-P_2 levels.[12] These effects of glucagon and insulin on Fru 2,6-P_2 levels correlate well with these hormones' ability to modulate cAMP levels and gluconeogenesis: glucagon raises and insulin lowers the rate of this process.

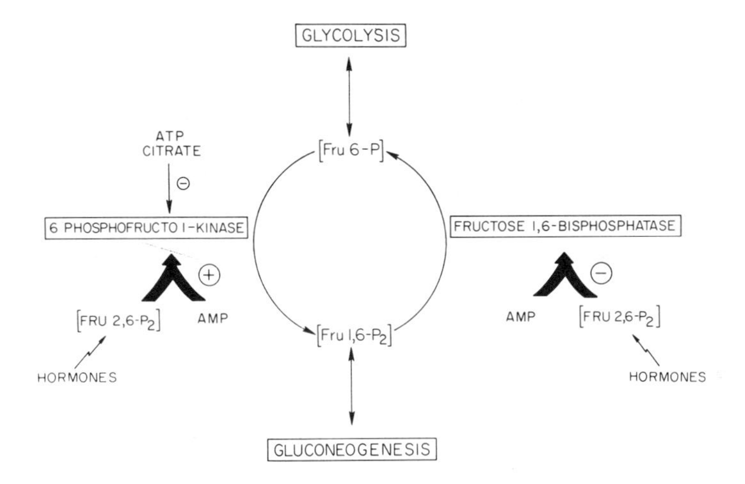

FIGURE 5. Regulation of the Fru 6-P/Fru 1,6-P_2 substrate cycle by Fru 2,6-P_2.

The enzyme responsible for the regulation of Fru 2,6-P_2 levels is bifunctional in nature as shown in schematic form in FIGURE 6. The enzyme possesses both a 6-phosphofructo-2-kinase (6PF-2K) activity that catalyzes the synthesis of Fru 2,6-P_2 from ATP and Fru 6-P as well as a Fru 2,6-P_2ase activity that catalyzes its degradation to Fru 6-P and P_i. (It was originally believed by Furuya *et al.*[13] that these enzyme activities were present in separate proteins, but this was definitely shown not to be the case when the enzyme was purified to homogeneity in our laboratory and its properties studied.[14,15]) The enzyme is regulated by cAMP-dependent protein kinase-catalyzed phosphorylation (FIG. 6) that inhibits the kinase and activates the bisphosphatase.[14–17] No other protein kinase has been shown to catalyze its phosphorylation. Furuya *et al.* reported that the enzyme was a substrate for phosphorylase kinase,[18] but this has been shown to be incorrect.[12] A phosphoprotein phosphatase in liver that dephosphorylates the enzyme has also been identified as the phosphoprotein 2A,[19] but there is no evidence that this phosphoprotein phosphatase is regulated by hormones.

6PF-2K/Fru 2,6-P_2ase consists of a dimer of two 55 000-dalton subunits as determined by SDS-gel electrophoresis.[14–17] As far as is known the two subunits are identical. Most evidence suggests that the two reactions of the enzyme are catalyzed at two discrete active sites on each subunit.[20–24] Kinetic studies of the kinase reaction have shown that ADP is a competitive inhibitor with respect to ATP[4,16,22,25] whereas Fru 2,6-P_2 is a noncompetitive inhibitor with respect to either substrate.[16,21,25,26] Studies on the bisphosphatase reaction indicated that Fru 6-P is a noncompetitive inhibitor[4,10,14,16,21,26–30] whereas P_i is a competitive inhibitor.[28] Recent studies have also revealed that the Fru 2,6-P_2ase has a very high affinity for Fru 2,6-P_2 ($K_m = 4$ nM) and undergoes substrate inhibition at concentrations of Fru 2,6-P_2 above 100 nM.[28] Both inorganic phosphate and α-glycerol phosphate overcome substrate inhibition and act as activators of the bisphosphatase at high substrate concentrations. It has been suggested that one or both reactions may proceed via one or more phosphorylated intermediate(s) because the enzyme catalyzes both ATP/ADP and Fru 2,6-P_2/Fru 6-P exchanges and is phosphorylated on a histidine residue from Fru 2,6-P_2.[26–28,31]

FIGURE 7 summarizes current knowledge of hormonal regulation of the gluconeogenic/glycolytic pathway in mammalian liver. Panel A shows the situation when cAMP levels are high or in the presence of Ca^{2+}-linked hormones. In the case of elevated cAMP levels, PK and the bifunctional enzyme are in the phosphoform, as are 6PF-1K and Fru 1,6-P_2ase. This results in inhibition of glycolytic flux and increased gluconeogenic flux as indicated. In the case of Ca^{2+}-linked hormones, the phosphorylation state of the bifunctional enzyme is not altered, but PK is phosphorylated on both seryl and threonyl residues resulting in inactivation of the enzyme and decreased recycling of PEP to PYR.[7] The effect of Ca^{2+}-linked hormones on rates of gluconeogenesis is small and correlates well with the small effect on PK phosphorylation state and activity.[1–3] The work summarized above is taken from studies on rat liver, and it is of course possible that the relative significance of cAMP- and Ca^{2+}-linked hormones on these processes in livers from other species may be quite different. Panel B illustrates the situation where cAMP levels are low, for example, when insulin/glucagon ratios are high. In this situation all of the key regulatory enzymes are dephosphorylated, Fru 2,6-P_2 and glycolytic flux are high, and gluconeogenic flux is inhibited. It seems reasonable to postulate that all hormonal regulation of pathway flux can be explained by changes in the phosphorylation state of PK and the bifunctional enzyme. Although both Fru 1,6-P_2ase[32] and 6PF-1K[33] are substrates *in vitro* and *in vivo* for the cAMP-dependent protein kinase, the evidence for physiologically relevant regulation of their activity by phosphorylation is both contradictory and controversial.[4,32–44]

IDENTIFICATION OF THE BIFUNCTIONAL ENZYME IN EXTRAHEPATIC TISSUES

Although the central role of the bifunctional enzyme in hormonal regulation of the glycolytic/gluconeogenic pathway in rat liver has been clearly shown, the role, if any, of these activities in other tissues is less certain. Either bisphosphatase or kinase activity has been detected and/or partially purified from other tissue sources including plants,[45,46] yeast,[47] rat and beef heart,[48,49] rat kidney, and bovine neutrophils (T. Chrisman and S. J. Pilkis, unpublished data). In most of these cases the bifunctionality of the protein has not been definitively established. In any case, liver is a tissue where a mechanism for switching between glycolysis and gluconeogenesis is necessary, and changes in Fru 2,6-P_2, mediated by the bifunctional enzyme, provides a signal for

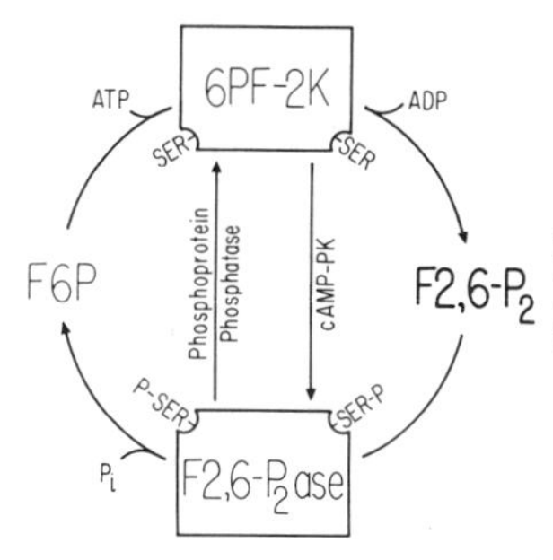

FIGURE 6. Regulation of the bifunctional catalyst 6PF-2K/Fru 2,6-P_2ase by phosphorylation/dephosphorylation reactions.

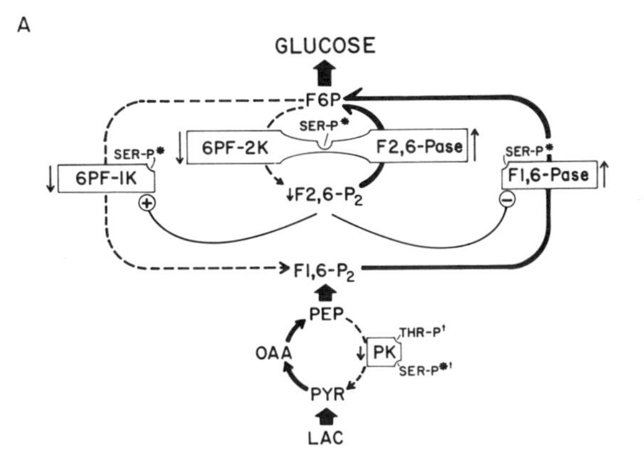

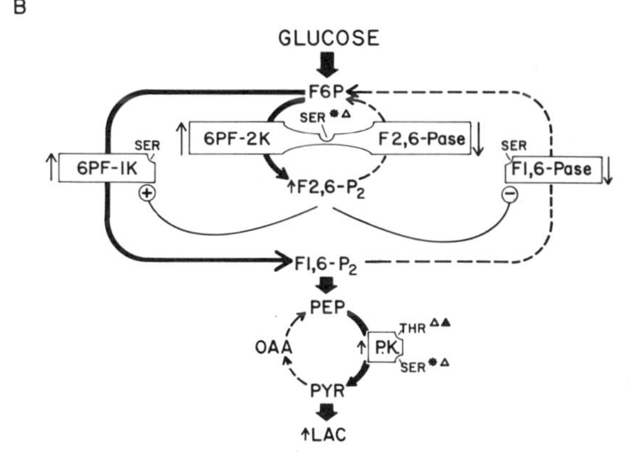

FIGURE 7. Regulation of the hepatic glycolytic/gluconegenic pathway by phosphorylation reactions. **(A)** Elevation of cAMP levels leads to phosphorylation of pyruvate kinase, 6PF-1K, Fru 1,6-P_2ase (in rat liver), and 6PF-2K/Fru 2,6-P_2ase on seryl residues (as indicated by ∗). PK can also be phosphorylated by Ca^{2+}- and calmodulin-dependent protein kinase on a threonyl residue (as indicated by +). These phosphorylations result in inhibition of glycolytic enzyme activities (PK, 6PF-2K, and, at least indirectly via the decrease in Fru 2,6-P_2 levels, 6PF-1K) and to activation of enzymes favoring gluconeogenesis (Fru 2,6-P_2ase and Fru 1,6-P_2ase). The final result is enhanced lactate to glucose flux. **(B)** In states where cAMP is low, such as with high insulin to glucagon ratios, the phosphorylations in **A** are all reversed leading to enhanced glycolysis.

such switching. In contrast, in heart the opposing gluconeogenic enzymes are very low or lacking, and such a bifunctional enzyme and its reciprocal regulation by phosphorylation may not be needed.

We decided to take advantage of some basic discoveries about the enzyme's reaction mechanism as well as an antibody to the liver enzyme to determine whether a similar bifunctional enzyme is present in extrahepatic tissues. It has been shown that Fru 2,6-P_2 interacts with a histidyl residue at the active site of the bisphosphatase to form a phosphoenzyme intermediate.[31] Although the N-P bond formed is acid labile, it is possible to trap this intermediate under certain conditions. A simple assay for determining the amount of bifunctional enzyme protein merely involves incubating Fru 2,6-[2-^{32}P]P_2 with tissue extract in the cold, spotting an aliquot on phosphocellulose paper, washing, and determining the amount of ^{32}P radioactivity incorporated into the enzyme.[50] The amount of ^{32}P radioactivity incorporated is a linear function of enzyme concentration over a wide range. A standard curve with known amounts of enzyme can be constructed, and nanogram quantities of bifunctional enzyme can be detected. This is particularly important for estimating Fru 2,6-P_2ase activity in extrahepatic tissues, where such activity is difficult if not impossible to measure. This also provides a specific test for ascertaining whether similar activities to those found in liver exist in other tissues.[50]

FIGURE 8 shows that it is possible to detect phosphoenzyme formation throughout the purification scheme for the liver enzyme as well as in a partially purified kidney extract. A band of radioactivity is seen in each lane, including the lane corresponding to the crude liver extract that comigrated with the pure liver enzyme during all the steps of the purification scheme. The kidney contains a Fru 2,6-P_2ase activity that forms a phosphoenzyme and is present in a peptide with the same relative molecular mass as the liver enzyme. Additional evidence for bifunctionality is seen in FIGURE 9, which shows DEAE-Sephadex chromatography of 6PF-2K activity (open symbols) and Fru 2,6-P_2ase activity determined by the phosphoenzyme method (closed symbols) from various tissues after partial purification by polyethylene glycol fractionation. Both activities copurified in liver, kidney, and skeletal muscle and eluted at the same salt concentration. In the case of heart, however, little or no phosphoenzyme could be detected, and the enzyme eluted at a slightly lower salt concentration, suggesting that it may not be bifunctional. We have not been able to detect any significant Fru 2,6-P_2ase activity associated with the heart enzyme by the traditional assay procedure.

TABLE 1 gives a summary of the properties of the Fru 2,6-P_2-metabolizing enzyme, from various tissues, including its ability to form a phosphoenzyme intermediate and to interact with rabbit antibody prepared against the rat liver enzyme. 6PF-2K and Fru 2,6-P_2ase activities are expressed relative to rat liver, the activity of which is taken as 100. Only liver (from rat, dog, and pig) contained substantial amounts of activity. All other tissues tested, including brain, lung, testis, kidney, skeletal muscle, and heart, contained only 1-10% that found in rat liver. Interestingly, both kinase and bisphosphatase activities were detected in bovine neutrophils. Both activities and phosphoenzyme formation from neutrophils copurified through polyethylene glycol fractionation, DEAE-Sephadex and Blue-Sepharose chromatography, and gel filtration steps (data not shown). The enzyme was bifunctional in all tissues tested on the basis of phosphoenzyme formation and coelution of kinase and bisphosphatase activity except for the heart. In addition, the heart enzyme did not react with the antibody to the rat liver enzyme. Although the heart enzyme appears to differ from the bifunctional enzyme from liver in a number of respects, its native molecular mass (Stokes radius: 45 A; M_r: 110 000 daltons) appears to be the same as judged by gel filtration on Sephadex G-100. However, Rider *et al.*[49] have reported that partially purified 6PF-2K from beef heart has a lower molecular mass than the rat liver enzyme. These

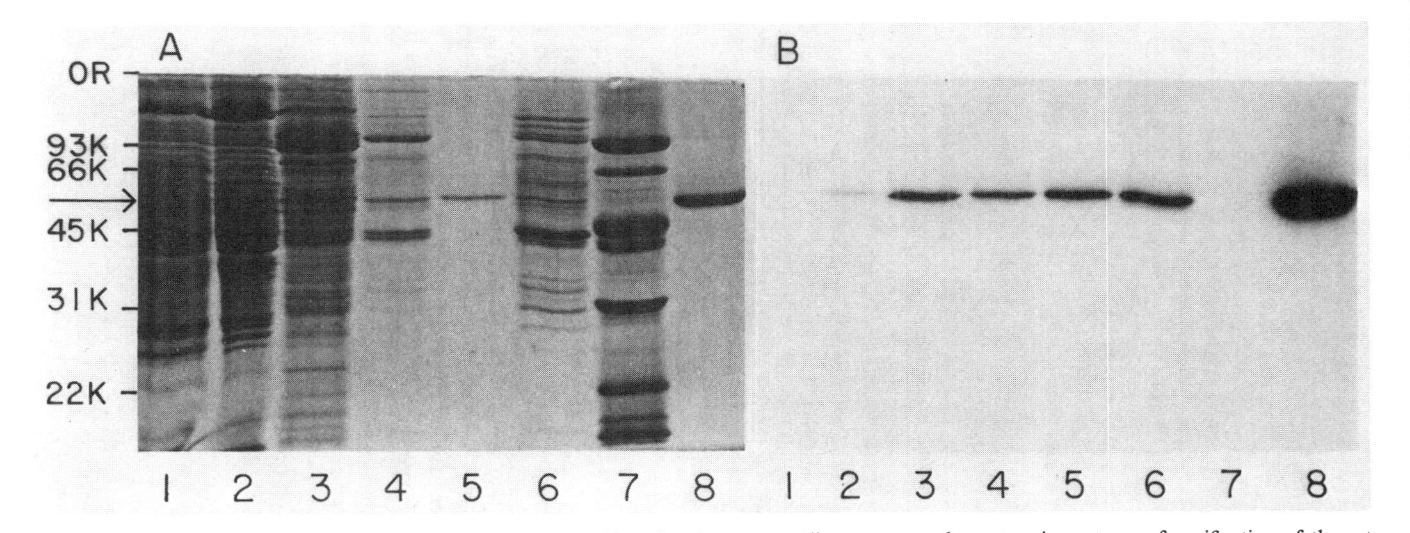

FIGURE 8. Detection of a phosphoenzyme intermediate of the bifunctional enzyme. Aliquots were taken at various stages of purification of the rat liver 6PF-2K/Fru 2,6-P_2ase and from a partially purified kidney preparation and incubated with Fru 2,6-[2-^{32}P]P_2 at 0 °C for 5 sec and then denatured with SDS and applied to a slab SDS-gel. Electrophoresis was then performed. **A** represents a Coomassie blue-stained gel and **B** an autoradiogram of the same gel. Lane 1: rat liver homogenate supernatant; lane 2: polyethylene glycol (6-12%) pellet; lane 3: DEAE-Sephadex pool; lane 4: Blue-Sepharose pool; lane 5: Fru 6-P eluted enzyme from P-Cellulose; lane 6: partially purified kidney extract; lane 7: molecular weight standards; lane 8: homogeneous rat liver 6PF-2K/Fru 2,6-P_2ase. The purification is that reported previously.[22]

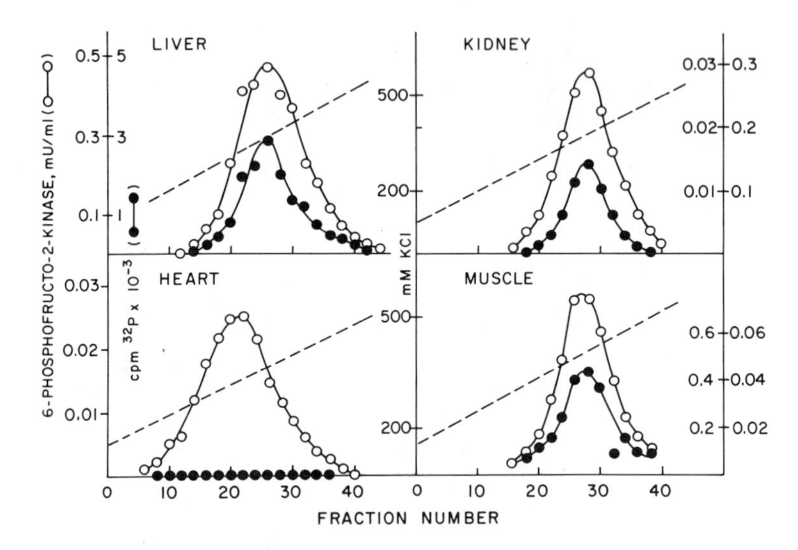

FIGURE 9. DEAE-Sephadex chromatography of 6PF-2K and Fru 2,6-P_2ase and activities from rat liver, kidney, skeletal muscle, and heart. Tissue extracts were prepared, subjected to polyethylene glycol fractionation, and applied to DEAE-Sephadex columns and eluted with linear KCl gradients.[22] 6PF-2K activity was assayed with the 6PF-1K activation assay and Fru 2,6-P_2ase was estimated with the phosphoenzyme assay.

workers also reported that beef heart enzyme possessed very low Fru 2,6-P_2ase activity compared to its 6PF-2K activity.

Clearly the gene for the bifunctional enzyme is only minimally expressed in tissues other than liver, and the heart enzyme may have a different form. This result is perhaps not unexpected because a finely tuned mechanism for switching between glycolysis and gluconeogenesis is unnecessary in extrahepatic tissues. Whether the difference between the heart and liver forms reflects different gene products or as yet unidentified posttranslational modifications remains to be determined. The lack of immunocross-reactivity supports the former theory. In order to answer this question it will be necessary to purify the heart enzyme to homogeneity and to characterize it in more detail.

LONG-TERM HORMONAL CONTROL OF HEPATIC 6PF-2K/FRU 2,6-P_2ASE ACTIVITIES

In order to obtain some idea of those factors controlling the expression of the gene for the bifunctional enzyme, the effect of various hormonal states on the expression of the hepatic bifunctional enzyme's activities was studied. Three different hormonal states were examined: diabetes, hyperthyroidism, and adrenal insufficiency with glucocorticoid replacement. TABLE 2 shows that in the diabetic state hepatic Fru 2,6-P_2 levels are decreased to 10% of those in the control group. In addition, both activities of the enzyme were decreased to about 25-30% of the control value and were restored to normal upon insulin administration for 24 hr. As also shown, the pattern seen was identical to that observed for glucokinase, an enzyme whose synthesis is controlled by insulin and cAMP.[51] The very low level of Fru 2,6-P_2 in livers of diabetic rats correlates with the low rate of glycolysis and high rate of gluconeogenesis in that state. The reduced level of the sugar diphosphate probably results from low glucokinase activity, which limits the amount of Fru 6-P available to the bifunctional enzyme, whose 6PF-2K activity is also decreased. If glucokinase activity were not low, a similar decrease in both 6PF-2K activity and Fru 2,6-P_2ase activity would not be expected to affect the steady state concentration of Fru 2,6-P_2. Coordinate inductions of both glucokinase and 6PF-2K appear to be necessary to restore Fru 2,6-P_2 to normal. It is also possible that the bifunctional enzyme is in the phosphoform in the diabetic state and that this contributes to the low Fru 2,6-P_2 concentration.

TABLE 3 shows that induction of hyperthyroidism by the administration of triiodothyronine (T_3) also decreased Fru 2,6-P_2 levels and 6PF-2K activity. In these same experiments there was no change in the total activity of or in the phosphorylation state of PK, as measured by the ratio of activity at 0.4 mM and 4 mM PEP, or in the activity of glucokinase (data not shown). This indicates that T_3 treatment did not have a general nonspecific effect on hepatic enzyme activity. Also shown is the lack of effect of T_3 on the activity ratio of the cAMP-dependent protein kinase. This result and the lack of effect on the activity ratio of PK tend to rule out phosphorylation as a mechanism for decreased 6PF-2K activity.

TABLE 1. Summary of Properties of 6PF-2K/Fru 2,6-P_2ase from Various Tissues

Source	Stokes Radius (A)	Reacts with Antibody to Rat Liver Enzyme Activity	Relative Amount of Kinase	Relative Amount of Fru 2,6-P_2ase
Liver				
Rat	45	Yes	100	100
Cow	45	Yes	40	40
Dog	45	Yes	40	30
Kidney, rat	45	Yes	3	0.5
Brain, rat			1	0.1
Spleen, rat			3	0.5
Skeletal muscle, rat		Yes	1	0.2
Neutrophil, cow	45			
Testis, rat	45		4	0.6
Adipose, rat			1	0.2
Lung, rat			6	0.9
Heart				
Rat	45	No	6	0.1
Dog		No	6	0.1
Pig		No	6	0.1

NOTE: The enzyme was partially purified from each tissue as previously described for the rat liver enzyme up to the Blue Sepharose step[22] using the kinase activity to monitor each step. Stokes radii were determined by gel filtration on Sephadex G-100 (superfine). Fru 2,6-P_2ase was determined by phosphoenzyme formation as described previously.[50] Equal amounts of 6PF-2K activity from each tissue were subjected to SDS-slab gel electrophoresis and a Western blot of the gel performed. As little as 2 ng of the rat liver bifunctional enzyme could be detected with a specific antibody to the enzyme raised in rabbits.[50]

TABLE 2. Effect of Diabetes and Insulin Treatment on 6PF-2K/Fru 2,6-P_2ase and Glucokinase Activities and on the Level of Fru 2,6-P_2ase

	N	Fru 2,6-P_2 (nmol/g)	6PF-2K (mU/g)	Fru 2,6-P_2ase (mU/g)	6PF-2K/ Fru 2,6-P_2ase	Glucokinase (mU/mg protein)
Normal, fed	12	10 ± 1.1	20.3 ± 4.0	9.3 ± 1.5	2.2	18.5 ± 1.5
Diabetic	11	1.1 ± 0.3	6.2 ± 2.0	2.2 ± 0.5	2.8	4.6 ± 1.6
Diabetic, treated with insulin	5	8.9 ± 2.0	34.6 ± 6.2	6.7 ± 1.3	5.2	22.1 ± 1.7

NOTE: Male Sprague-Dawley rats (175-240 g) were used. Diabetes was induced by the intravenous injection of alloxan (60 mg/kg) after starvation for 24 hr. Rats were used only if the blood glucose level was greater than 375 mg/100 ml of blood. Insulin (4 U) was given subcutaneously every 8 hr. Fru 2,6-P_2 levels and 6PF-2K, Fru 2,6-P_2ase, and glucokinase activities were determined as previously described.[54]

The effect of adrenalectomy and subsequent glucocorticoid replacement is shown in TABLE 4. Again there was no effect of adrenalectomy or steroid treatment on the activity ratio of PK, suggesting that changes in phosphorylation state are not responsible for any observed changes. Fru 2,6-P_2 levels and both activities of the enzyme were decreased by adrenalectomy, and the enzyme activities were restored, albeit slowly, by administration of glucocorticoids. Fru 2,6-P_2 levels were restored to normal within 24 hr even though the enzyme activities were still depressed. The reason for this is unknown, but may reflect effects of the steroid on the phosphorylation state of the bifunctional enzyme and/or effects on glucokinase. It should also be noted that the same decrease in total amount of enzyme was obtained using 6PF-2K activity (−80%) and phosphoenzyme formation (−77%), which measures the amount of enzyme protein. It would appear that phosphoenzyme formation will be a useful method for determining the extent of modulation of the actual amount of enzyme protein in various dietary and hormonal states.

We concluded from these studies that in addition to regulation by covalent modification and low-molecular-weight effectors, the bifunctional enzyme probably undergoes long-term multihormonal modulation of expression of its activities. Clearly, this enzyme presents a number of interesting problems, including those of structure/function and regulation of gene expression. It should be noted that other key regulatory

TABLE 3. Effect of T_3 Administration on PK and 6PF-2K Activity and the cAMP-dependent Protein Kinase Activity Ratio

Condition	N	Protein Kinase Activity Ratio (−cAMP/+cAMP)	PK		6PF-2K (mU/g)
			U/g	v/v	
Control	4	0.19 ± 0.20	15.0 ± 1.1	0.32 ± 0.03	0.6 ± 0.1
Treated with T_3	6	0.18 ± 0.10	14.0 ± 2.0	0.31 ± 0.01	0.20 ± 0.05

NOTE: Male Sprague-Dawley rats (175-240 g) were used. T_3 (2.5 μg/rat) treatment was begun 6 days after the operation and continued for 3 days, after which the animals and appropriate controls were sacrificed and the enzyme activities determined.

enzymes in the pathway such as glucokinase, PK, and PEPCK also undergo long-term hormonal regulation of their expression.[51-53]

ISOLATION OF A cDNA CLONE FOR 6PF-2K/FRU 2,6-P_2ASE

It was decided to attempt to obtain a cDNA clone for the enzyme that would provide primary sequence information as well as a probe to investigate the possible hormonal control of the gene's expression. Because the bifunctional enzyme and other key regulatory enzymes are relatively scarce proteins, a method for selecting the small proportion of enzyme clones from a large pool of recombinants was needed. Because antisera to all the regulatory hepatic enzymes, including the bifunctional protein, were available, immunological screening of a cDNA library from rat liver RNA in λgt11

TABLE 4. Effect of Glucocorticoid Administration on the Bifunctional Enzyme Activities and PK Activity

Condition	*N*	PK (v/v)	Fru 2,6-P_2 (nmol/g)	6PF-2K (mU/g)	Fru 2,6-P_2ase (μg/g)
Normal, fed	5	0.3 ± 0.1	5.0 ± 0.2	0.5 ± 0.2	9.45 ± 1.70
Adrenalectomized	6	0.3 ± 0.1	2.4 ± 0.5	0.1 ± 0.3	2.0 ± 0.1
Treated with hydrocortisone					
24 hr	5	0.30 ± 0.05	5.5 ± 0.4	0.10 ± 0.05	3.1 ± 1.0
48 hr	5	0.3 ± 0.1	5.1 ± 0.6	0.3 ± 0.1	3.6 ± 0.8
72 hr	5	0.3 ± 0.1	5.3 ± 1.0	0.4 ± 0.1	6.2 ± 2.0

NOTE: Male Sprague-Dawley rats (175-240 g) were used. Adrenalectomized rats were obtained from Harlan Sprague-Dawley (Indianapolis, IN). Hydrocortisone treatment (2.5 mg/rat/day) was begun 7 days after operation and continued for 1-3 days. The animals and appropriate controls were then sacrificed, and Fru 2,6-P_2 levels and enzyme activities were determined as described in TABLE 1. Fru 2,6-P_2ase protein, expressed as μg enzyme/g liver, was determined by the phosphoenzyme assay.[50]

was chosen.[52] The rat liver cDNA expression library was obtained from Dr. Jean Schwarzbauer at the Massachusetts Institute of Technology. Using the bifunctional enzyme antibody it was possible to detect 500-600 pg of enzyme protein as shown in FIGURE 10. Initial plaque screening of 250 000 independent recombinants identified a number of putative bifunctional enzyme clones as shown on the right panel. In results not shown here, a putative PK clone was detected using similar methods. The clones were taken through several rounds of purification and rescreening with antibodies and signals produced by a tertiary immunochemical screen for both enzymes as shown in FIGURE 11. The clones for PK and the bifunctional enzyme were analyzed, as shown in FIGURE 12, by restriction mapping in which *Eco*R1 and *Acc*1 digests were performed. The PK clone was estimated to comprise about 650 bases after *Eco*R1 and *Acc*1 digests. The bifunctional enzyme clone was estimated to comprise about 1.8 kilobases by *Acc*1 digest, but appeared to have an internal *Eco*R1 site(s) because several fragments were formed by digestion with this enzyme. The results suggest that this

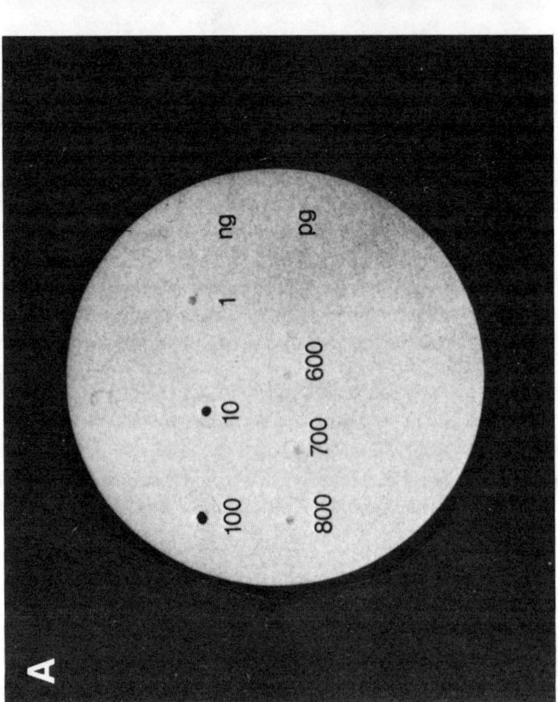

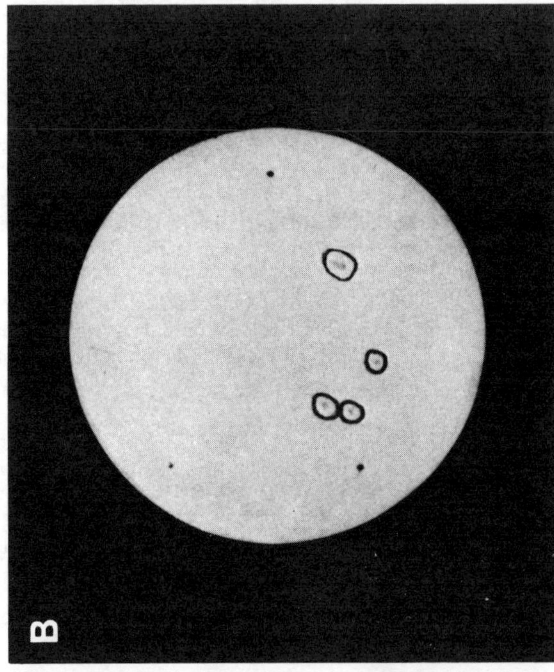

FIGURE 10. Immunological screening of a cDNA rat liver expression library in λgt11. **(A)** Sensitivity of the immunologic screening method for 6PF-2K/Fru 2,6-P_2ase. A rabbit antiserum to the rat liver bifunctional enzyme was diluted 5000-fold and used with a peroxidase-coupled double antibody in the screening procedure essentially by the method of Young and Davis.[51] **(B)** Immunopositive clones.

may represent a full-length or nearly full-length cDNA clone for the bifunctional enzyme.

With these clones we hope to approach the problems of structure/function and control of gene expression that regulatory enzymes like PK and 6PK-2K/Fru 2,6-P_2ase pose. We are particularly interested in the structure/function studies of the bifunctional enzyme; identification of the two catalytic centers; determination of whether there is any homology between the two sites; determination of whether the gene for the heart enzyme is indeed different from the gene for the rat liver enzyme; and isolation and identification of those gene elements responsible for the regulation of expression of the protein.

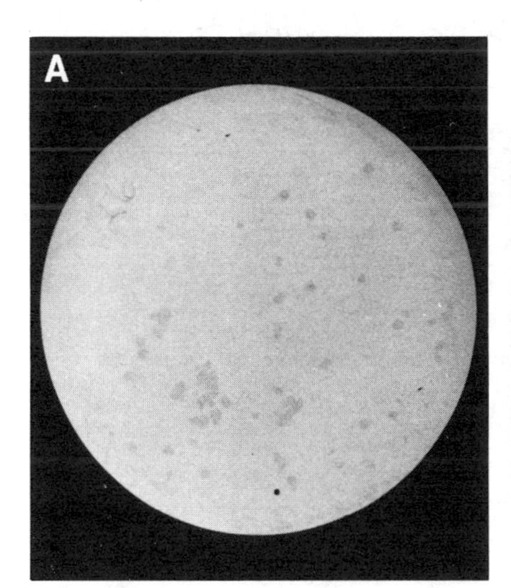

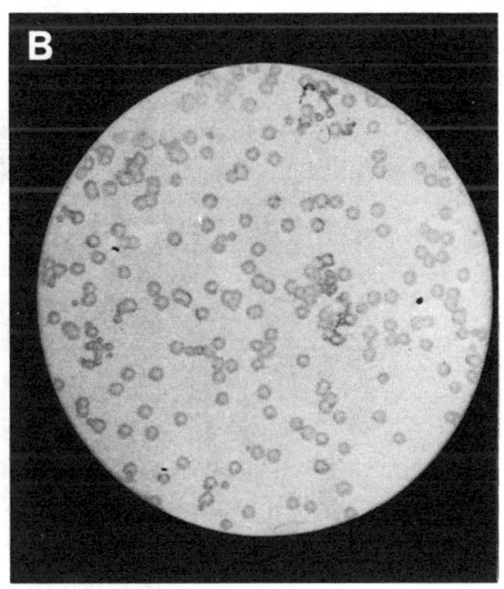

FIGURE 11. Plaque purification of clones for the liver bifunctional enzyme and PK. (**A**) Liver 6PF-2K/Fru 2,6-P_2ase. (**B**) PK.

SUMMARY

Short- and long-term regulation of the hepatic glycolytic/gluconeogenic pathway is exerted by several key regulatory enzymes. The most important acute effects are exerted at PK and on the steady state level of Fru 2,6-P_2, which is controlled by the activity of 6PF-2K/Fru 2,6-P_2ase. It appears reasonable to postulate that regulation by phosphorylation/dephosphorylation reactions of this bifunctional enzyme and PK can account in large part for the physiologically relevant acute actions of glucagon, insulin, and β-adrenergic agonists. With regard to gluconeogenesis, Ca^{2+}-linked hormones act only at the PK step. Long-term effects of hormones on key hepatic regulatory enzymes are a result of modulation of gene expression.[53] Such regulation occurs at the level of translation and/or transcription. These mechanisms, like short-term control, may also be mediated by phosphorylation/dephosphorylation reactions that modulate the activity of proteins and/or factors that interact with control elements at the gene level. Although long-term hormonal control of enzymes such as glucokinase, PEPCK, and PK has been studied intensively for decades, 6PF-2K/Fru 2,6-P_2ase may represent an additional enzyme whose expression is hormonally regulated.[54] The availability of a cDNA clone should permit studies to determine whether regulation occurs at the level of transcription.

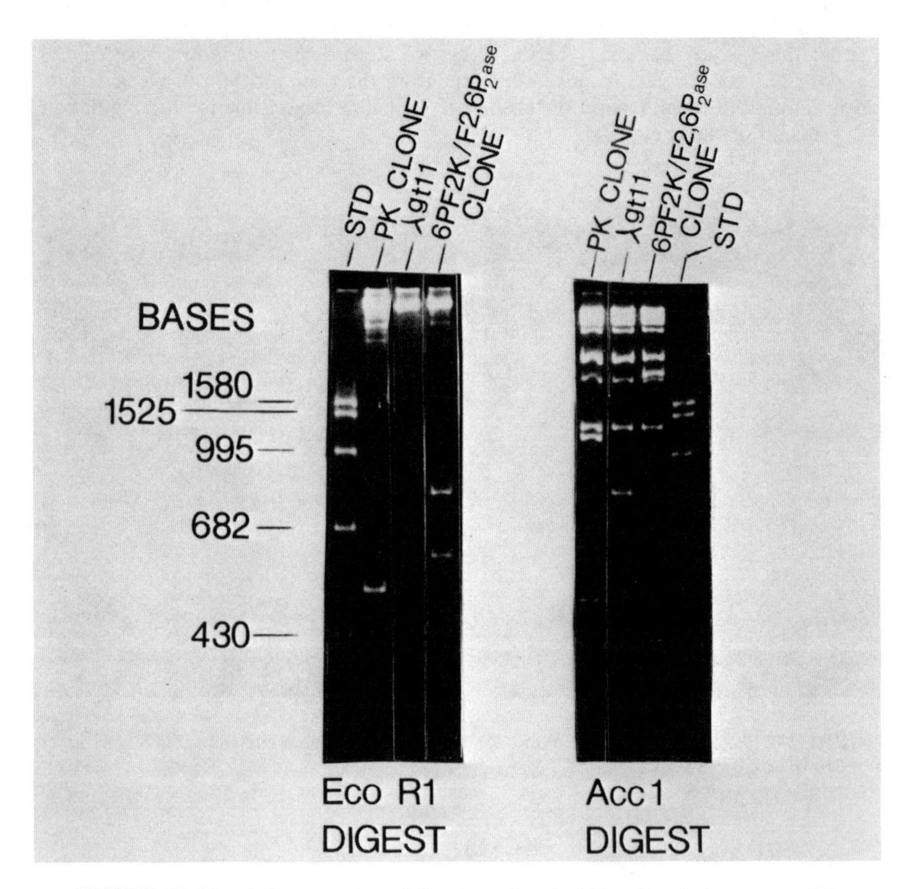

FIGURE 12. Restriction mapping of the clones for the bifunctional enzyme and PK.

ACKNOWLEDGMENTS

The authors would like to thank Ms. Jo Pilkis and Ms. Tina Pate for their skilled technical assistance.

REFERENCES

1. PILKIS, S. J., C. R. PARK & T. H. CLAUS. 1978. Hormonal control of hepatic gluconeogenesis. Vitam. Horm. **36:** 383-460.

2. CLAUS, T. H. & S. J. PILKIS. 1981. Hormonal control of hepatic gluconeogenesis. Biochem. Actions Horm. **8:** 209-271.
3. CLAUS, T. H., C. R. PARK & S. J. PILKIS. 1983. Glucagon and gluconeogenesis. *In* Handbook of Experimental Pharmacology. P. J. Lefebre, Ed. Vol. **66:** 315-360. Springer-Verlag. Berlin and New York.
4. CLAUS, T. H., M. R. EL-MAGHRABI, D. M. REGEN, H. B. STEWART, M. MCGRANE, P. D. KOUNTZ, F. NYFELER, J. PILKIS & S. J. PILKIS. 1984. The role of fructose 2,6-bisphosphate in the regulation of carbohydrate metabolism. Curr. Top. Cell. Regul. **23:** 57-86.
5. EXTON, J. H. 1985. Mechanism of action of Ca^{2+}-mobilizing hormones. Am. J. Physiol. **248:** E633-E647.
6. PILKIS, S. J., M. R. EL-MAGHRABI & T. H. CLAUS. 1986. Mechanism of action of insulin on the hepatic gluconeogenic/glycolytic pathway. *In* Symposium on the Mechanism of Action of Insulin. Elsevier. In press.
7. SCHWORER, C., M. R. EL-MAGHRABI, S. J. PILKIS & T. R. SODERLING. 1985. Phosphorylation of L-type pyruvate kinase by a Ca^{2+}/calmodulin-dependent protein kinase. J. Biol. Chem. **260:** 13018-13022.
8. GARRISON, J. & J. D. WAGNER. 1982. Glucagon and the Ca^{2+}-linked hormones angiotensin II, norepinephrine and vasopressin stimulate the phosphorylation of distinct substrates in intact hepatocytes. J. Biol. Chem. **256:** 13135-13142.
9. PILKIS, S. J., M. R. EL-MAGHRABI, M. MCGRANE, J. PILKIS, E. FOX & T. H. CLAUS. 1982. Fructose 2,6-bisphosphate: A mediator of hormone action at the fructose 6-phosphate/fructose 1,6-bisphosphate substrate cycle. Mol. Cell. Endocrinol. **25:** 245-266.
10. HERS, H.-G. & E. VAN SCHAFTINGEN. 1982. Fructose 2,6-bisphosphate 2 years after its discovery. Biochem. J. **206:** 1-12.
11. UYEDA, K., E. FURUYA, C. S. RICHARDS & M. YOKOYAMA. 1982. Fructose 2,6-bisphosphate: A new regulator of phosphofructokinase. Mol. Cell. Biochem. **48:** 97-120.
12. PILKIS, S. J., T. D. CHRISMAN, M. R. EL-MAGHRABI, A. COLOSIA, E. FOX, J. PILKIS & T. CLAUS. 1983. The action of insulin on hepatic fructose 2,6-bisphosphate metabolism. J. Biol. Chem. **258:** 1495-1503.
13. FURUYA, E., M. YOKOYAMA & K. UYEDA. 1982. An enzyme that catalyzes hydrolysis of fructose 2,6-bisphosphate. Biochem. Biophys. Res. Commun. **105:** 264-270.
14. EL-MAGHRABI, M. R., T. H. CLAUS, J. PILKIS, E. FOX & S. J. PILKIS. 1982. Regulation of rat liver fructose 2,6-bisphosphatase. J. Biol. Chem. **257:** 7603-7607.
15. EL-MAGHRABI, M. R., E. FOX, J. PILKIS & S. J. PILKIS. 1982. Cyclic AMP-dependent phosphorylation of rat liver 6-phosphofructo-2-kinase/fructose 2,6-bisphosphatase. Biochem. Biophys. Res. Commun. **106:** 794-802.
16. PILKIS, S. J., D. M. REGEN, H. B. STEWART, T. CHRISMAN, J. PILKIS, P. KOUNTZ, M. MCGRANE, M. R. EL-MAGHRABI & T. H. CLAUS. 1983. Rat liver 6-phosphofructo-2-kinase/fructose 2,6-bisphosphatase: A unique bifunctional enzyme regulated by cyclic AMP-dependent phosphorylation. Mol. Aspects Cell. Regul. **3:** 95-122.
17. MURRAY, K., M. R. EL-MAGHRABI, P. KOUNTZ, Y. LUKAS, T. R. SODERLING & S. J. PILKIS. 1984. Amino acid sequence of the phosphorylation site of rat liver 6-phosphofructo-2-kinase/fructose 2,6-bisphosphatase. J. Biol. Chem. **259:** 7673-7681.
18. FURUYA, E., M. YOKOYAMA & K. UYEDA. 1982. Regulation of fructose 6-phosphate 2-kinase by phosphorylation and dephosphorylation: Possible mechanism for coordinated control of glycolysis and glycogenolysis. Proc. Natl. Acad. Sci. USA **79:** 325-329.
19. PELLECH, S., P. COHEN, M. J. FISHER, C. POGSON, M. R. EL-MAGHRABI & S. J. PILKIS. 1984. The protein phosphatases involved in cellular regulation: Glycolysis, gluconeogenesis and aromatic amino acid breakdown in rat liver. Eur. J. Biochem. **145:** 39-49.
20. EL-MAGHRABI, M. R. & S. J. PILKIS. 1984. Rat liver 6-phosphofructo-2-kinase/fructose 2,6-bisphosphatase: A review of relationships between the two activities of the enzyme. J. Cell. Biochem. **26:** 1-17.
21. EL-MAGHRABI, M. R., T. PATE & S. J. PILKIS. 1984. Effect of sulfhydryl modification on the activities of rat liver 6-phosphofructo-2-kinase/fructose 2,6-bisphosphatase. J. Biol. Chem. **259:** 13104-13110.
22. EL-MAGHRABI, M. R., T. PATE, K. MURRAY & S. J. PILKIS. 1984. Differential effects of proteolysis and protein modification on the activities of 6-phosphofructo-2-kinase/fructose 2,6-bisphosphatase. J. Biol. Chem. **259:** 13096-13103.

23. SAKAKIBARA, R., S. KITAJIMA, F. C. HARTMAN & K. UYEDA. 1984. Hexose phosphate binding sites of fructose 6-phosphate, 2-kinase: Fructose 2,6-bisphosphatase. J. Biol. Chem. **259:** 14023-14028.
24. PILKIS, S. J., J. PILKIS, M. R. EL-MAGHRABI & T. H. CLAUS. 1985. The sugar phosphate specificity of rat hepatic 6-phosphofructo-2-kinase/fructose 2,6-bisphosphatase. J. Biol. Chem. **260:** 7551-7556.
25. KITAJIMA, S., R. SAKAKIBARA & K. UYEDA. 1984. Kinetic studies of fructose 6-phosphate, 2-kinase and fructose 2,6-bisphosphatase. J. Biol. Chem. **259:** 6896-6903.
26. PILKIS, S. J., D. M. REGEN, H. B. STEWART, J. PILKIS, T. M. PATE & M. R. EL-MAGHRABI. 1984. Evidence for two catalytic sites on 6-phosphofructo-2-kinase/fructose 2,6-bisphosphatase. J. Biol. Chem. **259:** 949-958.
27. KOUNTZ, P., M. R. EL-MAGHRABI & S. J. PILKIS. 1985. Isolation and characterization of 6-phosphofructo-2-kinase/fructose 2,6-bisphosphatase from bovine liver. Arch. Biochem. Biophys. **238:** 531-543.
28. STEWART, H. B., M. R. EL-MAGHRABI & S. J. PILKIS. 1985. Evidence for a phosphoenzyme intermediate in the reaction pathway of rat hepatic fructose 2,6-bisphosphatase. J. Biol. Chem. **260:** 12935-12941.
29. EL-MAGHRABI, M. R., T. M. PATE & S. J. PILKIS. 1984. Characterization of the exchange reactions of rat liver 6-phosphofructo-2-kinase/fructose 2,6-bisphosphatase. Biochem. Biophys. Res. Commun. **123:** 749-756.
30. VAN SCHAFTINGEN, E., D. R. DAVIES & H.-G. HERS. 1982. Fructose 2,6-bisphosphatase from rat liver. Eur. J. Biochem. **124:** 143-149.
31. PILKIS, S. J., M. WALDERHAUG, K. MURRAY, A. BETH, S. D. VENKATARAMU, J. PILKIS & M. R. EL-MAGHRABI. 1983. 6-Phosphofructo-2-kinase/fructose 2,6-bisphosphatase from rat liver: Isolation and identification of a phosphorylated intermediate. J. Biol. Chem. **258:** 6135-6141.
32. RIOU, J. P., T. H. CLAUS, D. A. FLOCKHART, J. D. CORBIN & S. J. PILKIS. 1977. *In vivo* and *in vitro* phosphorylation of rat liver fructose 1,6-bisphosphatase. Proc. Natl. Acad. Sci. USA **74:** 4615-4619.
33. PILKIS, S. J., M. R. EL-MAGHRABI & T. H. CLAUS. 1982. Studies on the *in vitro* phosphorylation of 6-phosphofructo-1-kinase from rat liver. Arch. Biochem. Biophys. **215:** 379-389.
34. SAKAKIBARA, R. & K. UYEDA. 1983. Differences in the allosteric properties of pure low and high phosphate forms of phosphofructokinase from rat liver. J. Biol. Chem. **258:** 8656-8662.
35. CLAUS, T. H., J. R. SCHLUMPF, M. R. EL-MAGHRABI, J. PILKIS & S. J. PILKIS. 1980. Mechanism of action of glucagon on hepatocyte phosphofructokinase activity. Proc. Natl. Acad. Sci. USA **77:** 6501-6505.
36. EKMAN, P. & U. DAHLQVIST-EDBERG. 1981. Effect of cyclic AMP-dependent phosphorylation on liver fructose 1,6-bisphosphatase activity. Biochim. Biophys. Acta **662:** 265-272.
37. EKDAHL, K. N. & P. EKMAN. 1985. Fructose 1,6-bisphosphatase from rat liver: A comparison of the kinetics of the unphosphorylated enzyme and the enzyme phosphorylated by cyclic AMP-dependent protein kinase. J. Biol. Chem. **260:** 14173-14179.
38. CLAUS, T. H., J. R. SCHLUMPF, M. R. EL-MAGHRABI, M. MCGRANE & S. J. PILKIS. 1981. Glucagon stimulation of fructose 1,6-bisphosphatase phosphorylation in rat hepatocytes. Biochem. Biophys. Res. Commun. **100:** 716-723.
39. PILKIS, S. J., M. R. EL-MAGHRABI, T. H. CLAUS, H. S. TAGER, D. E. STEINER, P. KEIM & R. HENRIKSON. 1980. Phosphorylation of rat hepatic fructose 1,6-bisphosphatase and pyruvate kinase. J. Biol. Chem. **255:** 2770-2775.
40. TAUNTON, O. D., F. B. STIFEL, H. L. GREENE & R. H. HERMAN. 1972. Rapid reciprocal change of rat hepatic glycolytic enzymes and fructose 1,6-bisphosphatase following glucagon and insulin injection *in vivo.* Biochem. Biophys. Res. Commun. **48:** 1663-1670.
41. TAUNTON, O. D., F. B. STIFEL, H. L. GREENE & R. H. HERMAN. 1974. Rapid reciprocal changes in rat hepatic glycolytic enzyme and fructose diphosphatase activities following insulin and glucagon injection. J. Biol. Chem. **249:** 7228-7233.
42. CHATTERJEE, T. & A. G. DATTA. 1978. Effect of glucagon administration on mice liver fructose 1,6-bisphosphatase. Biochem. Biophys. Res. Commun. **84:** 950-956.

43. Morikofer-Zwez, S., F. B. Stoecklin & P. Walter. 1981. Fructose 1,6-bisphosphatase in rat liver cytosol: Activation after glucagon treatment *in vivo* and inhibition by fructose 2,6-bisphosphate *in vitro*. Biochem. Biophys. Res. Commun. **101:** 104-111.
44. Pilkis, S. J., T. H. Claus, P. D. Kountz & M. R. El-Maghrabi. Role of phosphorylation/dephosphorylation in the regulation of enzymes of the fructose 6-phosphate/fructose 1,6-phosphate substrate cycle. *In* The Enzymes. P. Boyer & E. Krebs, Eds. Academic Press. In press.
45. Cseke, C., M. Stitt, A. Balogh & B. B. Buchanan. 1983. A product-regulated fructose 2,6-bisphosphatase occurs in green leaves. FEBS Lett. **162:** 103-106.
46. Stitt, M., C. Cseke & B. B. Buchanan. 1984. 6-Phosphofructo-2-kinase and fructose 2,6-bisphosphatase activities in plant tissue. Eur. J. Biochem. **143:** 89-97.
47. Clifton, D. & D. G. Franenkel. 1983. Fructose 2,6-bisphosphate and fructose 6-phosphate, 2-kinase, in *Saccharomyces cerevisiae* in relation to metabolic state in wild-type and fructose 6-phosphate, 1-kinase mutant strains. J. Biol. Chem. **258:** 9245-9249.
48. Narabayshi, H., T. W. R. Lawson & K. Uyeda. 1985. Regulation of phosphofructokinase in perfused rat heart: Requirement for fructose 2,6-bisphosphate and a covalent modification. J. Biol. Chem. **260:** 9750-9756.
49. Rider, M. H., D. Foret & L. Hue. 1985. Comparison of purified bovine heart and rat liver 6-phosphofructo-2-kinase: Evidence for distinct isoenzymes. Biochem. J. **231:** 193-196.
50. El-Maghrabi, M. R., J. J. Correia, P. J. Heil, T. M. Pate, C. E. Cobb & S. J. Pilkis. 1986. Tissue distribution, immunoreactivity, and physical properties of 6-phosphofructo-2-kinase/fructose 2,6-bisphosphatase. Proc. Natl. Acad. Sci. USA **83:** in press.
51. Weinhouse, S. 1976. Rat liver glucokinase. Curr. Top. Cell. Regul. **11:** 1-50.
52. Young, R. A. & R. N. Davis. 1985. Yeast RNA polymerase II genes: Isolation with antibody probes. Science **222:** 778-782
53. Granner, D. K. & T. L. Andreone. 1985. Insulin modulation of gene expression. Diabetes/Metab. Rev. **1:** 139-170.
54. Neely, P., M. R. El-Maghrabi, S. J. Pilkis & T. Claus. 1981. Effect of diabetes, insulin, starvation, and refeeding on the level of rat hepatic fructose 2,6-bisphosphate. Diabetes **30:** 1062-1064.

Coordinate Regulation of Rat Liver Genes by Thyroid Hormone and Dietary Carbohydrate[a]

TEH-YI TAO AND HOWARD C. TOWLE[b]

Department of Biochemistry
University of Minnesota
Minneapolis, Minnesota 55455

INTRODUCTION

The expression of many gene products in the liver is coordinately regulated by thyroid hormones and dietary carbohydrate. For example, hepatic malic enzyme in the rat is induced by either injection of 3,5,3′-triiodo-L-thyronine (T_3) or feeding of a high-carbohydrate, fat-free (CHO) diet.[1] The regulation by these two stimuli is exerted at a pretranslational level.[2,3] Studies on isolated primary rat hepatocytes in culture indicate that both T_3 and glucose can act directly at the hepatocellular level to stimulate malic enzyme synthesis.[4] An examination of approximately 230 hepatic mRNA species by cell-free translation and two-dimensional gel electrophoresis has revealed an extensive degree of overlap between the T_3- and CHO-diet-responsive "domains."[5] Thus, switching rats from a standard chow diet to a CHO diet led to alterations in the relative levels of 10 mRNA translational products. All but one of these mRNA species also responded to elevated plasma T_3 levels, and in each case the direction of change (that is, positive or negative) was the same. Because both elevated T_3 and the CHO diet result in increased fatty acid synthesis, it is possible that a subset of these mRNA products, like malic enzyme, are involved in adaptive hyperlipogenesis. These products provide an interesting system to study coordinate gene regulation by hormonal and nutritional stimuli.

In the past few years, we have focused on the regulation of one of these hepatic gene products, designated spot 14. Spot 14 was first identified by the nature of its translational product, which has an M_r of 17 000 and a pI of about 4.9.[6] The expression of the spot 14 gene was of particular interest due to its rapid kinetics of induction. Following a single injection of T_3 into hypothyroid rats, spot 14 mRNA levels increased with a lag time between 10 and 20 min.[7] A similar lag time was observed following the intragastric administration of sucrose to normal rats.[8] In cultures of primary hepatocytes, spot 14 mRNA levels are increased by addition of T_3 to the media.[9] The rapidity of the response of this gene product suggests that it may be responding as a primary event to the hormonal or nutritional stimuli.

[a] This work was supported by Grant AM 26919 from the National Institutes of Health.

[b] Address for correspondence: 4-225 Millard Hall, Department of Biochemistry, Minneapolis, Minnesota 55455.

We have recently explored the cellular site of action of T_3 in mediating changes in spot 14 mRNA. The gene-encoding spot 14 mRNA is present in a single copy per haploid genome in the rat and contains only one intervening sequence that interrupts the gene in the 3′-untranslated region.[10] The nuclear precursor to spot 14 was detected as a 4700-nucleotide RNA species that increased in relative concentration as early as 10 min after T_3 treatment of hypothyroid rats, prior to the earliest detectable rise in mature mRNA.[11] The levels of spot 14 nuclear precursor RNA changed proportionally to mature mRNA levels in chronically treated rats of varying thyroidal states.[12] Together these data indicate that regulation occurs in the nucleus at a step preceding accumulation of the nuclear precursor RNA. This nuclear site of action is consistent with the nuclear localization of the thyroid hormone receptor.[13] Somewhat surprisingly, however, measurements of the relative rates of spot 14 gene transcription by the nuclear run-on assay indicated that transcription was not the major site of control.[12] The spot 14 gene was actively transcribed in hypothyroid animals. Injection of T_3 led to a transient 2- to 3-fold increase in the relative rate of transcription, which was not sustained at longer times of treatment. In euthyroid or hyperthyroid animals the transcription rate was only 1.5-fold that of hypothyroid rats. This degree of change could not account for the much greater changes observed in both nuclear precursor and mature mRNA levels. These data led us to suggest that regulation of this particular gene product by T_3 occurs primarily at a posttranscriptional step involving stabilization of the nuclear precursor to spot 14 mRNA.

Because of the somewhat novel nature of this regulation, we were interested in further exploring the control of spot 14 gene expression. In particular, we have now asked whether the dietary regulation of spot 14 mRNA occurs by a similar mechanism to that observed for T_3. In addition, we have examined the regulation of several other genes responsive to both T_3 and CHO diet to determine whether posttranscriptional control is a unique mechanism for spot 14 or a more commonly employed form of control by these two stimuli.

REGULATION OF SPOT 14 GENE EXPRESSION BY CARBOHYDRATE

Male Sprague-Dawley rats weighing 200 to 250 g were either maintained on standard laboratory chow (Ralston-Purina) or switched to the CHO diet (ICN Biochemicals) for 7 days. There are two major differences between these diets. First of all, the CHO diet contains no fat, as opposed to 22% fat by weight in chow diet. Second, the CHO diet has all of its carbohydrate (58% by weight) in a simple, readily metabolizable form (that is, sucrose). These diets were fed to groups of hypothyroid, euthyroid (untreated), and hyperthyroid rats. Hypothyroidism was achieved by addition of 0.025% (w/v) methimazole to the drinking H_2O for a period of 3 to 4 weeks and was marked by a complete cessation of growth; hyperthyroidism was achieved by intraperitoneal injection of 20 μg T_3 per 100 g body weight per day for 7 days. All animals were given free access to food and water and were killed 3 hr into the 12-hr light cycle.

After appropriate dietary and hormonal treatment, livers were removed and divided into three portions for analysis of spot 14 mRNA levels, nuclear precursor levels, and rates of gene transcription. Spot 14 nuclear precursor levels were measured using an RNA dot blot assay. Aliquots containing from 1 to 4 μg of total nuclear RNA, extracted as previously described,[11] were fixed onto a nitrocellulose filter. These filters were then hybridized to a ^{32}P-labeled probe corresponding to a portion of the single intervening sequence of the spot 14 gene. This 850-bp probe contained no repetitive DNA elements and hybridized only to the 4700-nucleotide precursor RNA to spot 14 mRNA on electrophoretic analysis of nuclear RNA.[14] It is, thus, specific for the unspliced form of spot 14 RNA. Similarly, relative levels of mature spot 14 mRNA were assessed by an RNA dot blot assay using total cellular RNA and a 450-bp cDNA probe to the spot 14 mRNA sequence.[11]

A comparison of the degree of change observed in spot 14 nuclear precursor and mature mRNA levels is shown in TABLE 1. Feeding of the high-carbohydrate diet resulted in a dramatic increase in spot 14 mRNA levels in each of the three thyroid states tested. The extent of change is particularly marked in hypothyroid animals, indicating that the presence of thyroid hormone is not necessary for induction. The degree of change in mRNA levels found in this study was almost identical to that reported in a previous study in which surgical thyroidectomy was used to achieve hypothyroidism. Thus, the methimazole-treated rat appears to be a comparable model for inducing hypothyroidism, at least with respect to this gene product.

The relative levels of spot 14 nuclear precursor RNA were also increased by feeding the CHO diet (TABLE 1). These increases were generally in proportion to the increases observed for the mature mRNA of spot 14. Hence, the induction of spot 14 mRNA levels in response to diet and that seen in response to T_3 appear to be due to some nuclear event leading to an increased accumulation of nuclear RNA. This observation eliminates changes in the rate of RNA splicing, nuclear-to-cytoplasmic transport, and cytoplasmic mRNA turnover as potential control points.

Nuclei isolated from the same animals were used for assessing the relative rate of spot 14 gene transcription in the nuclear run-on assay.[16,17] In this assay, nuclei are incubated in the presence of ^{32}P-labeled ribonucleotides to allow labeling of nascent RNA chains. Radiolabeled nuclear RNA is then isolated and hybridized to excess unlabeled spot 14 cDNA immobilized on a nitrocellulose filter. Under the conditions of incubation, little or no RNA reinitiation or processing occurs. Thus, the extent of hybridization should be proportional to the number of RNA polymerase II molecules actively engaged in specific gene transcription at the time of nuclear isolation. Previous

TABLE 1. Relative Nuclear Precursor and mRNA Levels for Spot 14 in Various Dietary and Hormonal States

State[a]	mRNA Level[b]	Fold Induction[c]	Nuclear Precursor Level[b]	Fold Induction[c]
Hypothyroid, chow	0.8 ± 0.2	1.0	0.7 ± 0.2	1.0
Hypothyroid, CHO	8.9 ± 2.3	11.5	9.7 ± 1.6	13.8
Euthyroid, chow	4.0 ± 0.6	5.2	3.8 ± 0.6	5.5
Euthyroid, CHO	10.5 ± 2.2	13.6	6.5 ± 1.0	9.3
Hyperthyroid, chow	13.1 ± 1.7	17.0	13.5 ± 4.6	19.3
Hyperthyroid, CHO	36.1 ± 2.2	46.9	35.5 ± 5.5	50.6

[a] Groups of rats, four rats to each group, were treated as described in the text.

[b] Autoradiograms from RNA dot blot assays were quantitated by videodensitometry.[15] Values are expressed as OD/μg RNA and represent means ± SEM. All values are statistically different ($p < .01$) from the hypothyroid, chow-fed group.

[c] The "fold induction" is expressed relative to the hypothyroid, chow-fed group that is normalized to 1.0.

TABLE 2. Measurement of Spot 14 Gene Transcription Rate in Response to Carbohydrate Diet and Thyroid Hormone

State	Rate of Gene Transcription[a]	Fold Induction
Hypothyroid, chow	99 ± 5	1.0
Hypothyroid, CHO	214 ± 32	2.2
Euthyroid, chow	170 ± 9	1.7
Euthyroid, CHO	149 ± 17	1.5
Hyperthyroid, chow	174 ± 14	1.8
Hyperthyroid, CHO	369 ± 28	3.7

[a] Each result is expressed as the mean ppm hybridized ± SEM. Parts per million hybridized was determined from the following formula: (cpm hybridized to spot 14 cDNA − cpm bound to pBR322 DNA)/(total input cpm $\times 10^{-6} \times$ hybridization efficiency). Spot 14 cDNA probes used were pS14-c1 and pS14-c2 (1 μg each per filter), which together represent 1050 bp of the spot 14 mRNA.[10] Typical input cpm ranged from 4 to 8 $\times 10^6$ and typical backgrounds from 60 to 80 cpm. The efficiency of hybridization varied from 0.40 to 0.60 between individual samples, but did not vary systematically between experimental groups.

work from this laboratory has demonstrated the validity of this assay for spot 14 gene transcription.[12] Thus, synthesis of RNA capable of hybridizing to spot 14 cDNA was inhibited greater than 90% by the addition of 1 μg/ml of α-amanitin, indicating transcription is being carried out by RNA polymerase II. Furthermore, the labeled nuclear RNA only hybridized to the template strand of the spot 14 gene; no hybridization to the nontemplate strand was detectable. Hybridization was linearly related to input RNA, and background hybridization assessed using pBR322 DNA or a probe to the rat insulin gene was less than 10 ppm. We have, therefore, concluded that this assay is valid for quantitating the level of spot 14 transcripts in the radiolabeled nuclear RNA.

As shown in TABLE 2, minor changes in the rate of spot 14 gene transcription can be seen following either hormonal or dietary stimulation. The 1.7-fold change from hypothyroid, chow-fed animals to euthyroid, chow-fed animals was statistically significant, but no further change was seen in the transition between euthyroid, chow-fed and hyperthyroid, chow-fed animals. These results are in fairly good agreement with our previous observations.[12] Switching from the chow diet to the CHO diet led to a 2.1- and 2.0-fold further increase of spot 14 gene transcription in hypothyroid and hyperthyroid rats, respectively. The transcription rate of the spot 14 gene in euthyroid rats was approximately the same on either chow or CHO diet. Thus, as we concluded earlier for T_3 induction of spot 14 mRNA, the induction by CHO diet appears to be due to at least two factors. First, a small change occurs in the rate of gene transcription. As summarized in FIGURE 1, these changes can only partially account for the overall extent of induction seen for spot 14 nuclear precursor and mRNA levels. For example, in going from the hypothyroid, chow-fed animals to the hyperthyroid, CHO-fed rat, a 45- to 50-fold increase in nuclear precursor and mRNA levels occurs. This is accompanied by only a 3.7-fold increase in the rate of gene transcription. Thus, the major site of regulation for CHO diet, as well as for T_3, appears to be posttranscriptional. Coupled with our earlier conclusion that regulation must precede accumulation of the nuclear precursor, these data suggest that the control of spot 14 gene expression by both T_3 and CHO diet occurs largely at the level of nuclear precursor RNA stability.

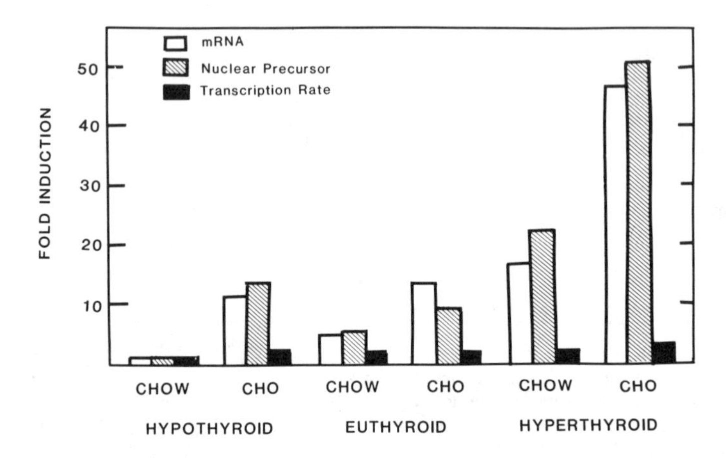

FIGURE 1. Summary of the effects of a high-carbohydrate, fat-free diet and T_3 on spot 14 gene expression. The "fold induction" is expressed relative to the value of the hypothyroid, chow-fed animals that was normalized to 1. For experimental details, see TABLES 1 & 2.

To further explore the effect of carbohydrate on spot 14 gene expression, we have examined the effect of force-feeding normal animals with sucrose. Recently, Mariash *et al.* have shown that spot 14 mRNA levels are induced rapidly by this treatment.[8] For these experiments, euthyroid animals were starved overnight and then fed with 1 ml of 60% (w/v) sucrose per 100 g body weight by intragastric tube. Spot 14 mRNA levels were induced within 30 min following this treatment, and, within 4 hr, had increased about 6-fold over controls gavaged with saline solution. Nuclei were isolated from these same animals for measurement of spot 14 nuclear precursor RNA levels and rates of gene transcription. As shown in FIGURE 2, the relative levels of spot 14 nuclear precursor RNA increased roughly in proportion to levels of mature mRNA. The rates of spot 14 gene transcription, however, did not change significantly during the first 2 hr of treatment. At 4 hr after sucrose gavage, a 1.6-fold increase ($p \leq .01$) in the rate of gene transcription was observed. Thus, the rate of spot 14 gene transcription following sucrose gavage does not correlate with the accumulation of spot 14 nuclear precursor and mRNA levels. As seen with chronically treated animals, it appears that posttranscriptional stabilization of nuclear precursor RNA plays a major role in regulation.

EFFECT OF T_3 AND CARBOHYDRATE ON TRANSCRIPTION OF OTHER RESPONSIVE HEPATIC GENES

Because both T_3 and CHO diet appear to regulate spot 14 gene expression primarily at a posttranscriptional level, it was of interest to examine other hepatic genes regulated by these two effectors to test whether this control mechanism was specific to the spot 14 gene or was of wider applicability. For this purpose, we have examined the expression of three other hepatic genes. Clones A2 and B3 represent two other functionally undefined hepatic mRNA sequences selected by the same differential screening technique used for isolating spot 14 cDNA (unpublished results). The B3 cDNA hybridizes to a hepatic mRNA of 1075 nucleotides and corresponds to a translational product that we have previously designated as spot 11 (M_r: ~25 000; pI: ~6.2).[6] The A2 cDNA hybridizes to a hepatic mRNA of 1700 nucleotides encoding a polypeptide with an M_r of about 50 000. A cDNA clone corresponding to 6-phosphogluconate dehydrogenase (6-PGD) mRNA, a member of the lipogenic enzyme family in rat liver, was previously cloned in this laboratory.[18] The relative levels of A2, B3, and 6-PGD mRNAs were each induced by both hormonal and dietary treatments, although the extent of induction varied for each product (FIG. 3). The maximally induced levels for A2, B3, and 6-PGD mRNAs observed in hyperthyroid rats fed CHO diet were 5.8-, 9.1- and 16.1-fold, respectively, that measured in hypothyroid animals on chow diet.

The relative rates of gene transcription for each of these three genes were measured in the same experimental groups as discussed above. For each gene, the rate of transcription was inhibited greater than 90% by addition of 1 μg/ml α-amanitin to the nuclear incubation. As seen for spot 14, changes in the transcription rate could only account for a minor portion of the much greater relative changes observed in the mRNA content for each gene (FIG. 3). The greatest change seen for any of these genes was a 1.8-fold increase in the rate of transcription of the 6-PGD gene in the hyperthyroid rats fed CHO diet. In this same group, relative mRNA levels increased 16.1-fold from the hypothyroid, chow-fed group. Thus, again the data indicate that posttranscriptional regulation plays a major role in control. For these genes, we have not quantitated relative nuclear precursor RNA levels and therefore cannot distinguish between nuclear and cytoplasmic sites of control.

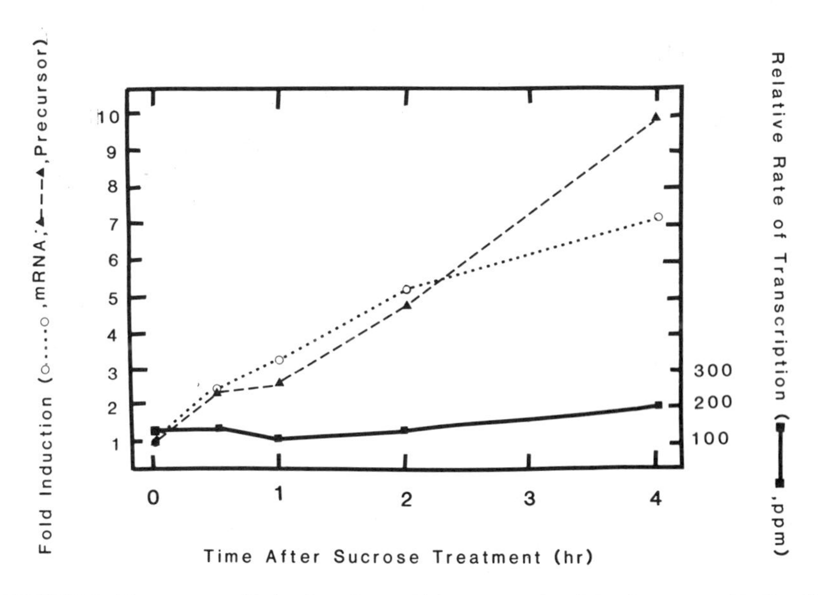

FIGURE 2. Time course of induction of spot 14 gene expression following sucrose feeding. The "fold induction" is expressed relative to the value of euthyroid rats subjected to an overnight fast and then gavaged with normal saline for 4 hr. The relative rate of gene transcription is expressed in ppm, as described in TABLE 2. Each value represents a mean of four values, one value from each of four animals in a group.

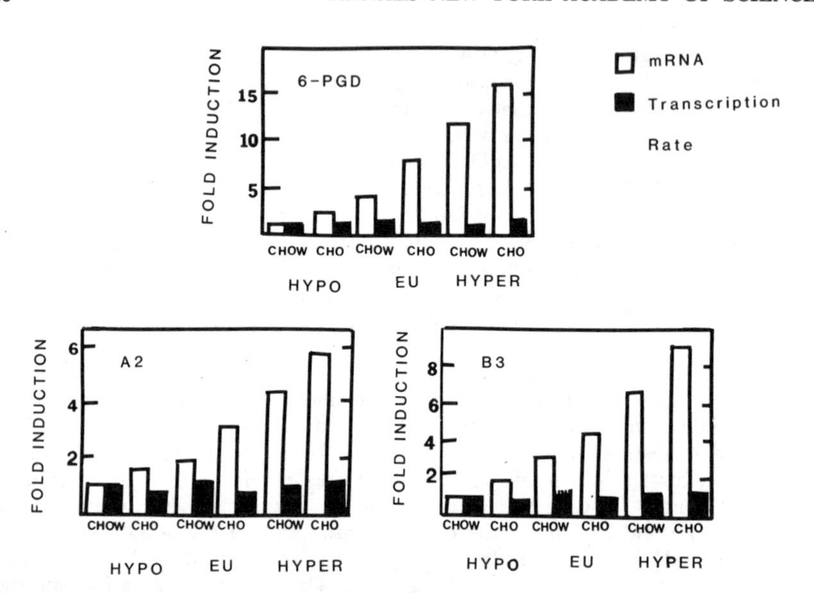

FIGURE 3. Effects of the CHO diet and T_3 on the expression of responsive hepatic genes. Relative levels of mRNA and rates of gene transcription were measured as described in the text. For the A2, B3, and 6-PGD genes, the relative rates of transcription were 45, 86, and 31 ppm, respectively, in hypothyroid, chow-fed rats. The cDNA inserts comprised 1150 (A2), 450 (B3), and 880 (6-PGD) bp, and no correction was made for the cDNA size relative to the size of the genes. All results are expressed relative to the hypothyroid, chow-fed animals and are the average of results from at least four animals per group.

By contrast, we have examined the regulation of the protein, α_{2U}-globulin. The expression of this multigene family in liver is dependent on the simultaneous presence of androgen, glucocorticoid, and thyroid and growth hormones in the rat.[19] Thus, in the transition between hypothyroidism and euthyroidism, the relative levels of α_{2U}-globulin mRNA increase over 50-fold (FIG. 4). This increase is most likely due in large part to the increased plasma levels of growth hormone, the production of which is highly dependent on the presence of T_3.[20,21] In hyperthyroid animals or animals fed CHO diet, α_{2U}-globulin mRNA levels are the same or slightly less than euthyroid, chow-fed animals. Recently, Kulkarni *et al.* have reported that the hormonal regulation of this gene is largely due to increased rates of transcription.[22] We have measured the rate of transcription of the α_{2U}-globulin genes in hypothyroid, euthyroid, and hyperthyroid rats fed the chow diet (FIG. 4). In this study, hypothyroidism was obtained by surgical thyroidectomy. Methimazole treatment only slightly depressed the mRNA levels for α_{2U}-globulin. The nature of the difference between these two treatments is not understood, but may be related to the multihormonal control of this gene. The relative rate of transcription of the α_{2U}-globulin genes was found to increase 9- to 10-fold in the transition from hypothyroid to either euthyroid or hyperthyroid states. A comparable result was obtained in animals maintained on the CHO diet (data not shown). Thus, for these genes, changes in the transcription rate were able to account for a major portion of the alterations in mRNA levels. Nevertheless, the 10-fold increase in the transcription rate could still not fully account for the greater than 50-fold change in mRNA levels. Thus, posttranscriptional mechanisms also may be operative in the control of this gene family.

CONCLUSIONS

For several hepatic genes regulated by thyroid hormone and dietary carbohydrate, the accumulation of mRNA following hormonal or nutritional stimuli did not correlate well with the rates of gene transcription as measured by the nuclear run-on assay. There are two possible explanations for these results. The first possibility is that the nuclear run-on assay is not accurately assessing the *in vivo* rate of gene transcription. For example, it is conceivable that a factor that normally inhibits transcription of these genes is lost during isolation of the rat liver nuclei. In this case, we would be measuring artificially high levels of gene transcription *in vitro* regardless of the *in vivo* transcription rate. To rule out this possibility, it will be necessary to measure transcription by an independent method, such as direct labeling of RNA with radiolabeled precursors. Although this approach is not feasible in whole animals, it could be attempted in cultured cells. Thus, a major thrust of future research in this laboratory will be to develop a cell system for studying regulation by thyroid hormone and carbohydrate. It should be mentioned, however, that results using the nuclear run-on assay have been shown to correlate well with the direct labeling of RNA to measure transcription rates in those studies in which the two methods have been directly compared.[16,17,23] In addition, for α_{2U}-globulin a change in the rate of gene transcription was detected in this study. Thus, the methods used for nuclear isolation and incubation appear to be capable of detecting bona fide changes in gene transcription at least for these genes.

The second possible explanation for the results is that the nuclear run-on assay is accurately reflecting the *in vivo* rate of gene transcription. In this case, the data suggest that thyroid hormone and dietary carbohydrate are acting similarly on several

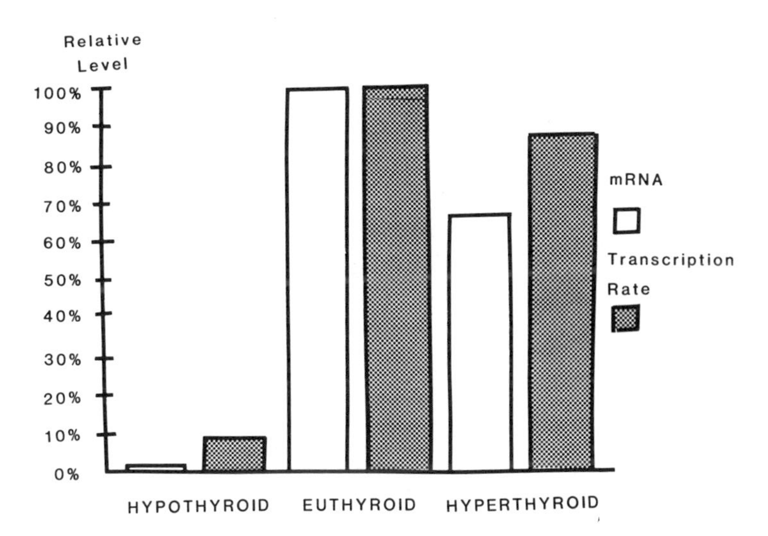

FIGURE 4. Effect of thyroid state on α_{2U}-globulin gene expression. Relative levels of mRNA and rates of gene transcription were measured as described in the text. Hypothyroidism was induced by surgical thyroidectomy and ^{131}I treatment. Results are expressed relative to the euthyroid, chow-fed group that was normalized to 100%. The relative rate of gene transcription was 703 ± 31 ppm in this group.

genes responsive to these stimuli, and that this regulation occurs at two distinct steps. One step is the rate of gene transcription. For each of the genes we have examined, a significant change in the rate of gene transcription was observed following hormonal and dietary treatment. In each case, however, this change could only account for a minor portion of the overall induction of mRNA. Thus, the major site of control would be posttranscriptional. For the A2, B3, and 6-PGD genes, we cannot distinguish whether this posttranscriptional control occurred at the level of nuclear or cytoplasmic RNA because nuclear RNA levels were not measured. For spot 14 gene expression, however, the control of gene expression would have to occur at a nuclear step preceding accumulation of the nuclear precursor. Thus, in the presence of T_3 or dietary carbohydrate, the spot 14 gene transcript appears to be more efficiently channeled into the pathway for processing and transport than in the absence of these stimuli.

It has been suggested for many years that only a portion of all nuclear transcripts are preserved and transported to the cytoplasm.[24] We would now suggest that this step may serve as a potential regulatory event in determining the levels of cytoplasmic spot 14 mRNA. Recently a few examples of regulation at this step in mRNA production have been presented, including the regulation of α_1-acid glycoprotein by glucocorticoid hormone,[25] prostatic binding protein by androgen,[26] and dihydrofolate reductase by growth state.[23,27] In each of these systems, RNA stabilization occurs at the level of the spliced, mature nuclear RNA species. For spot 14 gene expression, however, our data suggest that the control occurs at the level of the unspliced nuclear precursor RNA.

The mechanism by which the hormonal and dietary stimuli might act to alter stability of the spot 14 nuclear precursor is unclear. One possibility is that the effect is secondary to the induction of an intermediate protein stimulated via a transcriptional route. The extremely rapid kinetics of induction of the spot 14 nuclear precursor would seem to make this possibility unlikely. A second possibility is that these effectors might interact directly with the initial gene transcript to stabilize it. Thus, the T_3 receptor might bind to the spot 14 nuclear precursor to channel it into the appropriate pathway for processing and transport. No evidence of a direct binding of the T_3 receptor with nuclear RNA, however, has been reported. A third possibility is that RNA transcription and processing are directly coupled in the nucleus. In the absence of T_3 or carbohydrate, the spot 14 gene is actively transcribed, but the transcripts are not channeled into the processing pathway and are subject to rapid degradation. The action of T_3 or the carbohydrate-stimulated effector might alter the nuclear localization of the spot 14 gene into a domain more readily serviced by this processing pathway. Such a mechanism might account for the simultaneous effects on both the rates of gene transcription and nuclear RNA stability. Clearly much more work is necessary to explore these possibilities. The spot 14 gene would seem to be an advantageous model system in which to further study this interesting form of regulation.

ACKNOWLEDGMENTS

We wish to thank Julie Engle for valuable technical assistance, Dr. Cary Mariash for help in the sucrose gavage experiment, and Dr. Steve Seelig for the α_{2U}-globulin cDNA probe.

REFERENCES

1. MARIASH, C. N., F. E. KAISER, H. L. SCHWARTZ, H. C. TOWLE & J. H. OPPENHEIMER. 1980. Synergism of thyroid hormone and high-carbohydrate diet in the induction of lipogenic enzymes in the rat. J. Clin. Invest. **65:** 1126-1134.
2. TOWLE, H. C., C. N. MARIASH & J. H. OPPENHEIMER. 1980. Changes in the hepatic levels of messenger ribonucleic acid for malic enzyme during induction by thyroid hormone or diet. Biochemistry **19:** 579-585.
3. MAGNUSON, M. A. & V. A. NIKODEM. 1983. Molecular cloning of a cDNA sequence for rat malic enzyme. J. Biol. Chem. **258:** 1337-1342.
4. MARIASH, C. N., C. R. MCSWIGAN, H. C. TOWLE, H. L. SCHWARTZ & J. H. OPPENHEIMER. 1981. Glucose and triiodothyronine both induce malic enzyme in the rat hepatocyte culture. J. Clin. Invest. **68:** 1485-1490.
5. LIAW, C., S. SEELIG, C. N. MARIASH, J. H. OPPENHEIMER & H. C. TOWLE. 1983. Interactions of thyroid hormone, growth hormone, and high-carbohydrate, fat-free diet in regulating several rat liver messenger ribonucleic acid species. Biochemistry **22:** 213-222.
6. SEELIG, S., C. LIAW, H. C. TOWLE & J. H. OPPENHEIMER. 1981. Thyroid hormone attenuates and augments hepatic gene expression at a pretranslational level. Proc. Natl. Acad. Sci. USA **78:** 4733-4737.
7. JUMP, D. B., P. NARAYAN, H. TOWLE & J. H. OPPENHEIMER. 1984. Rapid effects of triiodothyronine on hepatic gene expression. J. Biol. Chem. **259:** 2789-2797.
8. MARIASH, C. N., S. SEELIG, H. L. SCHWARTZ & J. H. OPPENHEIMER. 1986. Rapid synergistic interaction between thyroid hormone and carbohydrate on mRNA-S14 induction. J. Biol. Chem. In press.
9. MARIASH, C. N., D. B. JUMP & J. H. OPPENHEIMER. 1984. T_3 stimulates the synthesis of a specific mRNA in primary hepatocyte culture. Biochem. Biophys. Res. Commun. **123:** 1122-1129.
10. LIAW, C. W. & H. C. TOWLE. 1984. Characterization of a thyroid hormone-responsive gene from rat. J. Biol. Chem. **259:** 7253-7260.
11. NARAYAN, P., C. W. LIAW & H. C. TOWLE. 1984. Rapid induction of a specific nuclear mRNA precursor by thyroid hormone. Proc. Natl. Acad. Sci. USA **81:** 4687-4691.
12. NARAYAN, P. & H. C. TOWLE. 1985. Stabilization of a specific nuclear mRNA precursor by thyroid hormone. Mol. Cell. Biol. **5:** 2642-2646.
13. OPPENHEIMER, J. H. 1979. Thyroid hormone action at the cellular level. Science **203:** 971-979.
14. TOWLE, H. C., P. NARAYAN, C. W. LIAW, J. A. ENGLE & T. TAO. 1985. Nuclear regulation of a thyroid hormone-responsive gene at a posttranscriptional level. *In* Nuclear Envelope Structure and RNA Maturation. E. A. Smuckler & G. A. Clawson, Eds.: 411-426. Alan R. Liss. New York, NY.
15. MARIASH, C. N., S. SEELIG & J. H. OPPENHEIMER. 1982. A rapid, inexpensive, quantitative technique for the analysis of two-dimensional electrophoretograms. Anal. Biochem. **121:** 388-394.
16. MCKNIGHT, G. S. & R. D. PALMITER. 1979. Transcriptional regulation of the ovalbumin and conalbumin genes by steroid hormones in chick oviduct. J. Biol. Chem. **254:** 9050-9058.
17. DERMAN, E., K. KRAUTER, L. WALLING, C. WEINBERGER, M. RAY & J. E. DARNELL. 1981. Transcriptional control in the production of liver-specific mRNAs. Cell **23:** 731-739.
18. MIKSICEK, R. J. & H. C. TOWLE. 1983. Use of a cloned cDNA sequence to measure changes in 6-phosphogluconate dehydrogenase mRNA levels caused by thyroid hormone and dietary carbohydrate. J. Biol. Chem. **258:** 9575-9579.
19. KURTZ, D. T. & P. FEIGELSON. 1978. Multihormonal control of the messenger RNA for the hepatic protein α_{2U}-globulin. *In* Biochemical Actions of Hormones. G. Litwack, Ed. Vol. **5:** 433-455. Academic Press. New York, NY.
20. ROY, A. K., B. CHATTERJEE, W. F. DEMYAN, T. S. NATH & N. M. MOTWANI. 1982.

Pretranslational regulation of α_{2U}-globulin in rat liver by growth hormone. J. Biol. Chem. **257:** 7834-7838.

21. LYNCH, K. R., K. P. DOLAN, H. L. NAKHASI, R. UNTERMAN & P. FEIGELSON. 1982. The role of growth hormone in α_{2U}-globulin synthesis: A reexamination. Cell **28:** 185-189.
22. KULKARNI, A. B., R. M. GUBITS & P. FEIGELSON. 1985. Developmental and hormonal regulation of α_{2U}-globulin gene transcription. Proc. Natl. Acad. Sci. USA **82:** 2579-2582.
23. LEYS, E. J., G. F. CROUSE & R. E. KELLEMS. 1984. Dihydrofolate reductase gene expression in cultured mouse cells is regulated by transcript stabilization in the nucleus. J. Cell Biol. **99:** 180-187.
24. DARNELL, J. E. 1979. Transcription units for mRNA production in eukaryotic cells and their DNA viruses. Prog. Nucleic Acid Res. Mol. Biol. **22:** 327-353.
25. VANNICE, J. L., J. M. TAYLOR & G. M. RINGOLD. 1984. Glucocorticoid-mediated induction of α_1-acid glycoprotein: Evidence for hormone-regulated RNA processing. Proc. Natl. Acad. Sci. USA **81:** 4241-4245.
26. PAGE, M. J. & M. G. PARKER. 1982. Effect of androgen on the transcription of rat prostatic-binding protein genes. Mol. Cell. Endocrinol. **27:** 343-355.
27. LEYS, E. J. & R. E. KELLEMS. 1981. Control of dihydrofolate reductase messenger ribonucleic acid production. Mol. Cell. Biol. **1:** 961-971.

Differential Expression of the Genes for the Mitochondrial and Cytosolic Forms of Phosphoenolpyruvate Carboxykinase[a]

YAACOV HOD, JONATHAN S. COOK,[b]
SHARON L. WELDON, JAY M. SHORT,[c] ANTHONY
WYNSHAW-BORIS,[d] AND RICHARD W. HANSON

Department of Biochemistry
School of Medicine
Case Western Reserve University
Cleveland, Ohio 44106

INTRODUCTION

In keeping with the theme of this volume on the application of recombinant DNA techniques to metabolism, we will present a brief review of the role of the isozymes of phosphoenolpyruvate carboxykinase (PEPCK) in the regulation of gluconeogenesis. We will attempt to integrate our current knowledge of both the metabolism and molecular biology of this enzyme, emphasizing the important role of PEPCK as a model for understanding the regulation of gene expression by peptide hormones. This article is dedicated to the memory of the late Merton F. Utter who discovered both PEPCK[1] and pyruvate carboxylase,[2] thereby establishing a mechanistic basis for the entry of carbon into the gluconeogenic pathway.

TWO FORMS OF PEPCK AND THEIR ROLE IN GLUCONEOGENESIS

Gluconeogenesis involves a series of reactions that requires the participation of enzymes in the mitochondria, the cytoplasm, and the microsomes of the cell. The

[a] The research carried out in our laboratory was supported in part by Grants AM 21859 and AM 24451 from the National Institutes of Health.

[b] Present address: Department of Physiological Chemistry, School of Medicine, Johns Hopkins University, Baltimore, Maryland 21205.

[c] Present address: Vector Cloning Systems, 3770 Tansy Street, San Diego, California 92121.

[d] A trainee on Metabolism Training Program AM 07319 of the National Institutes of Health.

mitochondria furnish the ATP and NADH required for gluconeogenesis, as well as phosphoenolpyruvate (PEP) or its cytosolic precursors, malate or aspartate. The reactions leading to the conversion of PEP to glucose 6-phosphate occur in the cytosol, whereas the final step in the pathway is catalyzed by glucose 6-phosphatase, an enzyme contained entirely in the endoplasmic reticulum. Previous reviews[3,4] have discussed the advantage of the compartmentation of gluconeogenesis in the regulation and coordination of this process with other metabolic pathways.

The initial reactions in gluconeogenesis involve the conversion of pyruvate to PEP and are catalyzed by two enzymes: pyruvate carboxylase (EC 6.4.1.1), which generates oxalacetate from pyruvate, and PEPCK (EC 4.1.1.32), which catalyzes the decarboxylation of oxalacetate in the presence of guanosine 5′-triphosphate or inosine 5′-triphosphate to produce PEP.[5,6] A simple reversal of the glycolytic reaction catalyzed by pyruvate kinase is thermodynamically unfavorable. Thus, the reactions of pyruvate carboxylase and PEPCK bypass this step. Oxalacetate can be utilized in several metabolic pathways, so it is PEPCK that catalyzes the first committed step in gluconeogenesis and is generally considered to be the pace-setting enzyme in the pathway.[7]

One important aspect in the intracellular compartmentation of gluconeogenesis is the presence of two isozyme forms of PEPCK, one located in the cytosol, the other in the mitochondria.[8] In most animals, including the human, cow, sheep, cat, dog, and guinea pig, the activity of the enzyme is evenly distributed between the two compartments; whereas the cytosolic enzyme is the major form in the rat, mouse, and hamster. Livers from the chicken, pigeon, and rabbit contain predominantly a mitochondrial form of the enzyme. In most species studied to date, with the exception of the chicken (see below), the intracellular distribution of the enzyme is the same in both liver and kidney, the two major gluconeogenic tissues.

Although it has been established that both forms of PEPCK are involved in glucose production,[9] the structural, functional, and genetic relationships between these two isozymes remain to be resolved. The chicken serves as an interesting model for understanding the tissue-specific development of PEPCK, since in this animal the tissue distribution of the enzyme varies from the general pattern described above. In the liver PEPCK is exclusively located within the mitochondria, whereas in the kidney the enzyme is found in both the mitochondria and cytosol.[9]

There are numerous differences in the regulation of gluconeogenesis between birds and mammals. Perhaps the most striking of these differences is the inability of avian liver to synthesize substantial amounts of glucose from pyruvate or amino acids.[10–15] Chickens maintain an elevated level of blood glucose (10 mM), which is supported by high rates of gluconeogenesis.[14] We have noted[9] that kidney tubules from the chicken can utilize both pyruvate and lactate with almost equal efficiency. In contrast, lactate is the major precursor for glucose synthesis in the liver. These differences in gluconeogenesis have been attributed, at least in part, to the unique intracellular distribution of the two forms of PEPCK in these tissues. Gluconeogenesis from pyruvate and amino acids such as alanine requires the transfer of reducing equivalents, as well as carbon, from the mitochondria to the cytoplasm to support the reductive synthesis of glucose in the cytosol. In tissues possessing a cytosolic form of PEPCK, this is accomplished by the movement of malate from the mitochondria to the cytosol,[16] followed by its conversion to oxalacetate thereby generating cytosolic NADH. In tissues lacking cytosolic PEPCK such as chicken liver, lactate is the major precursor for gluconeogenesis, and the required NADH is produced directly in the cytosol by the conversion of lactate to pyruvate. These findings suggest that the cytosolic form of PEPCK in the kidney participates in gluconeogenesis from amino acids derived from protein breakdown during starvation, whereas mitochondrial PEPCK is involved in the conversion of lactate to glucose as part of the Cori cycle.[9]

PROPERTIES OF THE MITOCHONDRIAL AND CYTOSOLIC ISOZYMES OF PEPCK

Earlier studies suggested that the cytosolic and mitochondrial forms PEPCK were identical proteins. Both enzymes have similar catalytic properties, have almost identical maximum velocities, and require the same cofactors.[17] Nowak *et al.*[18] have tested the inhibitory effect of the *E* and the *Z* isomers of phosphoenolbutyrate (substrate analogues of PEP) on the purified, hepatic enzymes from several species. Both cytosolic and mitochondrial PEPCK displayed the same stereospecificity toward the *E* and *Z* analogues, suggesting homology at the active site. The mitochondrial enzyme, however, is more susceptible to inhibition by these analogues than the cytosolic form, implying subtle differences in ligand interactions at the active sites of these two enzymes.

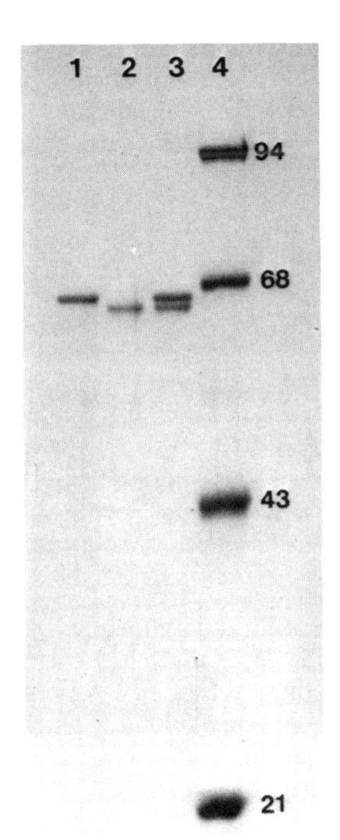

FIGURE 1. Different molecular masses of the mitochondrial and cytosolic forms of PEPCK. Mitochondrial and cytosolic forms of PEPCK from the liver and kidney of the chicken were purified to homogenity. Proteins (2 μg of each) were analyzed by electrophoresis on a sodium dodecyl sulfate-polyacrylamide (9%) gel. The Coomassie blue-stained gel is presented. Lane 1: mitochondrial PEPCK; lane 2: cytosolic PEPCK; lane 3: a mixture of both mitochondrial and cytosolic PEPCK; lane 4: molecular weight markers. The numbers denote the molecular mass $\times 10^{-3}$ of the proteins.

There are significant structural differences between cytosolic and mitochondrial PEPCK. Analysis of these two proteins by sodium dodecyl sulfate-polyacrylamide gel electrophoresis (FIG. 1) demonstrates that the two forms of PEPCK, both purified from the chicken, are not of identical size. There is also a difference in the isoelectric points of the two enzymes since it is possible to separate them by both ion-exchange chromatography[19] and isoelectric focusing.[20] The most convincing evidence that there are structural differences between the two forms of PEPCK is that they do not cross-react with the antibody to the counterpart enzyme. Earlier studies have shown that an antibody prepared against the cytosolic form of the enzyme from the rat did not interact with the mitochondrial form of PEPCK from the same species.[21] Similarly, antibody directed against the mitochondrial enzyme from chicken liver did not react with cytosolic PEPCK.[9] Similar results have been observed with the enzymes from monkey[18,22] and the enzymes from guinea pig (P.J. Markovitz and M. F. Utter, unpublished results). Some degree of cross-reactivity of the antibody, however, was found if the antigen was first denatured with sodium dodecyl sulfate and the reaction was carried out with a large excess of the counteracting antibody.[20] These results suggest that although the two forms of PEPCK are structurally different, they share some sequence homology.

To determine whether one form of the enzyme results from a posttranslational modification of the other, we have characterized the primary translational forms of the chicken PEPCK. When poly(A)$^+$ RNA isolated from the chicken liver (a tissue that contains only the mitochondrial form) was used to prime a cell-free system of protein synthesis, only one translation form of PEPCK was identified.[20] This product is approximately 1000-2000 daltons larger than the mitochondrial protein. Mitochondria isolated from the liver processed this larger translation form of PEPCK to its mature size.[20] Thus, mitochondrial PEPCK is initially synthesized as a precursor containing a signal peptide and, similar to many other mitochondrial proteins,[23] is excised when the precursor is transported across the mitochondrial membrane. In contrast, when poly(A)$^+$ RNA from the chicken kidney was used in an *in vitro* protein synthesis assay, two translation products were identified. One of the proteins behaved identically to the precursor for the mitochondrial form, whereas the second co-purified with PEPCK isolated from the cytosol fraction of chicken kidneys.[20] These results strongly indicate that the mitochondrial and cytosolic forms of PEPCK are synthesized independently of each other. This conclusion is further supported by the fact that mRNA species of different sizes code for the two forms of PEPCK in the chicken.[20]

STRUCTURAL DIFFERENCES BETWEEN THE CYTOSOLIC AND MITOCHONDRIAL FORMS OF PEPCK

To further elucidate the relationship between the cytosolic and mitochondrial forms of PEPCK, a cDNA library was constructed using chicken kidney RNA as a template.[24] Plasmids carrying sequences complementary to the mRNA for cytosolic PEPCK were identified by hybrid-selected translation.[24] Two such clones, which together span over 90% of the 2.8-kb mRNA, have been sequenced.[25] The remaining nucleotide sequences were obtained through sequencing the appropriate regions in the gene encoding the enzyme.[26] From the complete nucleotide sequence, we have deduced the amino acid structure of cytosolic PEPCK from the chicken.[25] This protein has a molecular mass of 69 522 daltons and contains 622 amino acids. Based on the amino acid sequence, we can predict that 36% of the enzyme's structure is organized as an α helix; 18%, as a β sheet; and 46%, as a random coil.[25] Putative binding domains in the enzyme for guanosine 5′-triphosphate (or inosine 5′-triphosphate) and PEP were postulated based on comparison with other proteins that interact with these cofactors.[25]

In our attempt to clone the mitochondrial form of PEPCK, we have sequenced several tryptic peptides derived from the protein. Sequence comparison shows an average of 56% homology between the 136 amino acids available for the mitochondrial protein and cytosolic PEPCK (FIG. 2). Of the amino acids between positions 92 to 110, only 8 out of 19 have been conserved; however, the amino acids between positions

448 and 467 are almost identical to those in the mitochondrial enzyme. It is evident from this analysis that the cytosolic and mitochondrial forms of PEPCK, although related proteins, are encoded by distinct genes. Because the two proteins are not structurally identical, the cytosolic and mitochondrial forms of PEPCK may offer a system to study the relationship between structure and function in two structurally different enzymes that catalyze the same reaction.

In contrast to the structural differences between the cytosolic and mitochondrial forms of the enzyme from the same species, there are many similarities between the cytosolic isozymes from different species. The cDNA for cytosolic PEPCK from the rat[27] hybridizes to the same size (2.8 kb) mRNA for the enzyme from such diverse animals as the frog, chicken, and human,[24,28] indicating conservation of nucleotide sequences. Direct comparison between the amino acid sequences of the enzyme from the chicken[25] and the rat[29] demonstrates an 85% conservation of the amino acid sequence between the two proteins. Both proteins are composed of 622 amino acids, and more than 58% of the 96 nonconserved amino acids are localized at the termini of both proteins. Thus, there are only minor changes in the primary structure of PEPCK from the chicken and the rat.

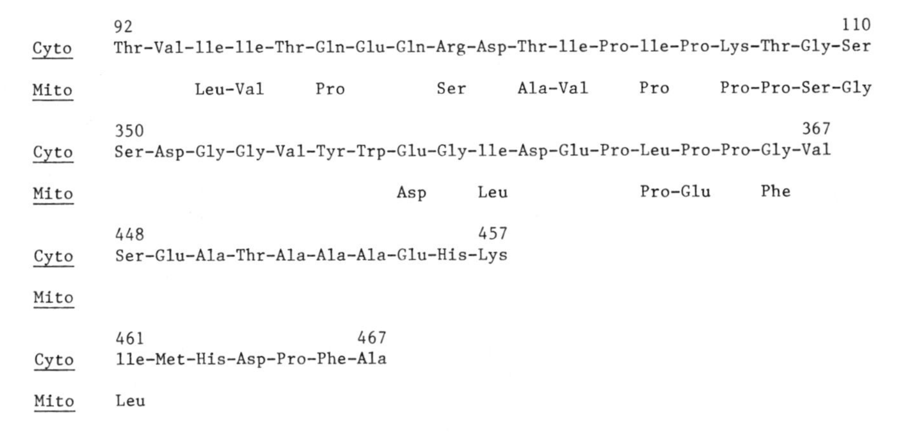

FIGURE 2. Comparison between the amino acid sequences of the mitochondrial and cytosolic forms of PEPCK from the chicken. The amino acid sequence for the cytosolic enzyme was deduced from the nucleotide sequence of the cDNA or the gene for this enzyme. The amino acid sequence for the mitochondrial enzyme was determined by Edman degradation of HPLC-isolated tryptic peptides of the purified protein. Cyto refers to the cytosolic form of PEPCK and Mito to the mitochondrial form of PEPCK.

REGULATION OF PEPCK GENE EXPRESSION

The significant homology in amino acid sequence between the mitochondrial and cytosolic PEPCK suggests that the two proteins are encoded by separate but related genes. The manner in which the expression of these two genes is controlled, however, is markedly different. The activity of the cytosolic form of the enzyme in rat liver is significantly increased during starvation and diabetes.[30] The synthesis of this isozyme of PEPCK is subject to complex regulation by a variety of hormones.[31] Hormones that will enhance the synthesis of the enzyme include glucagon and epinephrine (both acting through the intracellular mediation of cAMP); glucocorticoids; and the thyroid hormones.[30–38] On the other hand, insulin will counteract the effect of these hormones, resulting in a rapid deinduction of enzyme synthesis.[39–41] Little information is available concerning the hormonal control of gene expression for the mitochondrial enzyme. Most studies on the regulation of PEPCK synthesis have used the rat, a species that has only marginal levels of the mitochondrial enzyme (less than 10% of the total). One report indicated a lack of induction of mitochondrial PEPCK in guinea pig liver during starvation or diabetes.[42] Elliot and Pogson,[43] however, noted that starvation of guinea pigs increased the activity of both the mitochondrial and cytosolic forms of the enzyme. The reason for the difference between the two reports is not clear. In the chicken, starvation does not affect the activity of the mitochondrial enzyme in either the liver or the kidney, whereas cytosolic PEPCK is induced in the kidney.[9] Thus, the mitochondrial enzyme is constitutively expressed in the chicken, and only the synthesis of the cytosolic form is hormonally regulated. Conclusions concerning the regulation of mitochondrial PEPCK, however, are based on measurements of enzyme activity, so the mechanism of regulation of this isoenzyme remains to be established.

There is also a differential expression of the two forms of PEPCK during development. Livers from fetal rats contain only the mitochondrial enzyme whereas the cytosolic form is not synthesized until birth.[44,45] Experimentally induced diabetes[46–48] or Bt_2cAMP administration to late-gestation fetuses[49] results in the appearance of the mRNA and synthesis of the cytosolic enzyme. In the chicken, where a cytosolic form of PEPCK is absent in the adult liver, there is a temporary expression of the gene for that enzyme during the period just before and immediately after hatching.[50,51] Starting at the 14th day of embryonic life through the 7th day after hatching, the activity of the mitochondrial enzyme remains constant.[50] In the chicken kidney, however, both cytosolic and mitochondrial forms of PEPCK develop in parallel from the eighth day before hatching, so that the activity of the enzyme at the time of hatching is comparable to that observed in the kidney of an adult chicken.[52]

In marked contrast to the differences in the hormonal control of gene expression between the two forms of PEPCK from the same species, there are several similarities in the regulation of the cytosolic enzymes from a wide variety of species. Because the hormonal control of cytosolic PEPCK from the rat has been reviewed in detail elsewhere,[53–55] we will limit our discussion to the effect of cAMP on PEPCK gene expression. FIGURE 3 shows a time course of induction of cytosolic PEPCK mRNA by Bt_2cAMP in a human hepatoma cell line (a kind gift from Barbara Knowles, Wistar Institute). Within 2 hr after adding Bt_2cAMP to the medium, the level of PEPCK mRNA reaches a new steady state. Kinetic analysis of the turnover time[56] shows that the half-life of the mRNA for human PEPCK is about 1 hr. The significant increase in the level of the mRNA noted within 20 min after the addition of Bt_2cAMP demonstrates the rapid effect of this cyclic nucleotide on PEPCK gene expression in hepatoma cells. Such an effect can also be demonstrated in the intact animals. For example, 40 min after the administration of Bt_2cAMP to a fed chicken,[24] there is an increase in the mRNA coding for cytosolic PEPCK in both the liver and kidney. Because this mRNA is completely absent in the liver of fed chickens, we concluded that the hepatic gene is functional and can be activated by the proper hormonal stimuli.

The most detailed analysis of hormonal control of PEPCK gene expression has been carried out using the rat. Lamers *et al.*[36] have shown that immediately following the injection of Bt_2cAMP to a re-fed rat, the marked accumulation of mature mRNA for PEPCK parallels changes in the nuclear precursors for the enzyme mRNA. These changes in PEPCK mRNA are due, at least in part, to a 7-10-fold induction in the transcription rate of the gene, which occurs as early as 15 min after Bt_2cAMP administration to the animal.[36] These results clearly establish that cAMP can induce the transcription rate of a gene in eukaryotic cells.

IDENTIFICATION OF HORMONE-RESPONSIVE ELEMENTS IN THE PEPCK GENE

A series of chimeric genes containing the promoter-regulator region of the PEPCK gene were constructed, in order to identify specific nucleotide sequences that confer cAMP responsiveness to a linked gene.

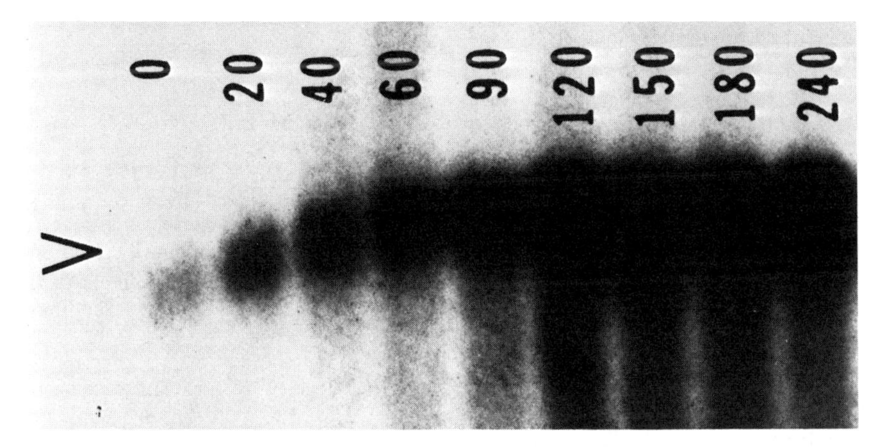

FIGURE 3. Induction of human PEPCK mRNA by Bt_2cAMP. Northern blot analysis of RNA extracted from human hepatoma cells incubated in media containing 0.5 mM Bt_2cAMP and 1 mM theophylline, for the time (in minutes) indicated at the top of each lane. ^{32}P-labeled plasmid containing a sequence complementary to the PEPCK mRNA from the rat (pPCK10rc)[27,57] was used as a probe for hybridization.

The gene for cytosolic PEPCK from the rat is approximately 6 kb in length, contains nine introns (FIG. 4),[29,57] and is a single-copy gene.[26] Recently, the complete cDNA as well as the gene have been sequenced and the amino acid sequence of the enzyme deduced.[29] The promoter-regulator region of the PEPCK gene has been localized within a 621-bp *Bam*Hl/*Bgl*II fragment (FIG. 4) present at the 5′ end of the gene.[29,58] The hormonal-control regions of the gene were studied by linking the 621-bp fragment to the structural segment of the gene encoding the herpes simplex virus thymidine kinase (TK) (lacking its own promoter) to generate the chimeric gene in the vector pPCTK-6A (FIG. 5B).[58] The 621-bp fragment of the PEPCK promoter contains 548 bp upstream from transcription start site of the PEPCK gene, as well as the first 73 bp of transcribed sequence contained in the portion of the mRNA sequence not translated into protein. This vector has been used to transfect cells of

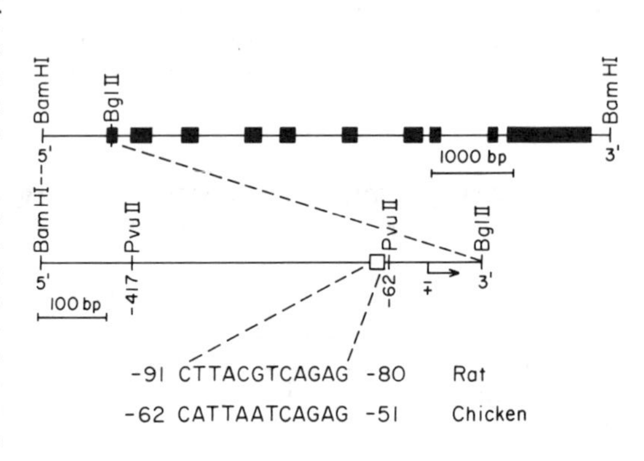

FIGURE 4. The structure of the gene for cytosolic PEPCK from the rat. Schematic representation of the rat gene coding for cytosolic PEPCK. For a detailed description see Beale *et al.*[29] and Yoo-Warren *et al.*[57] The top line shows the organization of the exons in the gene. The filled boxes represent the exons; the solid lines, the introns. The promoter region of the gene that is located between a *Bgl*II site at position +73 and a *Bam*HI site at position −548 relative to the transcription start site is presented at the bottom. A putative cAMP-regulatory sequence located at the promoter regions of the PEPCK gene from both the rat and the chicken is indicated. See the text for details.

the FTO-2B line, a TK-deficient rat hepatoma cell line[58] that is hormonally responsive and expresses the PEPCK gene. Cells containing active TK derived from pPCTK-6A have been selected in hypoxanthine, aminopterin, and thymidine (HAT) medium.[58] The addition of Bt_2cAMP to several clonally isolated cells transfected with pPCTK-6A resulted in a four- to six-fold induction of both the activity of TK and the level of its mRNA.[58] Because the parental TK gene did not respond to Bt_2cAMP, it was concluded that the promoter region contained in the *Bam*HI/*Bgl*II fragment of the PEPCK gene provided the sequences that confer cAMP responsiveness to the linked gene.[58]

In order to differentiate between promoter elements required for proper transcription and the cAMP-regulatory domain(s), chimeric genes were constructed using the complete TK gene, including its own promoter and transcription start site. A 355-bp *Pvu*II/*Pvu*II DNA fragment taken from the PEPCK and containing the region between positions −61 and −416 (FIG. 5) of the promoter was fused to the TK gene.[59] Enhancement in TK mRNA synthesis in response to Bt_2cAMP was observed when the *Pvu*II/*Pvu*II fragment was ligated either at position −109 or +1800 relative to the transcription start site of TK (which is defined as +1). The *Pvu*II/*Pvu*II fragment could also be "flipped" in the opposite orientation at any of the above sites and still confer responsiveness to Bt_2cAMP. In this chimeric gene, the *Pvu*II/*Pvu*II fragment serves only as a regulatory element because no portion of the sequence is transcribed into RNA. This demonstrates that the induction by Bt_2cAMP of TK mRNA is the result of an increase in the transcription rate of the chimeric gene and is not the result of posttranscriptional stabilization of the mRNA formed from the fused gene. Also, the positional flexibility, demonstrated by the cAMP-regulatory domain, is similar to that reported for viral enhancer elements.[60,61] Enhancer-like properties have been shown for the glucocorticoid-response elements of the murine mammary tumor virus,[62,63] the human methallothionein II gene,[64] the mouse methallothionein I gene,[65,66] the regulatory element of the β-interferon gene,[67] and the c-fos oncogene.[68]

To define the sequences in the PEPCK promoter that are required for the enhancement by cAMP of PEPCK gene transcription, a series of graded deletions through the PEPCK gene promoter-regulator region were constructed.[69] The transfected chimeric genes retained their responsiveness to cAMP as long as they included the sequence encompassing nucleotides −61 to −108 from the PEPCK promoter (FIG. 5). Deletion of this region completely eliminated cAMP inducibility. When this 47-bp fragment was inserted in front of the complete TK gene containing its own promoter, the synthesis of the TK mRNA was induced by Bt_2cAMP.[69] In contrast, the region between −109 and −416 did not confer cAMP responsiveness. However, because the 47-bp region alone did not restore the full magnitude of induction noted with the 621-bp PEPCK promoter fragment, it is possible that this region contains only the core element of the cAMP domain and that other sequences, ones closer to the 5′ end of the *Bam*HI/*Bgl*II fragment, are required to restore the full cAMP inducibility of the chimeric gene.

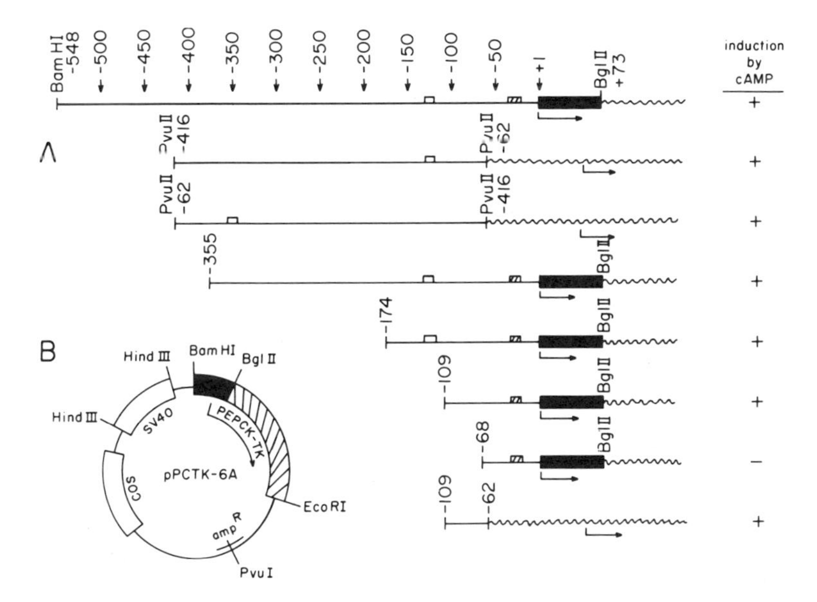

FIGURE 5. Identification of the cAMP-regulatory domain in the 5′ promoter region of the gene for cytosolic PEPCK from the rat. Various chimeric genes containing portions of the PEPCK promoter region fused to the gene encoding herpes simplex virus thymidine kinase (TK) were constructed. These chimeric genes were transfected into rat hepatoma cells (FTO-2B) to test for responsiveness to Bt_2cAMP. (**A**) Promoter regions of the PEPCK gene used in constructing chimeric genes. The solid line represents sequences of the PEPCK promoter, the filled box is part of the first 5′ exon (73 bp), and the wavy line is the TK gene. The hatched or clear boxes at the promoter region are the "TATA" and "CAAT" consensus sequences, respectively. The transcription start site and direction of transcription is indicated by an arrow, and the numbers at the top of the figure refer to the distance, in bp, from the transcription start site. TK mRNA in transfected FTO-2B cells was determined by S1 nuclease protection assay.[58-60] Enhancement in the level of transfected TK mRNA in FTO-2B cells upon incubation in the presence of 0.5 mM Bt_2cAMP and 1 mM theophylline is indicated by a plus sign. (**B**) Structure of the vector containing the PEPCK-TK chimeric gene (pPCTK-6A). The PEPCK promoter-regulator region is shown as filled box, and the TK gene is shown as a hatched box. The plasmid also contains 1 kb of a *Hind*III fragment from SV40 that contains the early and late promoters, the origin of replication, and the 72-bp enhancer repeats. COS refers to the phage λ cohesive ends, and amp refers to the β-lactamase gene of pBR322 (for details see Wynshaw-Boris *et al.*[58]).

Comparison of sequences in the promoter region of the chicken and rat PEPCK genes[26] shows two regions of over 80% homology in the first 200 nucleotides preceding the start site of mRNA synthesis. The orientation of these sequences relative to the transcription start site, as well as the distance separating them, are similar in both the chicken and the rat PEPCK genes, suggesting a possible role of these sequences in controlling PEPCK gene expression. One of these regions contains a sequence, CTTACGTCAGAG (FIG. 4), that is also conserved among other cAMP-regulated genes such as the rat preprosomatostatin,[70] porcine plasminogen activator,[71] and human vasoactive intestinal peptide.[72] This sequence, located in the rat PEPCK gene between positions −80 and −91, has been suggested as the core element in the cAMP-regulatory domain.[69]

POSSIBLE MECHANISMS OF cAMP STIMULATION OF PEPCK GENE EXPRESSION

DNA sequences of the promoter-regulator region of the PEPCK gene are undoubtedly only one element in the transcription complex. In prokaryotes, interaction between the genome and sequence-specific DNA-binding proteins has been shown to play an essential part in the control of gene expression. For example, cAMP is known to be involved in catabolite regulation in which the expression of several enzymes responsible for the catabolism of lactose, arabinose, maltose, and other sugars are induced.[73,74] The main element in the catabolite regulation in prokaryotes is the catabolite gene-activator protein (CAP). This protein, when complexed with cAMP, binds to a specific domain in the promoter thereby inducing transcription of the genes controlled by that promoter.[75] CAP is a dimer composed of two identical subunits. The complete amino acid sequence of CAP has been deduced from the nucleotide sequence of the cloned CAP gene.[76,77] The NH_2-terminal region that extends from residue 1 to 135 contains the cAMP-binding activity, and the COOH-terminal domain with the DNA-binding region is between residues 136 and 210.[78,79] Comparative analysis of sequences of several prokaryotic, cAMP-regulated genes emphasizes the involvement of the consensus sequence AA-TGTGA--T----C in binding cAMP-activated CAP.[80] The mechanism by which this interaction enhances transcription is not known. Two models, however, have been proposed.[75] First, cAMP-activated CAP could induce change(s) in the conformation of the DNA, thereby enhancing the rate of active initiation complex formation. These induced changes in DNA structure may also increase the fractional occupancy of the *lac* main promoter (PI) by RNA polymerase, probably due to inhibition of polymerase binding to accessory promoters. A second possible mechanism is through a direct interaction between cAMP-activated CAP and RNA polymerase.

There is also evidence that in eukaryotes selective and accurate initiation of transcription is dependent on promoter-specific factors.[81–83] Perturbation of chromatin structure at 5′ flanking sequences has been demonstrated for several actively transcribed genes. Such "open" chromatin conformation is typified by regions of greatly increased accessibility to cleavage by endonucleases such as DNase I. Detailed analysis of the lysozyme gene from various tissues of the chicken[84] shows different patterns of sites that are hypersensitive to DNase I. Recent studies by Reshef and colleagues (personal communication) have also demonstrated sites of greatly increased sensitivity to DNase I in the PEPCK gene. One such site has been identified in a region that correlates with the cAMP-regulatory domain. Specific factors that confer nuclease hypersensitivity at the 5′ end of the chicken adult β-globin gene have been described.[85,86] The elegant studies by Wu[87–89] have identified a heat shock-dependent activating protein factor that binds to upstream control sequences in the gene for the heat shock protein (hsp) 82 of *Drosophila.* Other work[90] has shown specific modulation of the transcription of a cloned avian vitellogenin gene by estradiol-receptor complex *in vitro.* The activation of gene transcription is a complex event and requires several types of chromatin modification. For example, ubiquitin was found to be covalently linked to histone 2H[91] in transcribed *Drosophila* genes. Benvenisty *et al.*[92,93] have correlated the DNA methylation pattern of the PEPCK gene in cells that express the gene and those that do not, and have shown sequential hypomethylation during the conversion of the inactive gene in the fetal liver to the fully active gene in the adult.

The molecular mechanism and the components that mediate the cAMP regulation of gene transcription are as yet unknown.[94] The only known cAMP receptor in eukaryotes is the cAMP-dependent protein kinase. This protein exists as an inactive tetramer of a regulatory subunit dimer and two catalytic subunits. Activation of protein kinase occurs by binding cAMP to the regulatory subunits, which causes their dissociation from the catalytic subunits. The physiological role of phosphorylation of a number of important regulatory enzymes by the catalytic subunit has been established.[95,96] There is to date only limited evidence, however, for an effect of protein phosphorylation on the control of gene expression. Murdoch *et al.*[97] have found that cAMP treatment stimulates the phosphorylation of a specific nuclear protein in GH3 cells and simultaneously enhances transcription of the prolactin gene. An alternative mechanism is that the regulatory subunit of protein kinase plays a major role in the regulation of gene expression. Several earlier studies have correlated the appearance of the regulatory subunit in the nucleus with transcriptional activity.[98,99] Another supportive evidence has been brought up by Weber *et al.*,[100] who have noted significant homology between the amino acid sequences of the regulatory subunit (RII) of protein kinase type II from bovine cardiac muscle and a segment of the NH_2-terminal domain of the prokaryotic CAP (residues 30-89). Because this region contains the cAMP-binding domain, it was proposed that conservation of this sequence in both prokaryotic and eukaryotic cAMP-responsive proteins may have functional significance.[100] In bacteria the cAMP-binding region is linked to a DNA-binding domain on the same polypeptide chain; in eukaryotes it combines with the catalytic subunits of protein kinase through protein-protein interaction. A recent report by Constantinou *et al.*,[101] however, demonstrated that the rat RII from the liver can directly bind to purified DNA. Incubation of the purified, phosphorylated cAMP-linked RII with supercoiled DNA (pBR322, SV40, M13, and ϕX174) resulted in the relaxation of the superhelical DNA in a manner similar to that seen with type I topoisomerases. This activity of RII is dependent on both cAMP binding and the phosphorylation of RII and can be abolished by either reassociation of RII with the catalytic subunits of protein kinase or by antibody against RII. Recent experiments by Jungmann *et al.*[102] have demonstrated that activated RII has a higher affinity for the PEPCK promoter-regulator region than for other control DNA segments. Although it is yet to be determined whether the regulatory subunit has higher affinity for the cAMP core domain at the PEPCK promoter than to other regions of the DNA, it is also possible that two separate activities are required for gene activation by cAMP. One involves the interaction at the cAMP core domain that is restricted to this specific sequence and carried out by a protein yet to be identified. The second is a topoisomerase I-like activity, mediated by the regulatory subunit of protein kinase. Due to the central role that cAMP plays in metabolic, hormonal, and developmental regulation, resolution of the molecular mechanism of cAMP action is of great importance.

CONCLUSIONS

Despite the progress in delineating the metabolic roles of the two isozymes of PEPCK and in mapping the hormonal-regulatory elements in the gene for the cytosolic form of the enzyme, several important questions remain to be answered. The gene for mitochondrial PEPCK must be isolated and characterized and its cDNA sequenced before we can begin to understand the evolutionary relationship between the two forms of the enzyme. Of particular interest in this regard is whether the gene for one isozyme is a modified form of the other. Because the pattern of regulation of gene expression of the mitochondrial and cytosolic enzyme is very different, we assume that the promoter-regulator regions of the two genes also differ. On the other hand, we expect some sequence homology in specific exons of both genes because the partial amino acid sequence indicates identical regions in the two enzymes. Another question of interest is the structural basis for the differences in the half-lives of the mRNAs for the mitochondrial and cytosolic forms of PEPCK. The half-life of mRNA for the cytosolic enzyme is between 30 and 60 min, whereas the mRNA for the mitochondrial isozyme is far more stable. We assume that this is due, in part, to structural variations in each mRNA, and we are currently carrying out a series of deletions in the mRNA for the cytosolic enzyme in order to determine the specific sequences responsible for the remarkably short half-life of the mRNA for this isozyme. Once identified, it may be possible to construct chimeric genes including sequences that alter the half-life of a heterologous gene, thereby altering its level of expression. Finally, only by introducing a modified form of the PEPCK regulatory element from either of the isozymes into cells will it be possible to test the functional significance of the putative control regions in each of the isozyme forms.

REFERENCES

1. Utter, M. F. & K. Kurahashi. 1954. J. Biol. Chem. **207:** 787-802.
2. Utter, M. F. & D. B. Keech. 1960. J. Biol. Chem. **235:** 17-18.
3. Garber, A. J., F. J. Ballard & R. W. Hanson. 1972. *In* Metabolism and the Regulation of Metabolic Processes in Mitochondria. M. A. Mehlman & R. W. Hanson, Eds.: 109-135. Academic Press. New York, NY.
4. Arinze, I. J. & R. W. Hanson. 1980. *In* The Nature of Metabolic Control in Mammals. R. H. Herman, F. B. Stifel & R. M. Cohn, Eds.: 495-534. Plenum. New York, NY.
5. Utter, M. F. & D. B. Keech. 1963. J. Biol. Chem. **238:** 2603-2608.
6. Keech, D. B. & M. F. Utter. 1963. J. Biol. Chem. **238:** 2609-2614.
7. Rognstad, R. 1979. J. Biol. Chem. **254:** 1875-1878.
8. Soling, H.-D. & J. Kleineke. 1976. *In* Gluconeogenesis: Its Regulation in Mammalian Species. R. W. Hanson & M. A. Mehlman, Eds.: 369-462. John Wiley & Sons. New York, NY.
9. Watford, M., Y. Hod, Y.-B. Chiao, M. F. Utter & R. W. Hanson. 1981. J. Biol. Chem. **256:** 10023-10027.
10. Soling, H.-D., J. Kleineke, B. Willms, G. Janson & A. Kuhn. 1973. Eur. J. Biochem. **37:** 233-243.
11. Dickson, A. J. & D. R. Langslow. 1978. Mol. Cell. Biochem. **22:** 167-181.
12. Langslow, D. R. 1978. Biochem. Soc. Trans. **6:** 1148-1152.
13. Mapes, J. P. & H. A. Krebs. 1978. Biochem. J. **172:** 193-203.
14. Brady, L. J., D. R. Romsos, P. J. Brady, W. G. Bergen & G. A. Leveille. 1978. J. Nutr. **108:** 648-657.
15. Brady, L. J., D. R. Romsos & G. A. Leveille. 1979. Comp. Biochem. Physiol. **63B:** 193-198.

16. Williamson, J. R. 1976. *In* Gluconeogenesis: Its Regulation in Mammalian Species. R. W. Hanson & M. A. Mehlman, Eds.: 165-220. John Wiley & Sons. New York, NY.
17. Utter, M. F. & H. M. Kolenbrander. 1972. *In* The Enzymes. P. P. Boyer, Ed. Vol. **6:** 117-168. Academic Press. New York, NY.
18. Duffy, T. H., P. J. Markovitz, D. T. Chuang, M. F. Utter & T. Nowak. 1981. Proc. Natl. Acad. Sci. USA **78:** 6680-6683.
19. Ballard, F. J. 1971. Biochim. Biophys. Acta **242:** 470-472.
20. Hod, Y., M. F. Utter & R. W. Hanson. 1982. J. Biol. Chem. **257:** 13787-13794.
21. Ballard, F. J. & R. W. Hanson. 1969. J. Biol. Chem. **244:** 5625-5630.
22. Utter, M. F. & D. T. Chung. 1978. Biochem. Soc. Trans. **6:** 11-16.
23. Hay, R., P. Bohni & S. Gasser. 1984. Biochim. Biophys. Acta **779:** 65-87.
24. Hod, Y., S. M. Morris & R. W. Hanson. 1984. J. Biol. Chem. **259:** 15603-15608.
25. Cook, J. S., S. L. Weldon, R. Garcia-Ruiz, Y. Hod & R. W. Hanson. 1986. Proc. Natl. Acad. Sci. USA. In press.
26. Hod, Y., H. Yoo-Warren & R. W. Hanson. 1984. J. Biol. Chem. **259:** 15609-15614.
27. Yoo-Warren, H., M. A. Cimbala, K. Felz, J. E. Monahan, J. P. Leis & R. W. Hanson. 1981. J. Biol. Chem. **256:** 10224-10227.
28. Cohen, H., B. Gidoni, D. Shouval, N. Benvenisty, D. Mencher, O. Meyuhas & L. Reshef. 1985. FEBS Lett. **180:** 175-180.
29. Beale, E. G., N. B. Chrapkiewicz, S. A. Hubert, R. J. Metz, D. P. Quick, R. L. Noble, J. E. Donelson, K. Biemann & D. K. Granner. 1985. J. Biol. Chem. **260:** 10748-10760.
30. Shargo, E., H. A. Lardy, R. C. Nordlie & D. O. Foster. 1963. J. Biol. Chem. **238:** 3188-3192.
31. Tilghman, S. M., R. W. Hanson & F. J. Ballard. 1976. *In* Gluconeogenesis: Its Regulation in Mammalian Species. R. W. Hanson & M. A. Mehlman, Eds.: 47-91. John Wiley & Sons. New York, NY.
32. Iynedjian, P. B. & R. W. Hanson. 1977. J. Biol. Chem. **252:** 655-662.
33. Kioussis, D., L. Reshef, H. Cohen, S. M. Tilghman, P. B. Iynedjian, F. J. Ballard & R. W. Hanson. 1978. J. Biol. Chem. **253:** 4327-4332.
34. Cimabala, M. A., W. H. Lamers, K. Nelson, J. E. Monahan, H. Yoo-Warren & R. W. Hanson. 1982. J. Biol. Chem. **257:** 7629-7636.
35. Beale, E. G., J. Hartley & D. Granner. 1982. J. Biol. Chem. **257:** 2022-2028.
36. Lamers, W. H., R. W. Hanson & H. M. Meisner. 1982. Proc. Natl. Acad. Sci. USA **79:** 5137-5141.
37. Meisner, H., D. S. Loose & R. W. Hanson. 1985. Biochemistry **24:** 421-425.
38. Loose, D. S., D. K. Cameron, H. P. Short & R. W. Hanson. 1985. Biochemistry **24:** 4509-4512.
39. Tilghman, S. M., R. W. Hanson, L. Reshef, M. F. Hopgood & F. J. Ballard. 1974. Proc. Natl. Acad. Sci. USA **71:** 1304-1308.
40. Granner, D., T. Andreone, K. Sasaki & E. Beale. 1983. Nature (London) **305:** 549-551.
41. Sasaki, K., T. P. Cripe, S. R. Koch, T. L. Andreone, D. D. Petersen, E. G. Beale & D. K. Granner. 1984. J. Biol. Chem. **259:** 15242-15251.
42. Nordlie, R. C., F. E. Varrichio & D. D. Holten. 1965. Biochim. Biophys. Acta **97:** 214-221.
43. Elliott, K. R. F. & C. I. Pogson. 1977. Biochem. J. **164:** 357-361.
44. Ballard, F. J. & R. W. Hanson. 1967. Biochem. J. **104:** 866-871.
45. Philippidis, H., R. W. Hanson, L. Reshef, M. F. Hopgood & F. J. Ballard. 1972. Biochem. J. **126:** 1127-1134.
46. Benvenisty, N., H. Cohen, B. Gidoni, D. Mencher, O. Meyuhas, D. Shouval & L. Reshef. 1984. *In* Lesions in Animal Diabetes. E. Shafrir & A. E. Renold, Eds.: 717-733. John Libbey & Co. London.
47. Mencher, D., D. Shouval & L. Reshef. 1979. Eur. J. Biochem. **102:** 489-495.
48. Mencher, D., H. Cohen, N. Benvenisty, O. Meyuhas & L. Reshef. 1984. Eur. J. Biochem. **141:** 199-203.
49. Hanson, R. W., L. Fisher, F. J. Ballard & L. Reshef. 1973. Enzyme **15:** 97-110.
50. Hamada, T. & M. Matsumoto. 1984. Comp. Biochem. Physiol. **77B:** 547-550.

51. FELICIOLI, R. A., F. GABRIELLI & C. A. ROSSI. 1967. Eur. J. Biochem. **3:** 19-24.
52. SHEN, C. A. & S. P. MISTRY. 1979. Poult. Sci. **58:** 663-667.
53. WYNSHAW-BORIS, A. & R. W. HANSON. 1983. Horizons Biochem. Biophys. **7:** 171-203.
54. LOOSE, D. S., A. WYNSHAW-BORIS, H. M. MEISNER, Y. HOD & R. W. HANSON. 1985. *In* Molecular Basis of Insulin Action. M. P. Czech, Ed.: 347-368. Plenum. New York, NY.
55. GRANNER, D. & E. BEALE. 1985. *In* Biochemical Actions of the Hormones. G. Litwack, Ed. Vol. **12:** 89-138. Academic Press. New York, NY.
56. SCHIMKE, R. T. & D. DOYLE. 1970. Annu. Rev. Biochem. **39:** 929-976.
57. YOO-WARREN, H., J. MONAHAN, J. SHORT, H. SHORT, A. BRUZEL, A. WYNSHAW-BORIS, H. MEISNER, D. SAMOLS & R. W. HANSON. 1983. Proc. Natl. Acad. Sci. USA **80:** 3656-3660.
58. WYNSHAW-BORIS, A., T. G. LUGO, J. M. SHORT, R. E. K. FOURNIER & R. W. HANSON. 1984. J. Biol. Chem. **259:** 12161-12170.
59. WYNSHAW-BORIS, A., J. M. SHORT, D. S. LOOSE & R. W. HANSON. 1986. J. Biol. Chem. **261:** 9714-9720.
60. KHOURY, G. & P. GRUSS. 1983. Cell **33:** 313-314.
61. GRUSS, P. 1984. DNA **3:** 1-5.
62. CHANDLER, V. L., B. A. MALER & K. R. YAMAMOTO. 1983. Cell **33:** 489-499.
63. PONTA, K., N. KENNEDY, P. SKROCH, N. HYNES & B. GRONER. 1985. Proc. Natl. Acad. Sci. USA **82:** 1020-1024.
64. KARIN, M., A. HASLINGER, H. HOLTGREVE, G. CATHALA, E. SLATER & R. D. PLAMITER. 1984. Cell **36:** 371-379.
65. STUART, G. W., P. F. SEARLE, H. Y. CHEN, R. L. BRINSTER & R. D. PALMITER. 1984. Proc. Natl. Acad. Sci. USA **81:** 7318-7322.
66. SEARLE, P. F., G. W. STUART & R. D. PALMITER. 1985. Mol. Cell. Biol. **5:** 1480-1489.
67. GOODBOURN, S., K. ZINN & T. MANIATIS. 1985. Cell **41:** 509-520.
68. TREISMAN, R. 1985. Cell **42:** 889-902.
69. SHORT, J. M., A. WYNSHAW-BORIS, H. P. SHORT & R. W. HANSON. 1986. J. Biol. Chem. **261:** 9721-9726.
70. MONTMINY, M. R., R. H. GOODMAN, S. J. HOROVITCH & J. F. HABENER. 1984. Proc. Natl. Acad. Sci. USA **81:** 3337-3340.
71. NAAMINE, Y., D. PEARSON, M. S. ALTUS & E. REICH. 1984. Nucleic Acids Res. **12:** 9525-9541.
72. TSUKADA, T., S. J. HOROVITCH, M. R. MONTMINY, G. MANDEL & R. H. GOODMAN. 1985. DNA **4:** 293-300.
73. PETERKOFSKY, A. 1976. Adv. Cyclic Nucleotide Res. **7:** 1-48.
74. PASTAN, I. & S. ADHYA. 1976. Bacteriol. Rev. **40:** 527-551.
75. DE CROMBRUGGHE, B., S. BUSBY & H. BUC. 1984. Science **224:** 831-838.
76. AIBA, H., S. FUJIMOTO, N. OZAKI. 1982. Nucleic Acids Res. **10:** 1345-1361.
77. P. COSSART & B. GICQUEL-SANZEY. 1982. Nucleic Acids Res. **10:** 1363-1378.
78. MCKAY, D. & T. STEITZ. 1981. Nature (London) **290:** 744-749.
79. MCKAY, D., I. WEBER & T. STEITZ. 1982. J. Biol. Chem. **257:** 9518-9524.
80. EBRIGHT, R. H. 1982. *In* Molecular Structure and Biological Activity. J. E. Griffin & W. L. Duax, Eds: 91-102. Elsevier. New York, NY.
81. DYNAN, W. S. & R. TJIAN. 1985. Nature **316:** 774-777.
82. JONES, K. A., K. R. YAMAMOTO & R. TJIAN. 1985. Cell **42:** 559-572.
83. GIDONI, D., W. S. DYNAN & R. TJIAN. 1984. Nature **312:** 409-413.
84. FRITTON, H. P., T. IGO-KEMENES, J. NOWOCK, U. STRECH-JURK, M. THEISEN & A. E. SIPPEL. 1984. Nature **311:** 163-165.
85. EMERSON, B. M. & G. FELSENFELD. 1984. Proc. Natl. Acad. Sci. USA **81:** 96-99.
86. EMERSON, B. M., C. D. LEWIS & G. FELSENFELD. 1985. Cell **41:** 21-30.
87. WU, C. 1985. Nature **317:** 84-87.
88. WU, C. 1984. Nature **309:** 229-234.
89. WU, C. 1984. Nature **311:** 81-84.
90. JOST, J.-P., M. GEISER & M. SELDRAN. 1985. Proc. Natl. Acad. Sci. USA **82:** 988-991.

91. LEVINGER, L. & A. VARSHAVSKY. 1982. Cell **28:** 375-385.
92. BENVENISTY, N., D. MENCHER, O. MEYUHAS, A. RAZIN & L. RESHEF. 1985. Proc. Natl. Acad. Sci. USA **82:** 267-271.
93. BENVENISTY, N., M. SZYF, D. MENCHER, A. RAZIN & L. RESHEF. 1985. Biochemistry **24:** 5015-5019.
94. KONDRASHIN, A. A. 1985. Trends Biol. Sci. **10:** 97-98.
95. KREBS, E. G. & J. A. BEAVO. 1979. Annu. Rev. Biochem. **48:** 923-959.
96. FLOCKHART, D. A. & J. D. CORBIN. 1982. Crit. Rev. Biochem. **12:** 133-186.
97. MURDOCH, G. H., R. FRANCO, R. M. EVANS & M. G. ROSENFELD. 1983. J. Biol. Chem. **258:** 15329-15335.
98. LAKS, M. S., J. J. HARRISON, G. SCHWOCH & R. A. JUNGMANN. 1981. J. Biol. Chem. **256:** 8775-8785.
99. SQUINTO, S. P., D. C. KELLEY-GERAGHTY, M. R. KUETTEL & R. A. JUNGMANN. 1985. J. Cycl. Nucleotide Protein Phosphoryl. Res. **10:** 65-73.
100. WEBER, I. T., K. TAKIO, K. TITANI & T. A. STEITZ. 1982. Proc. Natl. Acad. Sci. USA **79:** 7679-7683.
101. CONSTANTINOU, A. I., S. P. SQUINTO & R. A. JUNGMANN. 1985. Cell **42:** 429-437.
102. JUNGMANN, R. A., A. I. CONSTANTINOU, S. P. SQUINTO, J. KWAST-WELFELD & J. S. SCHWEPPE. 1986. Ann. N.Y. Acad. Sci. This volume.

Regulation of Genes for Enzymes Involved in Fatty Acid Synthesis[a]

ALAN G. GOODRIDGE, DONALD W. BACK,[b]
S. BRIAN WILSON,[c] AND MITCHELL J. GOLDMAN[d]

Departments of Pharmacology and Biochemistry
Case Western Reserve University
Cleveland, Ohio 44106

INTRODUCTION

Malic enzyme [L-malate-NADP$^+$ oxidoreductase (decarboxylating), EC 1.1.1.40] and fatty acid synthase are two of a set of "lipogenic" enzymes involved in the *de novo* synthesis of long-chain fatty acids. Malic enzyme is responsible for the generation of much of the NADPH that is required for the synthesis of saturated fatty acids, whereas the multifunctional enzyme, fatty acid synthase, catalyzes the polymerization and reduction of malonyl-CoA and acetyl-CoA to form long-chain saturated fatty acids (for reviews, see references 1-3). The levels and synthesis rates of malic enzyme and fatty acid synthase are correlated with the rate of fatty acid synthesis in the livers of animals in different nutritional and hormonal states. Thus, feeding a starved chick, duckling, or gosling causes large increases in the relative rates of synthesis of both malic enzyme and fatty acid synthase.[4-9] The levels of malic enzyme mRNA and fatty acid synthase mRNA correlate positively with the respective rates of synthesis of these enzymes, indicating that much of the nutritional regulation of the levels of these enzymes is pretranslational.[7-10]

Our long-term objectives are to identify plasma hormones that signal liver cells that the rate of alimentation has changed, and to understand, at the molecular level, each of the events that intervene between the binding of each hormone to its appropriate cellular receptor and the altered levels of specific mRNA. Our approach for identifying plasma hormones has been to test the effects of potential hormones on malic enzyme and fatty acid synthase levels in hepatocytes incubated in a serum-free medium[11] and to measure levels of effective hormones in the blood of intact animals during changes in alimentation. Our strategy for analyzing the intracellular components involved in

[a]The work described in this paper was supported in part by Grant AM 21594 from the National Institutes of Health.

[b]Postdoctoral Fellow of the Medical Research Council of Canada. Present address: Department of Biochemistry, Queen's University, Kingston, Ontario, Canada K7L 3N6.

[c]Supported as a predoctoral trainee by Grant GM 07382 from the National Institutes of Health.

[d]Supported as predoctoral trainee by Grant AM 07319 from the National Institutes of Health.

the regulation of the levels of malic enzyme and fatty acid synthase has been to start at the end of the pathway, that is, with altered enzyme levels, and then to work back towards the more proximal events. Maintaining a definitive cause and effect relationship at each step of the analysis is particularly important in a system, such as this one, where individual hormones have several different end-effects that may proceed by divergent or parallel pathways and where both malic enzyme and fatty acid synthase levels are regulated by several different hormones.

Using immunological techniques, we have established that nutritional and hormonal regulation of the levels of malic enzyme and fatty acid synthase involves regulation of enzyme synthesis, not enzyme degradation.[4,6,12] With cloned cDNAs, we have determined that enzyme synthesis is regulated primarily by controlling the cellular concentrations of malic enzyme mRNA and fatty acid synthase mRNA rather than by controlling the efficiency of translation of the specific mRNAs.[7-10] The next step, continuing with this strategy, is to identify the step(s) at which abundance of malic enzyme mRNA and fatty acid synthase mRNA are regulated by nutritional state *in vivo* and by hormones in hepatocytes in culture. Subsequent steps will involve identification of those sequences in the malic enzyme and fatty acid synthase genes that bestow selective hormone sensitivity on these genes and identification of the macromolecules that interact with those sequences to cause regulation of the expression of malic enzyme and fatty acid synthase. This presentation will emphasize identification of points of regulation in the nutritional and hormonal control of the production of the mRNAs for these two lipogenic enzymes.

ISOLATION OF CLONED cDNAs FOR MALIC ENZYME mRNA AND FATTY ACID SYNTHASE mRNA

Fatty acid synthase represents 20-30% of total protein in the uropygial gland of the domestic goose,[13] and goose uropygial gland is one of the most abundant known sources of malic enzyme.[8] Therefore, we isolated cDNAs for fatty acid synthase and malic enzyme from a cDNA library prepared from total poly(A)$^+$ RNA of the goose uropygial gland. The isolation and characterization of these cDNA clones has been described.[7,8] The goose malic enzyme and fatty acid synthase cDNAs cross-hybridized with their respective mRNAs in RNA prepared from tissues of chickens and ducks.[8-10] The goose clones, therefore, were used in hybridization assays to measure levels of these mRNAs in chicken and duck livers. For transcription studies, however, we used cDNAs which were homologous to duck or chicken malic enzyme and fatty acid synthase mRNAs. The homologous cDNAs were isolated from cDNA libraries constructed from total poly(A)$^+$ RNA of chick hepatocytes treated with insulin and triiodothyronine (S. Kawamoto, D. A. Fantozzi, and A. G. Goodridge, unpublished results) or of livers from fed ducklings.[14]

NUTRITIONAL REGULATION OF MALIC ENZYME AND FATTY ACID SYNTHASE mRNAs

The amount of malic enzyme mRNA in total RNA from liver increased rapidly when starved, newly hatched ducklings were fed a diet of high-carbohydrate mash

(FIG. 1, left panel), reaching 15 times the initial level at 9 hr and an apparent steady state, about 20 times the initial level, at 24 hr. The transient decrease observed at 12 hr was a consistent feature of the accumulation time course in each of the three experiments of this type that we performed. The half-life for malic enzyme mRNA in the livers of fed ducklings was estimated to be 3 hr (FIG. 1, left panel, inset) from the semilog plot of the differences between the apparent steady state level of mRNA in fed ducklings and mRNA levels during the accumulation phase versus time after the initiation of feeding.[15] In two additional experiments, estimates of the half-life of malic enzyme mRNA from fed ducklings were from 3 to 5 hr.

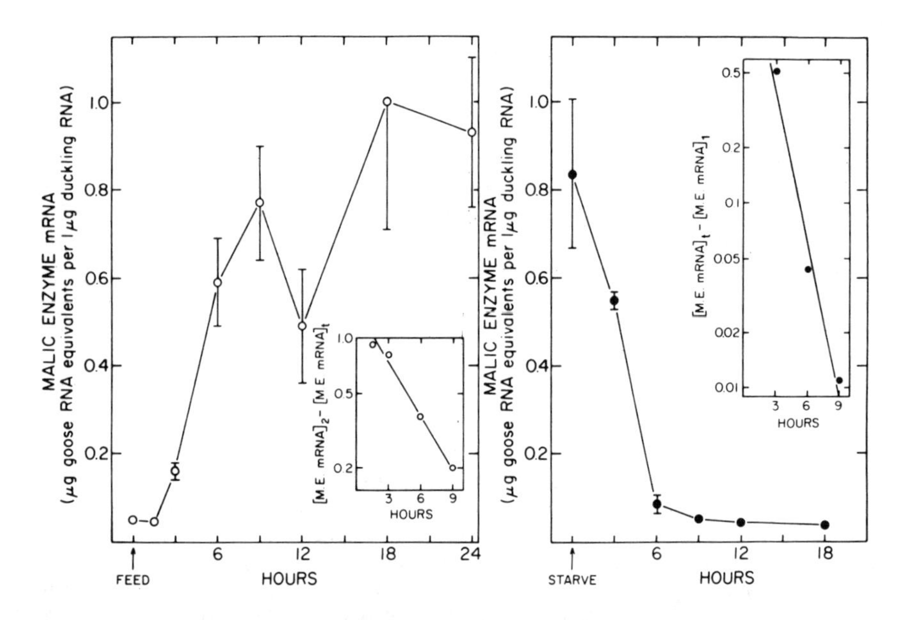

FIGURE 1. The effects of feeding (left panel) and starvation (right panel) on the abundance of malic enzyme mRNA in duckling in liver. At the beginning of the experiments shown in the left panel, the birds were 2 days old and unfed. For the experiments shown in the right panel, the birds were 11 days old and were fed 8 days, starved 2 days, and refed 1 day before beginning the experiment. Both sets of experiments were started at 10 A.M. This dietary regimen ensured a high level of malic enzyme mRNA at the beginning of the starvation experiment. Total RNA was extracted, fixed to GeneScreen (New England Nuclear), and hybridized to a single-stranded, ^{32}P-labeled malic enzyme cDNA as described by Goldman *et al.*[14] Five concentrations of total RNA from goose liver (0.2-10 μg) were dotted onto each sheet of GeneScreen. The hybridization signals for the goose RNA samples were used to construct a standard curve. The concentration of malic enzyme mRNA in the unknown samples is expressed as the equivalent of that amount of goose RNA giving the same hybridization signal. In the main plots each point is the mean ± SEM of three experiments (one duckling per experiment). The insets are semilog plots of the approaches of malic enzyme mRNA concentration to new steady states ($[\text{M.E. mRNA}]_1$, the steady state concentration of malic enzyme mRNA in the starved state; $[\text{ME mRNA}]_t$, the concentration at any time, t, during the approach to steady state; $[\text{ME mRNA}]_2$, the steady state concentration in the fed state). There were essentially no changes in malic enzyme mRNA concentrations when starved birds continued to starve or refed birds continued to feed. (Taken from Goldman *et al.*,[14] with permission of the *J. Biol. Chem.*)

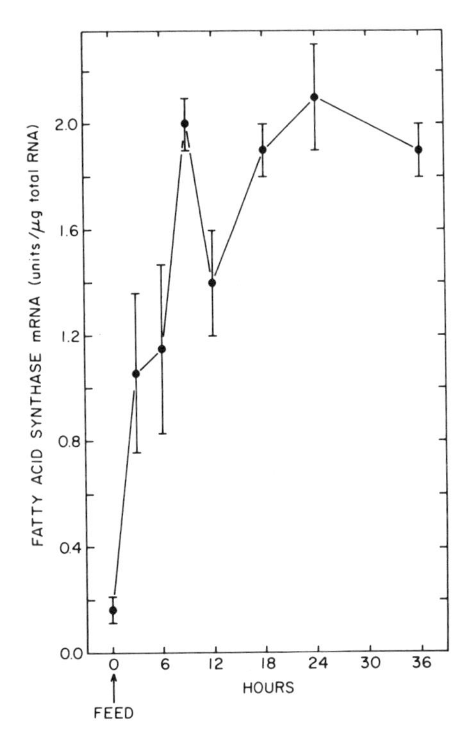

FIGURE 2. The effect of refeeding on the abundance of fatty acid synthase mRNA in liver of starved ducklings. At the beginning of the experiment the ducklings were 2 days old and unfed. The refeeding was started at 10 A.M. Total RNA was extracted, fixed to GeneScreen (New England Nuclear), and hybridized to a single-stranded, ^{32}P-labeled fatty acid synthase cDNA as described by Back *et al.*[16] Five amounts of total RNA from goose liver (0.2-10 μg) were dotted onto each sheet of GeneScreen. The hybridization signals from the goose RNA samples were used to construct a standard curve and to normalize the results from different hybridizations. The concentration of fatty acid synthase mRNA in the unknown samples is expressed as units per μg total RNA, where one unit is the hybridization signal equivalent to that observed with 1 μg of the goose RNA. Each point is the mean ± SEM of three experiments (one duckling per experiment). There were essentially no changes in fatty acid synthase mRNA concentrations when starved birds continued to starve. (Taken from Back *et al.*,[16] with permission of the *J. Biol. Chem.*)

Fatty acid synthase mRNA also was measured in these experiments. The level of fatty acid synthase mRNA increased rapidly, reached 10 times the initial level at 9 hr and, after a slight decrease at 12 hr, maintained the 10-fold increase (FIG. 2). The half-life for fatty acid synthase mRNA in fed ducklings, estimated from the time required to reach half its new steady state,[15] was about 4 hr. An independent but identical experiment, also involving three animals per time point, yielded a half-life of about 6 hr.

A similar set of measurements was carried out to determine the half-life of the malic enzyme and fatty acid synthase mRNAs in the livers of starved ducklings. The level of malic enzyme mRNA decreased rapidly when fed ducklings were starved (FIG. 1, right panel). The half-life of malic enzyme mRNA in starved ducklings was estimated from a semilog plot of the differences between mRNA levels during the decay phase and the steady state level of mRNA in starved ducklings versus time after removing food. A half-life of 1 hr was obtained (FIG. 1, right panel, inset). In the same experiment, the levels of fatty acid synthase mRNA decreased more slowly than the levels of malic enzyme mRNA (FIG. 3). The half-life of fatty acid synthase mRNA in starved ducklings was about 3 hr (FIG. 3, inset). The short half-life of malic enzyme mRNA in starved ducklings provides an explanation for the decrease in mRNA abundance at 12 hr after initiation of the feeding experiment (FIG. 1). This time point was taken during nighttime hours when the ducklings were not feeding, even though food was available. Because malic enzyme mRNA is degraded more rapidly than fatty acid synthase mRNA in starved ducklings, the decrease of fatty acid synthase mRNA during the nonfeeding nighttime hours was less than that for malic enzyme mRNA (*cf.* 12-hr time points and the steady state levels in FIGS. 1 & 2).

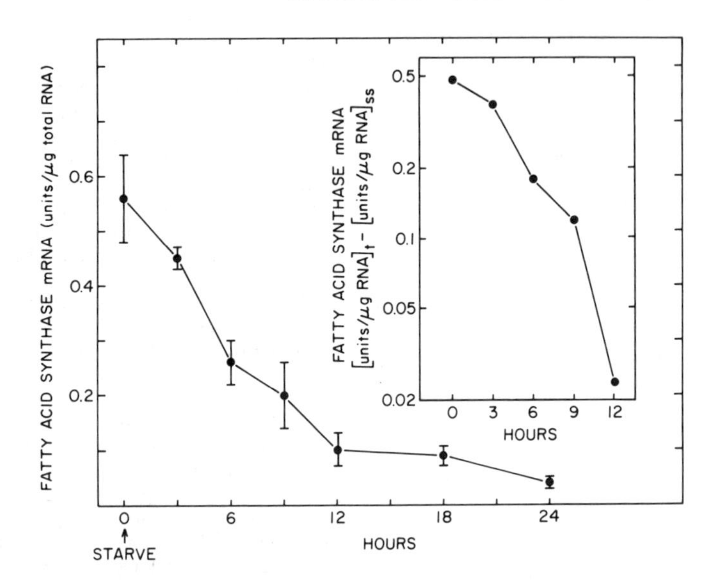

FIGURE 3. The effect of starvation on the abundance of fatty acid synthase mRNA in duckling liver. At the beginning of this experiment, the birds were 11 days old and were fed 8 days, starved 2 days, and refed 1 day before beginning the experiment. This dietary regimen ensured a high level of fatty acid synthase mRNA at the beginning of the starvation experiment. Other conditions and procedures were as described in the legend to FIGURE 2. The inset is a semilog plot of the approach of fatty acid synthase mRNA to the steady state ($[\text{units}/\mu\text{g RNA}]_{ss}$, the steady state concentration of fatty acid synthase mRNA in the starved state; $[\text{units}/\mu\text{g RNA}]_t$, the concentration at any time, *t,* during the approach to steady state). There were essentially no changes in fatty acid synthase mRNA concentration when starved birds continued to starve. (Taken from Back *et al.*,[16] with permission of the *J. Biol. Chem.*)

These indirect estimates of half-life suggested that starvation stimulated the degradation of malic enzyme mRNA, but not that of fatty acid synthase mRNA. Selectivity of the effects of starvation and feeding on the levels of malic enzyme mRNA and fatty acid synthase mRNA was indicated by the small increases or decreases of albumin mRNA abundance after feeding starved ducklings or starving fed ducklings, respectively.[14] The changes in albumin mRNA levels were not only small, they also were opposite in direction to those observed for malic enzyme mRNA and fatty acid synthase mRNA.

The 3- to 5-fold difference in the half-life of malic enzyme mRNA and the minimal change in the half-life of the fatty acid synthase mRNA were not sufficient to account for the 20- and 10-fold differences, respectively, in abundance between fed and starved ducklings. We therefore determined the effect of starvation and feeding on the rates of transcription of the malic enzyme, fatty acid synthase, and albumin genes.[14,16] Transcription was assessed by measuring the elongation of pre-existing malic enzyme, fatty acid synthase, or albumin transcripts in isolated nuclei,[17–19] a procedure which should assess primarily the rate at which initiation of transcription was occurring *in vivo* at the time the animals were killed.

Nutritional state had little or no effect on incorporation of ^{32}P-labeled uridine 5′-triphosphate into total RNA, a measure of total RNA synthesis.[14,16] Synthesis of malic enzyme mRNA increased threefold during the first 1.5 hr of feeding. Within 3 hr, there was further increase to five times the initial rate of unfed ducklings (FIG. 4). The transcription rate of the malic enzyme gene, however, also increased in ducklings that were starved rather than fed. The effect of feeding alone, therefore, was a two- to threefold increase. Starvation of previously fed ducklings reversed the effects of feeding on transcription of the malic enzyme gene. Maximum inhibition of the transcription of the malic enzyme gene was achieved within 3 to 6 hr after removing food (FIG. 4).

Transcription of the fatty acid synthase gene increased more than eightfold within 45 min of feeding. The slight decrease over the next 5 hr was not statistically significant (FIG. 5). After 24 hr of feeding, the transcription rate was still no higher than that achieved at 45 min. Between 24 and 30 hr of feeding, there appeared to be a further increase in the transcription rate. Variability and the small number of observations at each of these time points, however, makes the apparent increases statistically insignificant. To obtain a reliable indication of the "fold" increase in the transcription rate caused by refeeding starved ducklings, we compared the mean transcription rate of the fatty acid synthase gene in ducklings starved for 48 to 54 hr (0.13 ± 0.03 parts per thousand, mean ± SEM, $N = 18$) with that for ducklings fed for 24 to 30 hr (1.34 ± 0.12, $N = 16$). This 10-fold increase in transcription rate is equivalent in magnitude to the increase in fatty acid synthase mRNA level caused by feeding.

The changes in the rates of transcription of the malic enzyme and fatty acid synthase genes do not reflect changes in the rates of transcription of all RNAs. Transcription of the albumin gene was altered only sightly during these nutritional

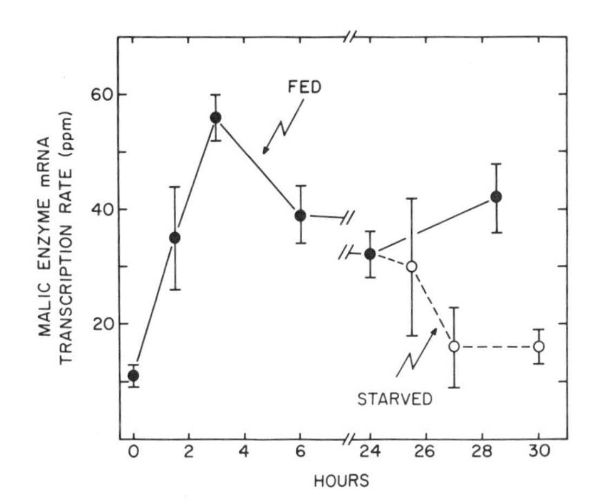

FIGURE 4. The effects of feeding and starvation on transcription of the malic enzyme gene. Ducklings were fed *ad libitum* for 10-14 days and starved for 48 hr before beginning these experiments. The ducklings were fed (closed circles) or starved (open circles) as indicated. Nuclei isolated from the treated ducklings were used for the measurements of the relative rates of transcription of the malic enzyme genes as described by Goldman *et al.*[14] The transcription rate for malic enzyme (expressed in parts per million) was determined using the following equation: ([cpm pDME1-cpm pBR322]/cpm in total RNA) × 100/hybridization efficiency × 2100/2000, where 2100 is the length of mature malic enzyme mRNA in nucleotides and 2000 is the length of pDME1 in base pairs. The efficiency of hybridization averaged about 50%. Hybridization to the paper containing pBR322 averaged about 120 cpm (5 ppm). The results are expressed as means ± SEM of four to eight experiments. Each experiment represents nuclei isolated from a different animal. (Replotted from data presented in Goldman *et al.*[14])

challenges, and always in a direction opposite to that of the malic enzyme and fatty acid synthase genes.[14] Furthermore, we have expressed the rate of transcription of specific genes relative to that of all DNA being transcribed in the isolated liver nuclei. This method of expressing transcription is consistent with the way mRNA is expressed, that is, relative to total liver RNA.

Thus, the 2- to 3-fold increase in transcription of the malic enzyme gene falls far short of accounting for the 20-fold increase in mRNA level that occurs when starved ducklings are fed. Given the difficulty of precise quantitation of either mRNA level, transcription rate, or mRNA half-life, the 2- to 3-fold change in transcription plus

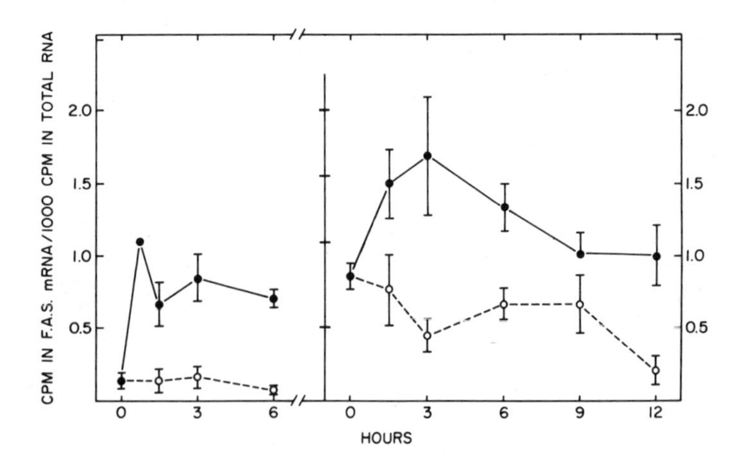

FIGURE 5. The effects of feeding and starvation on transcription of the fatty acid synthase gene. Ducklings were fed *ad libitum* for 10-14 days and starved for 48 hr before beginning these experiments. Left panel: ducklings were fed (closed circles) or starved (open circles) as indicated; right panel: ducklings were fed for 24 hr (starting with time zero of the left panel) and then fed (closed circles) or starved (open circles) as indicated. Nuclei isolated from the treated ducklings were used for the measurement of the relative rate of transcription of the fatty acid synthase gene as described by Back *et al.*[16] The transcription rate for fatty acid synthase (expressed as parts per thousand) was determined using the following equation: ([cpm pDFAS1-cpm pBR322]/cpm $\times$ 10^{-3} in total RNA) $\times$ 100/hybridization efficiency $\times$ 11 500/1350, where 11 500 is the average length of the two fatty acid synthase mRNAs in nucleotides and 1350 is the length of pDFAS1 in base pairs. The hybridization efficiency averaged about 50%. Hybridization to the paper containing pBR322 averaged about 100 cpm (0.005 ppt). The results are expressed as the means $\pm$ SEM of three to seven experiments; the lack of an error bar indicates that it was smaller than the symbol. In each experiment nuclei were isolated from a different animal. (Taken from Back *et al.*,[16] with permission of the *J. Biol. Chem.*)

the 3- to 5-fold change in stability may be sufficient to account for a 20-fold change in mRNA level. Regulation at other steps, however, cannot be eliminated. Nuclear malic enzyme transcripts were absent from the livers of starved ducklings and appeared rapidly after feeding was initiated.[14] This result is not consistent with an increase in processing of nuclear mRNA transcripts. Because transcription was increased only about threefold, this result suggests the possibility that a decrease in the degradation of nuclear malic enzyme mRNA transcripts may contribute to the increased production of cytoplasmic mRNA.

At their respective steady states, both fatty acid synthase mRNA level and rate of transcription of the fatty acid synthase gene were 10 times higher in the fed ducklings than in starved ones. This result suggests that, in contrast with regulation of the malic enzyme gene, nutritional regulation of the fatty acid synthase gene is primarily at a transcriptional level. Transcription is unlikely to be the only step involved in the regulation of fatty acid synthase mRNA concentration, however, because transcription of the fatty acid synthase gene decreased slowly when fed ducklings were starved (FIG. 5). Nine hours after the ducklings began to starve, the rate of transcription of the fatty acid synthase gene was still more than 50% of that in fed birds. If transcription was the only regulated step, then the decay in mRNA level should lag behind the decay in transcription. In fact, the decrease in the level of fatty acid synthase mRNA occurred faster than the decrease in its rate of synthesis. Thus, posttranscriptional mechanisms also may play a role in the regulation of fatty acid synthase mRNA by starvation. The posttranscriptional regulation must be transient, however, because 48 hr of starvation decreases both transcription and mRNA level to the same degree.

RELATIONSHIP BETWEEN ENZYME SYNTHESIS AND ENZYME mRNA LEVEL IN CONTROL AND HORMONE-TREATED HEPATOCYTES IN CULTURE

For these experiments, hepatocytes were isolated from the livers of 19-day-old embryos and incubated for 3 days in a serum-free medium, Waymouth MD 705/1.[20,21] Addition of insulin to these cells caused a greater than twofold increase in the relative synthesis of malic enzyme, but this increase was not significant statistically ($p > .05$) (FIG. 6). Malic enzyme mRNA levels were unaffected by insulin (FIG. 6). When triiodothyronine was added in the absence of insulin, synthesis of malic enzyme and malic enzyme mRNA level were stimulated 94-fold and 7-fold, respectively. In the presence of insulin, triiodothyronine stimulated enzyme synthesis by 43-fold and mRNA level by 11-fold relative to cells incubated with insulin alone. This result suggests that regulation of malic enzyme activity is exerted primarily at a pretranslational level. The discrepancy between stimulation of synthesis rate and stimulation of mRNA level may be due to the difficulty in measuring very low rates of enzyme synthesis or mRNA levels, especially in the absence of insulin. Alternatively, triiodothyronine may stimulate the efficiency with which malic enzyme mRNA is translated.

For fatty acid synthase, triiodothyronine alone caused a twofold increase in the rate of enzyme synthesis, and a similar increase in enzyme activity (FIG. 7). This contrasts with the large increase in malic enzyme activity elicited by triiodothyronine alone. The level of fatty acid synthase mRNA, however, was increased sevenfold, which is about the same as for malic enzyme mRNA. Insulin alone caused about a twofold increase in both synthesis rate and activity of fatty acid synthase, but a negligible change in the level of fatty acid synthase mRNA. When insulin was used in conjunction with triiodothyronine, activity and synthesis rate were increased by 8- and 17-fold, respectively, relative to cells treated with triiodothyronine alone, but mRNA was increased only about 2-fold. Taken together, these data indicate that triiodothyronine and insulin have similar effects on levels of the mRNAs for both fatty acid synthase and malic enzyme, with triiodothyronine causing the major increase. In the presence of insulin, however, fatty acid synthase mRNA is translated much more efficiently, resulting in a synergistic effect of insulin and triiodothyronine on the synthesis and activity of fatty acid synthase.

Glucagon also regulates the level of malic enzyme and fatty acid synthase in avian hepatocytes in culture.[6,12,21] Relative synthesis of malic enzyme and malic enzyme mRNA accumulation were inhibited by 98 and 93%, respectively (FIG. 6). The effects of glucagon on fatty acid synthase were much less dramatic, decreasing enzyme activity and enzyme synthesis by 80%, and mRNA level by 60% (FIG. 7).

Thus, triiodothyronine and glucagon appear to regulate the level of both malic enzyme and fatty acid synthase mRNA in a coordinated fashion, suggesting that common mechanisms may be involved. Insulin also stimulates accumulation of both mRNAs, at least when triiodothyronine is present. In addition to regulation at a pretranslational step, however, insulin appears to stimulate the translation of fatty acid synthase mRNA.

STEPS AT WHICH TRIIODOTHYRONINE AND GLUCAGON REGULATE THE LEVEL OF MALIC ENZYME mRNA

The rate of transcription of the malic enzyme gene was assessed in hepatocytes incubated in culture for 24 hr with triiodothyronine or triiodothyronine plus glucagon (TABLE 1). Triiodothyronine caused a twofold increase in transcription, whereas glucagon had no effect. Total transcription was unaffected by triiodothyronine and only slightly inhibited by glucagon (TABLE 1). Based on the results of the run-on assay, therefore, most of the regulation of malic enzyme mRNA level by triiodothyronine and all of the regulation by glucagon is posttranscriptional.

The effect of triiodothyronine on degradation of malic enzyme mRNA has not been assessed. Indirect evidence, however, suggests that glucagon significantly stimulates degradation of malic enzyme mRNA. Hepatocytes were incubated with triiodothyronine for 24 hr, resulting in a much increased level of malic enzyme mRNA.

TABLE 1. The Effects of Triiodothyronine and Glucagon on Transcription of the Malic Enzyme Gene in Hepatocytes in Culture[a]

Additions	Total Incorporation[b]	Malic Enzyme[c]
None	26×10^6	19
Triiodothyronine	19×10^6	50
Triiodothyronine plus glucagon	17×10^6	65

[a]Hepatocytes were isolated from the livers of 19-day-old chick embryos and incubated for 48 hr in Waymouth medium MD 705/1 containing insulin (300 ng/ml), as previously described,[12,20,21] except that pCME4, a nearly full-length chicken malic enzyme cDNA clone, was used to isolate malic enzyme mRNA. Hormones were present for 24 hr before harvesting the cells. Nuclei were prepared and transcription measured by a run-on assay as previously described.[14,17-19] One experiment, representative of three that were carried out, is shown.

[b]Total incorporation is expressed as cpm.

[c]Transcription of the malic enzyme gene (expressed as ppm) was determined using the following equation: ([cpm pCME1-cpm pBR322]/cpm $\times 10^6$ in total RNA) $\times$ 100/hybridization efficiency. Efficiency of hybridization varied from 30 to 46%. In all cases, cpm bound to the paper containing pBR322 DNA ($\sim$160 cpm) represented less than 50% of cpm bound to the paper containing pCME1.

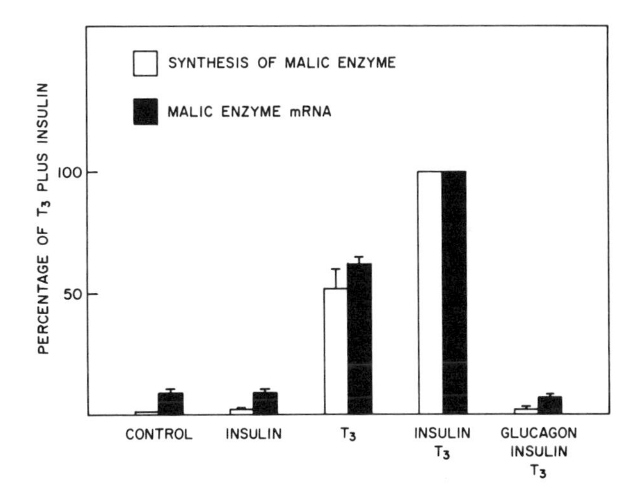

FIGURE 6. The effects of insulin, triiodothyronine, and glucagon on the relative synthesis and mRNA level for malic enzyme in hepatocytes in culture. Hepatocytes were isolated from the livers of 19-day-old chick embryos and incubated for 3 days in Waymouth medium MD 705/1.[12,20,21] Relative synthesis of malic enzyme from [^{3}H]leucine (open bars) and malic enzyme mRNA level (closed bars) were measured as previously described.[12,14] The synthesis measurements and the mRNA measurements were carried out at different times. Relative synthesis of malic enzyme was expressed as dpm incorporated into malic enzyme per 100 dpm incorporated into total soluble protein and then normalized to the rate in cells treated with insulin plus triiodothyronine (100%). Malic enzyme mRNA was measured as described in the legend to FIGURE 1. It was then expressed as units of malic enzyme mRNA per μg total RNA and normalized to the level in cells treated with insulin plus triiodothyronine. The results are expressed as means ± SEM of four to nine (synthesis rates) or three to seven (mRNA) experiments. Concentrations of the hormones were as follows: insulin, 300 ng/ml; triiodothyronine, 1 μg/ml; and glucagon, 1 μg/ml.

Different groups of cells were then incubated with glucagon, actinomycin D, α-amanitin, or no extra agent, all in the continued presence of triiodothyronine (FIG. 8). In the presence of the inhibitors of transcription, malic enzyme mRNA decayed with a half-life of 8 to 11 hr. In the presence of glucagon, malic enzyme decayed with a half-life of less than 2 hr. A 5-fold change in the half-life is not sufficient to account for the about 40-fold difference in malic enzyme mRNA level in untreated versus glucagon-treated cells. Other levels of posttranscriptional regulation must therefore be involved.

RELATIONSHIP OF HORMONAL REGULATION IN HEPATOCYTES IN CULTURE TO HORMONE LEVELS IN THE BLOOD DURING STARVATION AND REFEEDING

A large body of evidence suggests that insulin, triiodothyronine, and glucagon in the blood are important mediators of the effects of feeding and starvation on the concentrations of malic enzyme and fatty acid synthase in the liver. The effects of

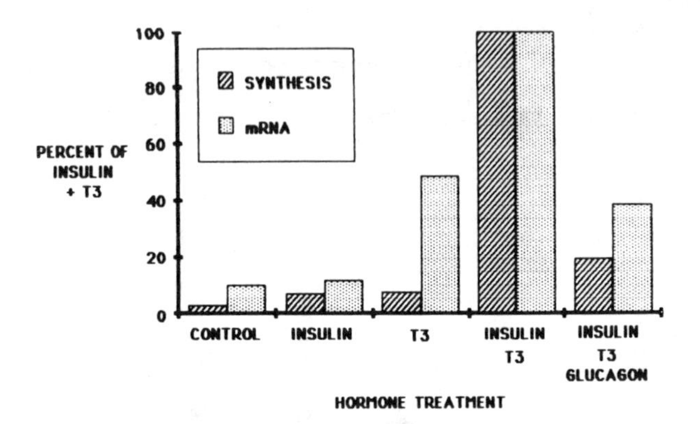

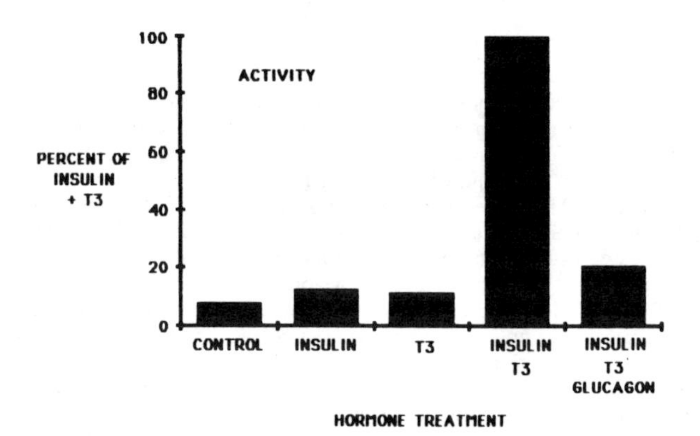

FIGURE 7. The effects of insulin, triiodothyronine, and glucagon on the activity, relative synthesis, and mRNA level for fatty acid synthase. Hepatocytes were isolated and incubated with hormones as indicated in the legend to FIGURE 6. Activity (solid bar, bottom panel) and relative synthesis (crosshatched bar, upper panel) of fatty acid synthase was measured as described by Fischer and Goodridge,[6] and the results were normalized to the rate in cells treated with insulin plus triiodothyronine. Fatty acid synthase mRNA (dark stippled bar, upper panel) was measured as described in the legend to FIGURE 2, expressed as units of fatty acid synthase mRNA per μg total RNA, and then normalized to the level in cells treated with insulin plus triiodothyronine. The results are expressed as the means of five to nine different experiments.

these hormones on the levels of these enzymes in hepatocytes in culture, as described in the previous section, are consistent with that interpretation. This hypothesis predicts that changes in the rates of processes that regulate malic enzyme and fatty acid synthase levels will correlate temporally with changes in level of these hormones in the blood. Because the actual changes in enzyme activity or level are several steps removed from the initiating events, and both enzymes have half-lives of 1 to 2 days,[4,5] steady state levels of enzyme would not be achieved until several days after a refeeding stimulus had been applied. Even mRNA levels might not correlate well because half-lives of malic enzyme and fatty acid synthase mRNAs are 4 to 6 hr in fed ducklings (FIGS. 1 & 2). To the extent that regulation of transcription is the step that controls enzyme level, transcription rates of these genes should change essentially in parallel with changes in the concentrations of the appropriate hormone(s) in blood. We have, therefore, measured the levels of glucose, insulin, glucagon, thyroxine, and triiodothyronine in the blood of ducklings during the first 6 hr after refeeding, a time that encompasses virtually all of the increases in transcription rates of malic enzyme and fatty acid synthase (*cf.* FIGS. 4 & 5).

As expected, there was a rapid increase in plasma glucose when starved ducklings were fed a high-carbohydrate mash diet. The maximum level was achieved 1.5 hr after feeding began (FIG. 9, upper panel). Plasma glucose remained elevated during the entire 6 hr of the experiment. The rate of secretion of insulin is stimulated and that of glucagon inhibited by glucose.[22,23] The responses of insulin and glucagon secretion to glucose in waterfowl are similar to those in mammals.[24,25] Plasma insulin increased and plasma glucagon decreased within 3 hr after refeeding starved ducklings and remained increased and decreased, respectively, at 6 hr (TABLE 2). The kinetics of the changes in insulin and glucagon levels are consistent with roles in the regulation of mRNA level.

The concentration of triiodothyronine in the plasma increased significantly within 45 min after starved ducklings were refed, and reached a maximum of more than two times the initial level within 1.5 hr (FIG. 9, middle panel). After 1.5 hr, the triiodothyronine level decreased slowly but remained significantly elevated for the entire 6 hr of this experiment. The timing of these changes is consistent with triiodothyronine playing a significant role in the increase in the rates of transcription of the malic enzyme and fatty acid synthase genes caused by refeeding starved ducklings.

Plasma thyroxine level was not significantly affected by refeeding starved ducklings (FIG. 9, bottom panel). Thus, the increase in plasma triiodothyronine caused by feeding was due to either increased conversion of thyroxine to triiodothyronine in the periphery or decreased clearance of triiodothyronine. Because refeeding starved animals is reported to stimulate the conversion of thyroxine to triiodothyronine in liver,[26,27] it seems probable that the increased level of triiodothyronine in refed ducklings was due to increased conversion of thyroxine to triiodothyronine. The liver is the major site of

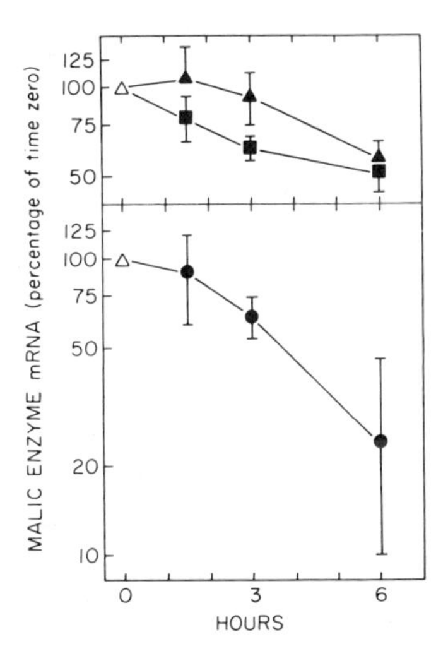

FIGURE 8. The effects of glucagon, α-amanitin, and actinomycin D on the level of malic enzyme mRNA in hepatocytes in culture incubated with triiodothyronine. Hepatocytes were prepared from the livers of 19-day-old chick embryos as described in the legend to FIGURE 6 and incubated in Waymouth medium MD 705/1 containing insulin (300 ng/ml) for 2 days. The medium was then changed to contain insulin plus triiodothyronine (1 μg/ml), and the incubation continued for an additional 24 hr. At 24 hr (open triangles), glucagon (closed circles, 1 μg/ml), actinomycin D (closed triangles, 2.5 μg/ml), or α-amanitin (closed squares, 5 μg/ml) was added to the indicated dishes, and incubation continued as indicated. RNA was extracted and assayed for malic enzyme mRNA sequences essentially as described in the legend to FIGURE 1 and normalized to the level of malic enzyme mRNA in cells incubated with triiodothyronine plus insulin for 24 hr (time zero). The results are expressed as means ± SEM of three to eight independent experiments. In the absence of glucagon or transcription inhibitors, malic enzyme mRNA continued to accumulate as indicated in FIGURE 10.

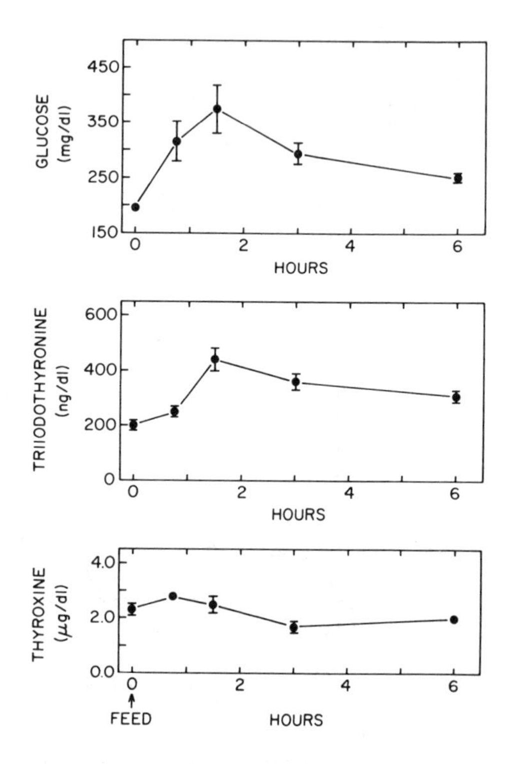

FIGURE 9. Glucose and thyroid hormones in the blood of starved/refed ducklings. Newly hatched ducklings were fed *ad libitum* for 7 or 8 days and then starved for 24 hr before starting the feeding experiments. Glucose (upper panel) was measured in serum by a standard clinical assay. Triiodothyronine (middle panel) and thyroxine (lower panel) were measured in serum by radioimmunoassay using standard clinical assay. The assays were carried out in the Clinical Biochemistry Laboratory, University Hospitals of Cleveland. The results are expressed as means ± SEM of 6 to 10 different experiments, except for the thyroxine measurements, which are means ± SEM 3 to 6 experiments.

conversion of thyroxine to triiodothyronine and of degradation of thyroid hormones in general. It is possible, therefore, that the change in concentration of triiodothyronine in liver cells was much greater than the twofold change in the plasma. In hepatocytes in culture, triiodothyronine caused a 2-fold increase in transcription and a greater than 10-fold increase in malic enzyme mRNA level. Therefore, the three- to fivefold increase in the rate of transcription of malic enzyme caused by feeding for 3 to 6 hr seems unlikely to have been due exclusively to a twofold change in hormone concentration. If, however, the local hepatic concentration had changed much more, as suggested in the discussion above, triiodothyronine may be the primary controller of malic enzyme transcription during nutritional changes in intact animals. It remains

TABLE 2. Insulin and Glucagon in the Blood of Starved and Refed Ducklings[a]

Time (hr)	Insulin (pg/ml)	Glucagon (pg/ml)	Ratio (insulin/glucagon)
0	480 ± 40	220 ± 80	2.4 ± 0.6
3	1180 ± 130	140 ± 10	8.3 ± 0.3
6	1210 ± 190	100 ± 40	13 ± 3

[a]Newly hatched ducklings were fed for 10-14 days and then starved for 2 days before refeeding. Serum insulin and plasma glucagon were measured by radioimmunoassay, using standard clinical assays. The results are expressed as means ± SEM of three different experiments (one duckling per experiment).

possible, however, that as yet undiscovered mediators play important roles in regulation of the expression of both the malic enzyme and the fatty acid synthase genes.

STIMULATION OF THE ACCUMULATION OF MALIC ENZYME AND FATTY ACID SYNTHASE mRNAs MAY INVOLVE A PEPTIDE INTERMEDIATE

When triiodothyronine was added to hepatocytes that had been incubated in culture for 2 days with insulin, both malic enzyme (FIG. 10) and fatty acid synthase mRNA (data not shown) accumulated with sigmoidal kinetics and took 48 hr or more to

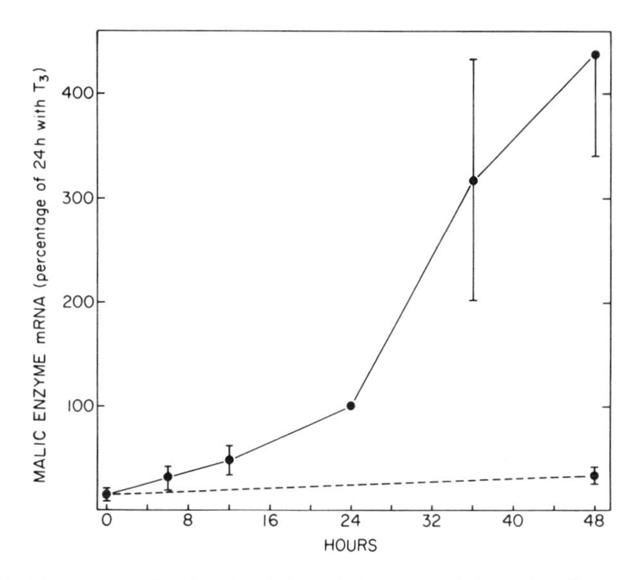

FIGURE 10. Time course for the stimulation of the accumulation of malic enzyme mRNA caused by triiodothyronine in hepatocytes in culture. Hepatocytes were isolated from the livers of 19-day-old chick embryos as described in the legend to FIGURE 6 and incubated in Waymouth medium MD 705/1 containing insulin (300 ng/ml). The medium was changed once after about 20 hr of incubation. After about 44 hr of incubation, triiodothyronine (1 μg/ml) was added without changing the medium. No triiodothyronine was added to the control cells. RNA was isolated and malic enzyme mRNA quantitated as described in the legend to FIGURE 1. The data were then normalized to the amount of malic enzyme mRNA in hepatocytes incubated for 24 hr with triiodothyronine. The results are expressed as means ± SEM of four or five independent experiments.

approach their new steady state levels. These observations suggested the possibility that triiodothyronine might stimulate the accumulation of an intermediate that is involved in the triiodothyronine-stimulated accumulation of mRNAs for malic enzyme and fatty acid synthase. We assessed the possible involvement of a peptide intermediate with the inhibitor of protein synthesis, puromycin. Puromycin caused a small and statistically insignificant decrease in the levels of malic enzyme or fatty acid synthase mRNAs in the absence of triiodothyronine but almost completely blocked the increases in mRNA accumulation in the presence of that hormone. (The malic enzyme mRNA results are shown in TABLE 3. The fatty acid synthase mRNA results are not shown but were essentially the same as those for malic enzyme.) Puromycin did not decrease the level of β-tubulin mRNA either with or without triiodothyronine in the medium (data not shown). β-Tubulin mRNA is degraded somewhat more rapidly than either malic enzyme mRNA or fatty acid synthase mRNA[28,29] (our unpublished results) and is unaffected by triiodothyronine in the hepatocytes in culture. Puromycin also inhibited further accumulation of malic enzyme or fatty acid synthase mRNAs when added to hepatocytes 24 hr after the triiodothyronine (data not shown). This putative intermediate may be a "true" intermediate in thyroid hormone action, its concentration increasing as a consequence of the addition of triiodothyronine. Alternatively, it could be a rapidly degraded peptide factor required for the increase in level of malic enzyme and fatty acid synthase mRNAs caused by triiodothyronine.

TABLE 3. The Effect of Puromycin on the Accumulation of Malic Enzyme mRNA Caused by Triiodothyronine[a]

Incubation Time (hr)	Insulin		Insulin and Triiodothyronine	
	Control	Puromycin	Control	Puromycin
1.5	35	36	52	55
3.0	36	14	100	44
6.0	19	8	177	36

[a]Hepatocytes were isolated from the livers of 19-day-old chick embryos and incubated in Waymouth medium MD 705/1 containing insulin (300 ng/ml), as previously described.[12,20,21] Triiodothyronine (1 μg/ml) was added, as indicated above, after about 48 hr of incubation. Puromycin (1 mM) was added, as indicated above, 30 min prior to adding the triiodothyronine. Malic enzyme mRNA was assayed by hybridization with cloned malic enzyme cDNA.[14] The results are expressed as a percentage of the mRNA level in cells incubated for 3 hr with triiodothyronine and without puromycin, and are the averages of two independent experiments.

SPECULATION

Rather than analyzing the effects of one hormone on one or more genes, we are attempting to analyze the mechanisms by which individual genes are regulated by several different hormones. Although the same cluster of agents regulates both malic enzyme and fatty acid synthase gene expression, the mechanisms of action of the individual hormones can be the same or be dramatically different. Thus, synthesis rates for these two "lipogenic" enzymes are regulated coordinately in intact animals by starvation or starvation followed by refeeding and in culture by triiodothyronine, possibly involving a common peptide intermediate. There are, however, striking differences in the mechanisms by which changes in enzyme synthesis are achieved. *In vivo*, regulation of fatty acid synthase level is exerted primarily at the level of transcription whereas for malic enzyme both transcriptional and posttranscriptional events are involved, with the latter likely dominating. In cells in culture, insulin affects the translational efficiency of fatty acid synthase mRNA but not malic enzyme mRNA. Furthermore, the accumulation of malic enzyme mRNA is much more sensitive to the inhibitory action of glucagon than is the accumulation of fatty acid synthase mRNA.

Why should malic enzyme and fatty acid synthase be regulated by different mechanisms despite their coordinated expression under most physiological conditions? Fatty acid synthesis is only one of several pathways in the cytoplasm that utilize NADPH. Both microsomal drug hydroxylation and maintenance of reduced glutathione levels, for example, require NADPH. Regulation of these pathways, both major consumers of NADPH, is independent of the regulation of fatty acid synthesis. Differences in the regulation of malic enzyme and fatty acid synthase may reflect a requirement for the organism to maintain regulatory properties for malic enzyme that are related to functions other than fatty acid synthesis. In the course of evolution, mechanisms for regulating the abundance of these two proteins may have evolved independently with acquisition of a role for malic enzyme in fatty acid synthesis being a relatively recent event.

SUMMARY

The levels of malic enzyme and fatty acid synthase are increased by feeding and decreased by starvation in liver *in vivo* and are increased by triiodothyronine and decreased by glucagon in hepatocytes in culture. Cloned malic enzyme and fatty acid synthase cDNAs are being used to analyze regulation of these unique genes. Dietary regulation of both enzymes occurs at pretranslational steps. Increased transcription and increased mRNA stability contribute about equally to a 20-fold increase in malic enzyme mRNA level when starved ducklings are refed. In contrast, a 10-fold increase in the level of fatty acid synthase mRNA is largely accounted for by increased transcription of this gene. In chick-embryo hepatocytes incubated in serum-free medium containing insulin, triiodothyronine causes a greater than 10-fold increase in levels of both malic enzyme and fatty acid synthase mRNAs. Kinetic and inhibitor experiments suggest a protein intermediate in the increases of malic enzyme and fatty acid synthase mRNAs caused by triiodothyronine. For malic enzyme, the stimulation by triiodothyronine is predominantly posttranscriptional. Glucagon decreases the level of malic enzyme mRNA by 90 to 95%, with regulation occurring at a posttranscriptional step. Inhibitor experiments suggest that stimulation of the degradation of malic enzyme mRNA is partially responsible. Glucagon inhibited fatty acid synthase mRNA level by less than 50%; the inhibited step has not been identified. Thus, the coordinated regulation of malic enzyme and fatty acid synthase proteins by nutritional state may involve different hormones regulating at different points. A surprisingly large component of the regulation is posttranscriptional.

ACKNOWLEDGMENTS

We are indebted to Dr. William R. Carpenter for critically reviewing this manuscript and to Timothy Uyeki for expert technical assistance with some of the experiments.

REFERENCES

1. WAKIL, S. J., J. K. STOOPS & V. C. JOSHI. 1983. Annu. Rev. Biochem. **52:** 537-579.
2. VOLPE, J. J. & P. R. VAGELOS. 1973. Annu. Rev. Biochem. **42:** 21-60.
3. GOODRIDGE, A. G. 1985. *In* Biochemistry of Lipids and Membranes. D. E. Vance & J. E. Vance, Eds.: 143-180. Benjamin/Cummings. Menlo Park, CA.
4. SILPANANTA, P. & A. G. GOODRIDGE. 1971. J. Biol. Chem. **246:** 5754-5761.
5. ZEHNER, Z. E., V. C. JOSHI & S. J. WAKIL. 1977. J. Biol. Chem. **252:** 7015-7022.
6. FISCHER, P. W. F. & A. G. GOODRIDGE. 1978. Arch. Biochem. Biophys. **190:** 332-344.
7. MORRIS, S. M., JR., J. H. NILSON, R. A. JENIK, L. K. WINBERRY, M. A. MCDEVITT & A. G. GOODRIDGE. 1982. J. Biol. Chem. **257:** 3225-3229.
8. WINBERRY, L. K., S. M. MORRIS, JR., J. E. FISCH, M. J. GLYNIAS, R. A. JENIK & A. G. GOODRIDGE. 1983. J. Biol. Chem. **258:** 1337-1342.
9. GOODRIDGE, A. G., R. A. JENIK, M. A. MCDEVITT, S. M. MORRIS, JR. & L. K. WINBERRY. 1984. Arch. Biochem. Biophys. **230:** 82-92.
10. MORRIS, S. M., JR., L. K. WINBERRY, J. E. FISCH, D. W. BACK & A. G. GOODRIDGE. 1984. Mol. Cell. Biochem. **64:** 63-68.
11. GOODRIDGE, A. G. 1986. Fed. Proc. **45:** in press.
12. GOODRIDGE, A. G. & T. G. ADELMAN. 1976. J. Biol. Chem. **251:** 3027-3032.
13. BUCKNER, J. S. & P. E. KOLATTUKUDY. 1976. Biochemistry **15:** 1948-1957.
14. GOLDMAN, M. J., D. W. BACK & A. G. GOODRIDGE. 1985. J. Biol. Chem. **260:** 4404-4408.
15. BERLIN, C. M. & R. T. SCHIMKE. 1965. Mol. Pharmacol. **1:** 149-156.
16. BACK, D. W., M. J. GOLDMAN, J. E. FISCH, R. S. OCHS & A. G. GOODRIDGE. 1986. J. Biol. Chem. **261:** 4190-4197.
17. MORY, Y. & M. GEFTER. 1978. Nucleic Acids Res. **5:** 3899-3912.
18. SPINDLER, S. R., S. H. MELLON & J. D. BAXTER. 1982. J. Biol. Chem. **257:** 11627-11632.
19. MCKNIGHT, G. S. & R. D. PALMITER. 1979. J. Biol. Chem. **254:** 9050-9058.
20. GOODRIDGE, A. G. 1973. J. Biol. Chem. **248:** 1924-1931.
21. GOODRIDGE, A. G., A. GARAY & P. SILPANANTA. 1974. J. Biol. Chem. **249:** 1469-1475.
22. VANCE, J. E., K. D. BUCHANAN, D. R. CHALLONER & R. H. WILLIAMS. 1968. Diabetes **17:** 187-193.
23. UNGER, R. H. & L. ORCI. 1981. N. Engl. J. Med. **304:** 1518-1524.
24. SITBON, G. & P. MIALHE. 1978. Horm. Metab. Res. **10:** 117-123.
25. SITBON, G. & P. MIALHE. 1979. Horm. Metab. Res. **11:** 123-129.
26. BALSAM, A. & S. H. INGBAR. 1978. J. Clin. Invest. **62:** 415-424.
27. GAVIN, L. A., F. A. MCMAHON & M. MOELLER. 1981. Endocrinology **109:** 530-536.
28. BEN-ZE'EV, A., S. R. FARMER & S. PENMAN. 1979. Cell **17:** 319-325.
29. STIMAC, E., V. E. GROPPI, JR. & P. COFFINO. 1984. Mol. Cell. Biol. **4:** 2082-2090.

Multihormonal Regulation of Milk Protein Gene Expression

J. M. ROSEN, J. R. RODGERS, C. H. COUCH,
C. A. BISBEE, Y. DAVID-INOUYE, S. M. CAMPBELL,
AND L.-Y. YU-LEE

Department of Cell Biology
Baylor College of Medicine
Houston, Texas 77030

INTRODUCTION

Milk protein gene expression in mammary epithelial cells is regulated by the complex interplay of several peptide and steriod hormones at both the transcriptional and posttranscriptional levels.[1] Developmental processes that affect both cell-cell and cell-substratum interactions also play a critical role in the expression of these hormonally regulated genes.[2] The overall objective of the studies to be described is to elucidate the mechanisms regulating the developmental and multihormonal control of milk protein gene expression. The general approach employed is first to define by DNA-mediated gene transfer *cis*-acting regulatory sequences required for the control of milk protein gene expression by stage- and tissue-specific hormone-inducible *trans*-acting factors. Once both wild-type and "mutant" genes have been characterized it may then be possible to identify these specific regulatory molecules. The principal problems encountered in our laboratory are not the result of limitations of recombinant DNA technology, but rather those of cell biology; that is, cell-cell and cell-substratum interactions as well as cell shape are critical factors regulating milk protein gene expression.

In contrast to most of the other systems involved in metabolic regulation discussed at this conference, the studies to be described from our laboratory will be concerned with the regulation of genes encoding proteins with nonenzymatic functions. Caseins, the principal milk proteins, are abundant dietary proteins important for the provision of essential amino acids and the transport of supersaturating concentrations of calcium and phosphate into milk. Therefore, there is a need for chronic regulation leading to a sustained high level of synthesis and secretion of these proteins, rather than an acute modulation of enzyme activity. In mammary epithelial cells this regulation is primarily posttranscriptional as a result of stabilization of the primary transcripts rather than by modulation of the rates of transcription. As with acutely regulated genes, such as the phosphoenolpyruvate carboxykinase (PEPCK) gene, there is no need for a strong promoter. Instead, a moderately active promoter generating extremely stable mRNAs leads to the accumulation of high levels of the milk protein mRNAs during lactation. The consequence of these effects is that 5′ flanking DNA sequences alone appear to be insufficient to elicit hormonal regulation.

Maximal levels of milk protein gene expression are observed in the presence of several peptide and steroid hormones, notably insulin, prolactin, and hydrocortisone. Basal levels of casein mRNA are observed in explant cultures maintained in insulin (I) and hydrocortisone (F) alone. Upon prolactin (M) addition, however, there is a marked induction, ranging from 10- to 250-fold, in casein mRNA levels observed after 24 hr. The mechanism of transmembrane signaling by prolactin that results in specific gene activation remains an enigma. In fact, this is also true for the other members of this family of peptide hormones, that is, growth hormone and placental lactogen. Although preliminary characterization of an approximately 40-kDa receptor for prolactin has been reported[3] and monoclonal antibodies to this receptor have been generated,[4] no well-documented effects of the prolactin-receptor complex on adenylyl cyclase or cAMP-dependent protein kinase, or intrinsic tyrosine or type C kinase activities have been reported. Other known activators of these latter kinases, epidermal growth factor and phorbol esters, have been reported to inhibit prolactin-mediated casein synthesis.[5,6] Of interest is the observation that colchicine administration will block prolactin induction of casein gene expression.[7] Although the specificity of this inhibitor remains to be established, it once again brings into question whether cell shape is of critical importance for hormone-mediated gene expression. These points will be briefly illustrated in the following presentation.

RESULTS AND DISCUSSION

Hormonal Regulation in Cell and Explant Cultures

One of the initial reasons for choosing to study the regulation of milk protein gene expression was the availability of a well-defined explant culture system in which hormonal concentrations could be manipulated in a chemically defined medium.[8] This system was useful for studying the regulation of steady state levels of the milk protein mRNAs, as well as for limited pulse-chase analyses. It was poorly suited, however, for experiments requiring large amounts of cells, for nuclear-cytoplasmic fractionation, and especially for gene-transfer experiments. Furthermore, retention of hormones in explants isolated from midpregnant rats precluded the definition of the precise role of hormones in this culture system.[9]

Despite these limitations it was possible to study the regulation of three mRNAs, which are members of the calcium-sensitive casein gene family, and a fourth mRNA encoding a novel whey protein in mice and rats, designated whey acidic protein (WAP). Both prolactin and hydrocortisone, in the presence of insulin, are required for the maximal induction of these mRNAs,[10] although there are differential effects of the peptide and steroid hormones on WAP and casein mRNA levels. Previous studies had suggested that the major effect of prolactin on casein mRNA accumulation was exerted at the posttranscriptional level, but it was not possible to distinguish in these experiments effects exerted on cytoplasmic mRNA turnover from those on nuclear RNA processing.[11]

In order to discriminate between these processes, additional pulse-chase analyses were performed in the explant cultures. Focusing just on the regulation of β-casein mRNA, pulse-chase studies were carried out using either a short, 30-min pulse-labeling period with [^{3}H]uridine, or a long, 24-hr steady state labeling period. Using the short labeling time, only a 2-fold increase in the rate of β-casein mRNA synthesis is observed when the [^{3}H]RNA is hybridized to an immobilized β-casein cDNA, whereas after a 24-hr labeling period, a 40-fold increase in β-casein mRNA is observed in cultures that had been treated with IFM versus IF alone for 24 hr.[11] When a rapid chase is elicited by the addition of actinomycin D, and the label remaining in β-casein mRNA is expressed relative to a stable control RNA, such as 18S ribosomal RNA (rRNA), the results obtained in FIGURE 1 for 30-min pulse-labeled RNA are obtained. In the absence of prolactin, the half-life of β-casein mRNA is approximately 10 min, whereas in the presence of prolactin it is equivalent to rRNA's, or greater than 18 hr (FIG. 1A). Under the same conditions prolactin has no effect on the stability of another nuclear RNA with rapid turnover, the external transcribed spacer (ETS) region of the ribosomal transcription unit. When similar pulse-chase studies were performed after the 24-hr labeling period, quite a different result was obtained. In this case the β-casein mRNA transcript stability is unchanged in either the presence or absence of prolactin, and in both cases a long half-life comparable to rRNA is observed (data not shown). These results suggest that the principal effect of prolactin is exerted on the turnover of nuclear casein mRNA transcripts, but once the mature mRNA has entered the cytoplasm it is stable in both the absence and presence of prolactin. Furthermore, only a 2-fold effect of prolactin is detected on the synthesis of β-casein mRNA, under conditions where mRNA accumulation increases 35- to 40-fold.[12]

In order to elucidate the precise mechanism of posttranscriptional regulation of nuclear casein mRNA processing, it was necessary to develop a cell culture model in which the endogenous β-casein gene could be correctly regulated, and into which exogenous DNA constructs could be introduced. Although there are several breast cancer cell lines, such as T-47D, which retain functional receptors for both peptide and steroid hormones, no clonal cell line is available to date in which casein gene regulation has been observed. Primary cultures of mammary epithelial cells derived from midpregnant animals have been shown to retain hormone-inducible casein gene expression (when these cells are grown on a floating collagen gel matrix).[13] Although these cells are suitable for transient expression assays, they are clearly inappropriate for the development of transfected cell lines in which hormonally regulated RNA processing could be studied.

Recently, a cell line derived from midpregnant Balb/c mice, designated COMMA-1D, has been reported to synthesize β-casein.[14] Using early passage cells, the hormonal responsiveness of these cells was characterized in greater detail. COMMA-1D cells were grown in 10% dextran-coated charcoal stripped horse serum in the presence of I or IFM for 74 hr and the level of the endogenous mouse β-casein and cytoplasmic actin mRNAs determined (FIG. 2). When the cells are cultured on a type I floating collagen gel matrix, a 150-fold induction of β-casein mRNA is observed in IFM versus I cultures after 74 hr. This induced level represents approximately 5% of the level of casein mRNA in a standard RNA isolated from 10-day lactating mouse mammary tissue. However, when the cells are cultured on a type I collagen gel, which remains attached to the dish, a 15-fold lower level of β-casein mRNA is observed. Furthermore, under these conditions no decrease is seen when the cultures are withdrawn into I alone for 24 hr. Under comparable conditions of hormone withdrawal the cells grown on the floating gel exhibited approximately a 75% decrease in β-casein mRNA levels. Cells grown on a plastic substratum either in the presence or absence of F and M had a barely detectable level of casein mRNA. Finally, the specificity of these changes for β-casein mRNA is indicated by the constant level of actin mRNA under any of these culture conditions. Similar effects of substratum on casein gene expression in COMMA-1D cells have been observed by Bissell and co-workers.[15]

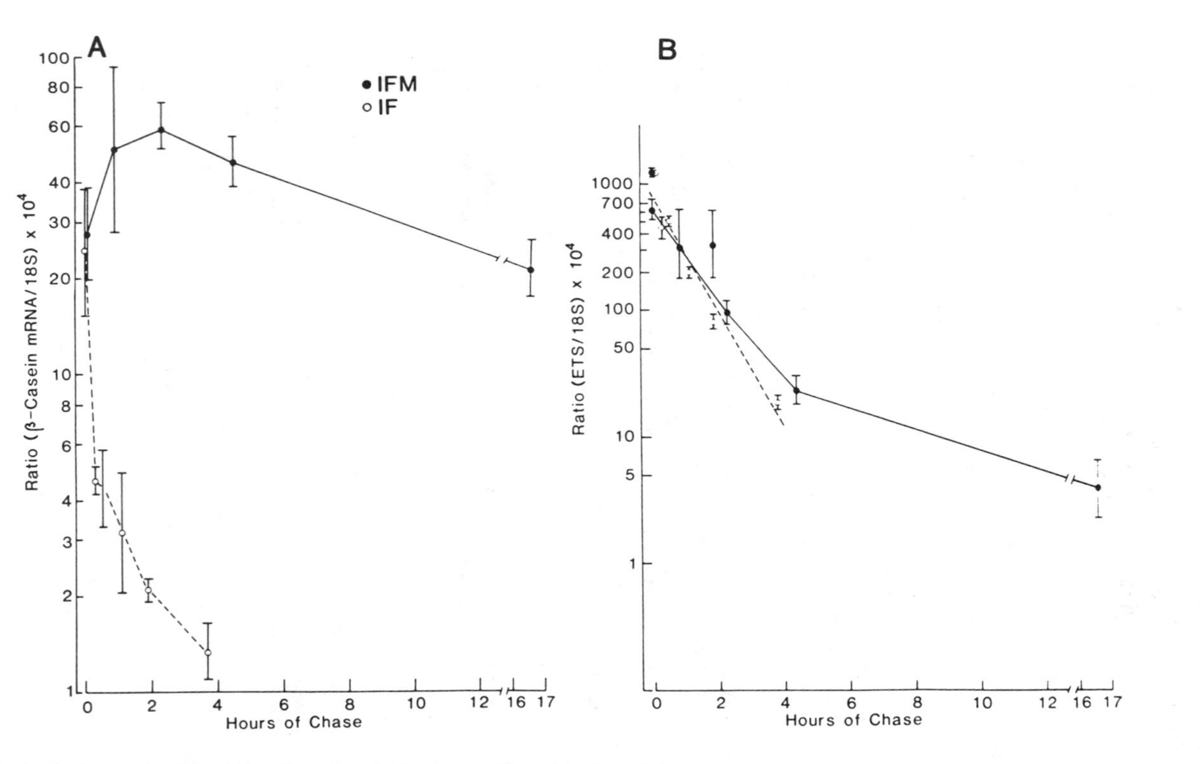

FIGURE 1. Prolactin-induced stabilization of casein mRNA. The relative labeling of β-casein mRNA and ribosomal 18S and ETS sequences during a chase with actinomycin D was analyzed as follows: 300 dishes of explants from 15-day midpregnant rats were incubated for 48 hr in IF medium, then switched to fresh IF or IFM medium for 24 hr.[11] After 1 hr of labeling with [^{3}H]uridine, the medium was replaced with IF or IFM medium containing 20 μg/ml of actinomycin D. RNA was extracted by phenol/$CHCl_3$/SDS at pH 8.0 followed by reprecipitation from 3 M NaOAc. The fraction of cpm hybridized to cloned β-casein cDNA (**A**) or to an ETS probe (**B**) was divided by the fraction hybridized to an 18S rRNA probe. These values are plotted semilogarithmically.

Additional experiments have been performed to characterize the hormonal responsiveness of COMMA-1D cells, as compared to mammary explant cultures. The synergistic effect of hydrocortisone and prolactin on β-casein mRNA levels described previously for explant cultures has also been seen using COMMA-1D cells (FIG. 3). Unlike explant cultures, in which steroid retention may complicate the interpretation of the roles of individual hormones, it was possible to examine directly the response to added F and M of cultures maintained in charcoal-stripped horse serum depleted of glucocorticoid and lactogenic hormones. No significant induction of β-casein mRNA levels is seen in cultures exposed to IF for 70 hr as compared to I alone, yet an approximately two- to threefold effect of IM versus I is apparent. IFM-treated cultures displayed a time-dependent increase in casein mRNA levels as depicted in FIGURE 3, and following withdrawal displayed a rapid deinduction. These results suggested that COMMA-1D cells would provide a suitable model system for analysis of transfected milk protein genes.

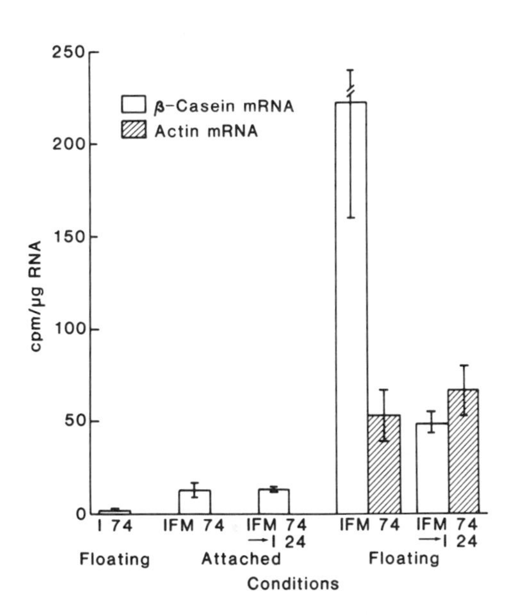

FIGURE 2. Substratum-dependent induction of β-casein mRNA in COMMA-1D cells. COMMA-1D cells (passage 23) were grown in DMEM containing 10% charcoal-stripped horse serum plus either I alone (5 μg/ml) or I and F (1 μg/ml) and M (200 ng/ml) for 74 hr on an attached or floating type I rat-tail collagen gel. Because the COMMA-1D cells are a mixed cell population we do not continuously passage the cells, but remove cells stored in liquid N_2 before each experiment. Confluent cells are usually passaged with 10X trypsin for 2 min to avoid selection of subpopulations of cells. In the two withdrawal experiments shown in the figure, the medium was changed to I alone and incubated for an additional 24 hr. Cytoplasmic RNA was extracted as described in FIGURE 1, and the levels of mouse β-casein and cytoplasmic actin mRNAs were determined by an RNA dot blot procedure using M13 single-stranded DNA probes. The results are expressed as the cpm hybridized per μg of RNA. The levels of actin mRNA were invariant as a function of hormonal or substratum conditions and are only shown for the IFM floating gel conditions.

When the levels of α- and γ-casein mRNAs are examined in COMMA-1D cells, similar responses are observed. Quite unexpectedly, however, no expression of the usually abundant WAP mRNA is observed under any of these culture conditions (data not shown). A similar loss of WAP gene expression has been observed in primary mouse mammary epithelial cell cultures, whereas casein mRNA levels were inducible in a hormone- and substratum-dependent fashion.[16] As discussed previously, WAP gene expression can be regulated in mammary explant cultures in a serum-free, chemically defined medium. Thus, cell-cell interactions, functional hormone-receptor systems, and, presumably, cell shape all play a role in the regulation of WAP gene expression. The only case in which WAP gene expression has been observed in cell culture in our laboratory is in cultures of primary dimethylbenz[*a*]anthracene (DMBA)-induced mammary tumor cells grown on attached collagen gels. Under these conditions, a small subpopulation of cells retains the ability to synthesize milk proteins in a hormone-inducible fashion. Presumably, the regulation of WAP gene expression in these tumor cells is analogous to the expression of certain liver-specific enzymes in several minimal deviation hepatoma cell lines discussed in this volume. The loss of WAP gene expression in dissociated mammary epithelial cells is analogous to that observed for many liver-specific enzymes in dissociated parenchymal cells, as compared to intact liver slices.[17] The difference between casein and WAP gene regulation was

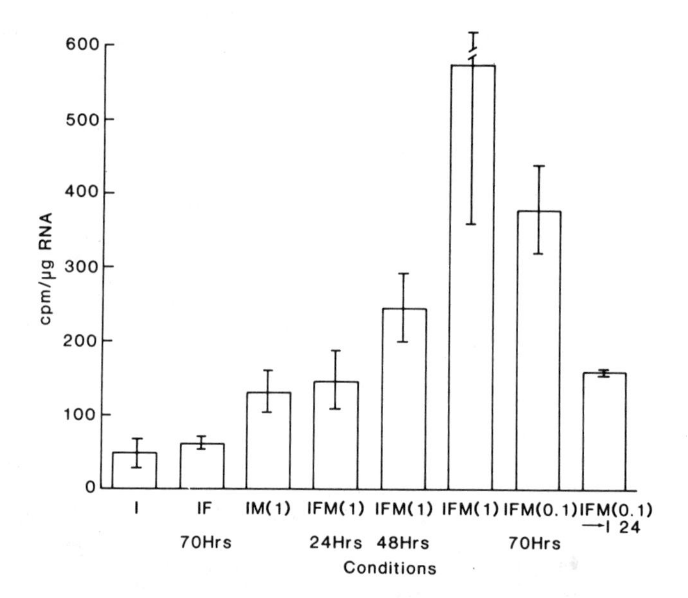

FIGURE 3. Multihormonal regulation of β-casein mRNA levels in COMMA-1D cells. COMMA-1D cells (passage 20) were grown in DMEM containing 10% charcoal-stripped horse serum on floating collagen gels as described in FIGURE 2. The concentration of M was either 1 or 0.1 μg/ml as shown in the figure. The time of exposure to the various hormonal conditions is as shown. Withdrawal to I alone for 24 hr is depicted by the arrow.

unexpected, however, because the hormonal induction of casein mRNA is observed in dissociated cells in a substratum-dependent manner.

One additional difference that is observed between the COMMA-1D cells and explant cultures is the initial lag time observed prior to the induction of β-casein mRNA following the addition of prolactin and hydrocortisone. In explant cultures mRNA levels increase as early as 1 hr following hormone addition,[8] but in cells grown on a floating collagen gel a 12- to 16-hr lag period is observed (data not shown). This lag is not dependent upon the time at which the gels are detached from the culture dish, but instead is related to the time after the addition of prolactin and hydrocortisone even if the gels had already been detached for 24 hr.

One of the well-documented changes that is observed in primary mammary epithelial cell cultures grown on a floating collagen gel substratum is the contraction of the gel accompanied by a marked change in cell shape.[13,15] In fact if collagen gels are cross-linked with glutaraldehyde to prevent contraction of the gel, no hormone-dependent induction of casein synthesis is observed in these primary cultures.[13] The role of hormones in eliciting these changes in cell shape is less well defined. One possible reason for the increased lag time observed in COMMA-1D cells as compared to explant cultures, however, might be the requirement for a hormone-dependent change in cell morphology.

In the course of screening several other cell lines, both clonally derived and mixed cell populations, which are also derived from midpregnant Balb/c mice, a striking example of a hormone-dependent change in cell morphology was observed. Using a clonal mammary epithelial cell line, designated M8A-4,[18] a hormone-dependent change in cell morphology was observed when cells grown on a floating collagen gel were transferred from a medium containing I and 5% fetal bovine serum to one containing IFM plus serum for 74 hr (FIG. 4). The M8A-4 cells are cuboidal in cultures with I alone, but appear to form ductule-like aggregates in the presence of prolactin and hydrocortisone. Interestingly, these cells failed to produce detectable β-casein mRNA even though they were clearly hormonally responsive.

In this regard analysis of the original COMMA-1D cell line has revealed that it is a mixed cell population with approximately 10% of the cells capable of synthesizing casein.[18] Repeated attempts to clone the cells synthesizing casein by either transfection with a dominant selectable marker or by limiting dilution have failed to generate a homogeneous clonal cell line. Instead, the selected cells are still heterogeneous, and only a small subpopulation synthesizes casein. These results support the concept that cell-cell interactions may be important for the expression of differentiated function in mammary epithelial cells.[2] The lack of a hormonally responsive clonal cell line presents a problem for the analysis of transfection experiments. One approach to generating such a line might be to target various oncogenes to the mammary gland using tissue-specific promoter and enhancer elements.[19,20] Cell lines derived from tumors induced by these oncogenes may overcome some of the cell-cell, cell-substratum, and cell-shape requirements necessary for the hormonal regulation of milk protein gene expression.

Milk Protein Gene Structure

As a necessary prerequisite for elucidating the mechanism of hormonal regulation of milk protein gene expression, the structure and organization of the three casein mRNAs and the WAP mRNA as well as the corresponding genomic sequences encoding these mRNAs have been studied. Extensive nucleotide sequence analysis of a number of casein mRNAs from rat,[21,11] mouse,[23] guinea pig,[24,25] and cow[26] has revealed a considerable divergence among the casein nucleotide sequences. This divergence was expected: the caseins represent one of the most rapidly diverging protein families yet studied.[27] Three regions of these mRNAs, however, are highly conserved. These include the 5′ noncoding, signal peptide, and casein kinase phosphorylation sequences. The unusual conservation of the 5′ noncoding region of the casein mRNAs may be related to the formation of a potentially stable secondary structure, which in turn may have a role in posttranscriptional regulation of casein synthesis.[11] The casein signal peptides are all 15 amino acids in length and contain an invariant lysine residue in position 2 and a cysteine residue in position 8. These residues may play a role in the efficient translocation, the recognition and removal, and the secretion into milk of the caseins. The conservation of the signal peptide sequences is also unexpected because they usually diverge more rapidly than the coding regions of their respective proteins.

Recent studies on the genomic structure and organization of the rat α-, β-, and γ-, and the bovine α-casein genes revealed that these three conserved structural gene regions are each encoded by separate exons.[28-30] The evolution of the casein gene family appears to have involved initially the recruitment of exons with distinct functional domains to generate a primordial casein gene, which was followed by intragenic and intergenic duplications to generate the different members of the multigene family.[28] Support for this hypothesis comes from genetic studies that show that the bovine genes occur as a gene cluster,[31] and that all three mouse calcium-sensitive casein genes are localized on a single chromosome.[32]

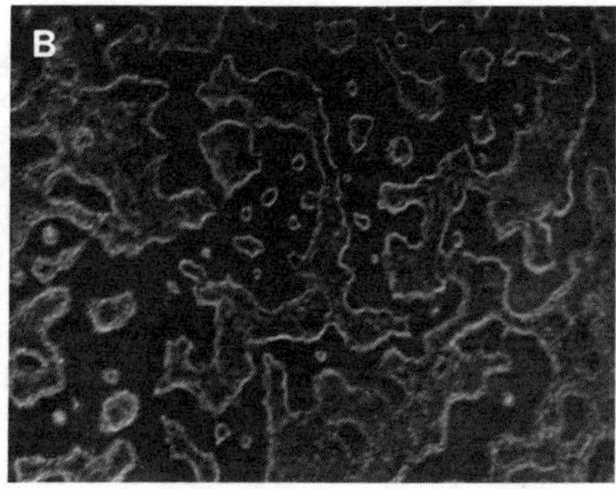

FIGURE 4. Hormonal regulation of mammary cell morphology in M8A-4 cells. Cells of the M8A-4 line (passage 36), a clonal mammary epithelial cell line derived from the MOMA-1 cell population,[18] were grown in DMEM containing 5% fetal bovine serum for 72 hr on a floating collagen gel in either **(A)** I or **(B)** IFM medium as described in FIGURE 3. No detectable β-casein mRNA was observed in these cells under either hormonal condition by an RNA dot blot analysis.

Analysis of the 5′ flanking DNA of the three rat casein and the bovine α-casein genes has revealed several regions that are more highly conserved than both the entire mature coding and intron regions of these genes.[30] These sequences reside within the first 200 base pairs of the 5′ flanking DNA and extend beyond the TATA- and CAAT-box sequences known to be important for promoter function in many other eukaryotic genes. In the rat and bovine α-casein genes significant homology is observed extending to 550 base pairs upstream of the site of initiation of transcription or CAP site. These conserved 5′ flanking sequences may contain potential *cis*-regulatory elements that are responsible for the coordinate expression of the functionally related casein genes during mammary gland development.

There are several noteworthy features of the 5′ flanking region of the casein genes. First, these sequences are particularly AT-rich. The 200-base-pair region proximal to the CAP site has a mean melting temperature of only 65° C, almost 10° C less than the comparable mammalian globin gene region and considerably less than many housekeeping genes with highly GC-rich promoter regions.[33] Second, both the α- and γ-casein genes have the unusual TATA sequence TTTAAAT. The transition of an

A to a C, T, or G in the second position of the TATA sequence has been shown to reduce the *in vitro* efficiency of the promoter sequence in other genes.[34] This same altered TATA sequence is also found in both the rat and mouse WAP gene promoter regions.[35] Third, the rat α-casein gene contains a repeating sequence, $(CA)_{39}$, at −650 base pairs. This sequence may act as an enhancer of gene transcription,[36] and may account for the higher basal level of α-casein gene expression in the absence of hormonal induction.

The structure of the rat and mouse WAP genes is considerably less complex than that of the casein genes.[35] WAP is a cysteine-rich protein, and is a member of an apparently functionally unrelated family of proteins, designated the four-disulfide core proteins.[37] Analysis of WAP mRNA and gene structure has confirmed the hypothesis that the protein has arisen by an intragenic duplication involving the cysteine domain. Analysis of somatic cell hybrids also has revealed that the WAP gene is on a different chromosome from the casein gene cluster.[32] Although the levels of WAP mRNA are regulated both developmentally and hormonally in an analogous, but not identical fashion to the casein mRNAs, a comparison of their 5′ flanking sequences does not reveal any unusual sequence conservation, except for the previously mentioned unusual TATA sequence.[35] Comparison of the casein or WAP 5′ flanking sequences with those of the rat α-lactalbumin gene has also failed to reveal any striking sequence similarities. Thus, there are no obvious consensus sequences present in the first 500 base pairs of 5′ flanking DNA of these genes that can account for their developmental and hormonal regulation. In addition to these structural studies the functional role of both the flanking and intragenic sequences of these genes needs to be established by transfection experiments before any definitive conclusions about their regulatory nature can be reached.

Transfection of the Milk Protein Genes

Two general strategies have been employed in our laboratory to study the regulatory sequences involved in milk protein gene expression by gene transfection. The first and least difficult approach is to assay fusion gene constructs in which flanking regions of the casein genes are linked to a readily assayable reporter gene, in this case the gene encoding the bacterial enzyme chloramphenicol acetyltransferase (CAT), which is not usually expressed in eukaryotic cells.[38] These experiments may employ either transient expression assays after the introduction of the DNA by a calcium phosphate-dimethylsulfoxide-shock transfection protocol,[39] or use co-selection with a dominant selectable marker. This latter approach is useful for the evaluation of hormone-dependent promoter and enhancer sequences, but would not necessarily detect sequences involved in posttranscriptional regulation.

The second strategy is to introduce the entire milk protein gene or a functional minigene in which the flanking and exon sequences are fused to the cDNA sequence allowing the generation of a mature mRNA. These constructs are cloned into a eukaryotic expression vector derived from bovine papilloma virus (BPV). These BPV vectors have been engineered as shuttle vectors to permit their replication in both bacterial and eukaryotic cells, and usually incorporate a dominant selectable marker.[40] They have several potential advantages for studying posttranscriptional regulation. These include the presence of amplified copies, the use of the transfected gene's rather than the viral promoter, and their usual maintenance as stable, unrearranged minichromosomes.[41]

In order to assay for hormone-dependent promoter and enhancer function, a series of constructions in which the 5′ flanking regions of the β- and γ-casein genes were cloned into the SV_0CAT vector has been generated.[42] These constructions contained either 500 or 1000 base pairs of 5′ flanking DNA upstream from the CAP site, or the flanking DNA, plus the first noncoding exon and 500 to 600 base pairs of the first intron. They have been tested by DNA-mediated gene transfer into several different cell lines, including primary cultures of midpregnant rat mammary epithelial cells, primary DMBA-induced tumor cells, COMMA-1D cells, and T-47D cells. For all the cells, except the T-47D cells, it was necessary to develop a transfection protocol in which cells were plated on an attached collagen gel and the DNA introduced by a calcium phosphate-dimethylsulfoxide-shock procedure.[38] For the COMMA-1D and midpregnant cells the gels were then released and hormones added for 48 or 72 hr. In all these experiments the parental SV_0CAT and SV_2CAT vectors were employed as controls, the latter containing the SV40 promoter and enhancer sequences. The maximal hormonal effect observed in any of these experiments was twofold comparing cells exposed to I alone versus IFM, and when triplicate samples were analyzed these changes were not statistically significant (data not shown).

In order to avoid problems of differential DNA uptake in the transient expression assays, the casein-CAT fusion genes were also stably introduced into COMMA-1D cell populations by co-transfection with a pSV_2*neo*-dominant selection vector.[43] The hormonal regulation of these fusion genes was then determined and normalized for the number of CAT copies quantitated by a DNA dot blot assay. As shown in FIGURE 5, a twofold effect at best of prolactin and hydrocortisone is observed using a variety of different constructions, even in the extreme situation of comparing cells cultured on an attached collagen gel with I alone to a floating gel with IFM. Under these same hormonal conditions the maximal change in endogenous β-casein mRNA levels (as great as 150-fold) is observed.

Several conclusions can be drawn from these experiments. First, the 5′ flanking regions of the β- and γ-casein genes are relatively weak promoters, as compared to the SV40 promoter and enhancer, usually yielding only 1 to 10% of the activity observed with SV_2CAT. Second, the activity of these promoters is orientation dependent; that is, they are considerably more active in the 5′-to-3′ orientation than in the opposite orientation.[42] Finally, constructions that contain the first noncoding exon of the casein genes are more active than those containing 5′ flanking DNA alone. This may have the trivial explanation that the 5′ noncoding exon is providing an efficient eukaryotic translational signal for the bacterial CAT gene. Alternatively, the unusual conservation of the 5′ noncoding region of the casein mRNAs suggests that the increased activity of these constructions might reflect a posttranscriptional role of these sequences.

The failure to observe any major hormonal effect on the expression of the casein fusion genes, even an effect of glucocorticoids as compared to insulin alone, suggests that the important *cis*-acting regulatory sequences lie either further 5′ to the CAP site of these genes, as has recently been observed for glucocorticoid regulation of tyrosine aminotransferase-CAT fusion genes,[44] or alternatively that sequences within, as well as 5′ flanking to, the genes may be important to permit the correct interaction of transcriptional and posttranscriptional regulatory signals.[45]

In order to test this possibility the entire rat β-casein gene containing 2300 base pairs of 5′ flanking DNA and 400 base pairs of 3′ flanking DNA,[12] an α-casein minigene containing 680 base pairs of 5′ flanking DNA, and the first two exons and two introns of the gene fused in frame to the α-casein cDNA (Yu-Lee and Rosen, unpublished observations) have been constructed in BPV-dominant selection vectors. These have been transfected into either COMMA-1D or T-47D cells, and either clonal or mixed cell populations have been assayed for the integrity of the genes as well as their expression. Amplified copies of the unrearranged genes have been identified in both cell lines using the β-casein and α-casein constructions, and in both cases strand-specific expression has been detected using SP 6 riboprobes. At present, however, hormonal regulation of the transfected β-casein gene and α-casein minigene has not been observed. Furthermore, analysis of the RNA transcripts suggests that very little of the mature mRNA of the correct size is present in either case. Instead, a series of

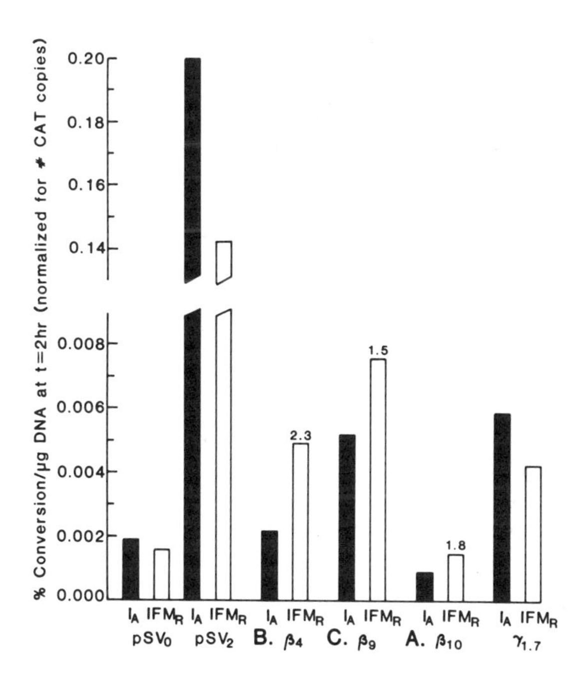

FIGURE 5. Assay of casein-CAT fusion genes in G418-selected COMMA-1D cell populations. COMMA-1D cells (passage 19) were co-transfected with 20 μg of pSV_0CAT-based casein-CAT fusion plasmids and 2 μg of pSV_2*neo* DNA[43] by calcium phosphate precipitation followed by a 20% dimethylsulfoxide shock.[39] After approximately 6 weeks of selection for G418 resistance (300 μg/ml), pooled cells from each co-transfection were assayed for CAT activity as previously described.[38] Casein promoters (base pairs numbered relative to the transcription start site): B.β_4: β-casein, -511 to -12; C.β_9: β-casein, -511 to $+535$ (contains first exon and 5′ part of first intron); A.β_{10}: β-casein, -1100 to -12; $\gamma_{1.7}$: γ-casein, -1100 to $+600$ (contains first exon and 5′ part of first intron). Abbreviations: I_A: attached collagen gel cultures in medium containing I (5 μg/ml) alone; IFM_R: floating collagen gel cell cultures in IFM (1 μg/ml of F and M).

higher molecular weight transcripts that may correspond to primary pre-mRNAs and their processed intermediates have been observed.[46] Additional mapping experiments are required to definitively establish if these transcripts are correctly initiated, and, in fact, if they truly represent unprocessed casein gene transcripts. At present, however, this unexpected result might suggest that there is a failure of the correct hormonally mediated RNA processing of the transfected genes. This is consistent with the observation that the major effect of prolactin appears to be exerted at the level of nuclear

RNA processing. It is conceivable that either the constructions analyzed to date are still lacking the correct genetic signals to elicit the correct hormonal regulation or that the transfected cells are lacking the necessary hormonal responsiveness. Both of these possibilities are currently being studied in our laboratory.

In contrast to these studies employing the rat casein genes, there has been a failure to detect the expression of the entire WAP gene containing 1000 base pairs of 5′ flanking and 1200 base pairs of 3′ flanking DNA in a BPV vector, in experiments in which this gene was transfected into the COMMA-1D, M8A-4, T-47D, and C127 cell lines. Once again there were amplified, unrearranged copies of the gene present in these cells, but only in a single clone containing 50 copies of the gene was any correctly sized WAP mRNA detected, and it was not hormonally regulated (Campbell and Rosen, unpublished observations). Thus, the correct signals for RNA transcription and processing appear to be present in the WAP-BPV construction. The lack of WAP gene expression is unexpected because the WAP gene was inserted into the BPV vector near to the viral enhancer, which has been shown recently to stimulate the expression of the insulin gene even in nonpancreatic cell lines.[47] This failure to observe WAP expression in a variety of cell lines suggests, therefore, that a negative regulatory factor, perhaps specific for WAP gene expression, may be overriding the positive effect of the BPV enhancer in these cell lines. Studies are currently underway to attempt to reactivate the expression of the transfected and endogenous WAP genes in the COMMA-1D cells by reimplanting the transfected cells in a cleared mammary fat pad.[14]

The preceding discussion illustrates some of the problems inherent in attempting to define the *cis*-acting sequences important for the multihormonal regulation of milk protein gene expression. These results suggest several approaches that should be taken in future experiments. First, the analysis of transfected genes will be performed in mixed cell populations rather than in clonal isolates because of the increasing evidence of the importance of cell-cell interactions for the induction of casein gene expression. Second, constructions employing additional 5′ flanking DNA and lacking any prokaryotic vector sequences will be employed. Third, these constructions may be screened more rapidly by employing transient expression assays utilizing primary cultures and a polyoma vector, which will be amplified 10- to 20-fold in mouse cells, and contains a histone reference gene.[45] Fourth, transgenic animals provide an alternate method to define stage-specific and tissue-specific *cis*-acting sequences, although they may not be useful for studies of multihormonal regulation of gene expression. Especially in the case of WAP gene expression, transgenic animals may provide a method to circumvent the problems encountered in cell culture, at least until the development of responsive cell lines capable of WAP gene expression.

REFERENCES

1. Rosen, J. M., R. J. Matusik, D. A. Richards, P. Gupta & J. R. Rodgers. 1980. Recent Prog. Horm. Res. **36:** 157-194.
2. Levine, J. F. & F. E. Stockdale. 1985. J. Cell Biol. **100:** 1415-1422.
3. Necessary, P. C., P. A. Humphrey, P. B. Mahajan & K. E. Ebner. 1984. J. Biol. Chem. **259:** 6942-6946.
4. Djiane, J., I. Dusanter-Fourt, M. Katoh & P. A. Kelly. 1985. J. Biol. Chem. **260:** 11430-11435.
5. Taketani, Y. & T. Oka. 1983. Proc. Natl. Acad. Sci. USA **80:** 1646-1649.
6. Taketani, Y. & T. Oka. 1983. Proc. Natl. Acad. Sci. USA **80:** 2647-2650.

7. HOUDEBINE, L. M. & J. DJIANE. 1980. Mol. Cell. Endocrinol. **17:** 1-15.
8. MATUSIK, R. J. & J. M. ROSEN. 1978. J. Biol. Chem. **253:** 2343-2347.
9. BOLANDER, F. F., JR., K. R. NICHOLAS & Y. J. TOPPER. Biochem. Biophys. Res. Commun. **91:** 247-252.
10. HOBBS, A. A., D. A. RICHARDS, D. J. KESSLER & J. M. ROSEN. 1982. J. Biol. Chem. **257:** 3598-3605.
11. ROSEN, J. M., W. K. JONES, J. R. RODGERS, J. R. COMPTON, C. A. BISBEE, Y. DAVID-INOUYE & L.-Y. YU-LEE. 1986. Ann. N. Y. Acad. Sci. **464:** 87-99.
12. GUYETTE, W. A., R. J. MATUSIK & J. M. ROSEN. 1979. Cell **17:** 1013-1023.
13. LEE, E. Y.-H., G. PARRY & M. J. BISSELL. 1984. J. Cell Biol. **98:** 146-155.
14. DANIELSON, K. G., C. J. OBORN, E. M. DURBAN, J. S. BUTEL & D. MEDINA. 1984. Proc. Natl. Acad. Sci. USA **81:** 3756-3760.
15. BISSELL, M. J., E. Y.-H. LEE, M.-L. LI, L.-H. CHEN & H. G. HALL. 1985. *In* The Second NIADDK Symposium on the Study of Benign Prostatic Hyperplasia. U.S. Government Printing Office. In press.
16. LEE, E. Y.-H., W.-H. LEE, C. S. KAETZEL, G. PARRY & M. J. BISSELL. 1985. Proc. Natl. Acad. Sci. USA **82:** 1419-1423.
17. CLAYTON, D. F., A. L. HARRELSON & J. E. DARNELL, JR. 1985. Mol. Cell. Biol. **5:** 2623-2632.
18. MEDINA, D., C. J. OBORN, F. S. KITTRELL & R. L. ULLRICH. 1986. J. Natl. Cancer Inst. **76:** 1143-1156.
19. STEWART, T. A., P. K. PATTENGALE & P. LEDER. 1984. Cell **38:** 627-637.
20. HANAHAN, D. 1985. Nature (London) **315:** 115-122.
21. BLACKBURN, D. E., A. A. HOBBS & J. M. ROSEN. 1982. Nucleic Acids Res. **10:** 2295-2307.
22. HOBBS, A. A. & J. M. ROSEN. 1982. Nucleic Acids Res. **24:** 8079-8098.
23. HENNIGHAUSEN, L. G. & A. E. SIPPEL. 1982. Eur. J. Biochem. **125:** 131-141.
24. HALL, L., J. E. LAIRD & R. K. CRAIG. 1984. Biochem. J. **222:** 561-570.
25. HALL, L., J. E. LAIRD, J. C. PASCALL & R. K. CRAIG. 1984. Eur. J. Biochem. **138:** 585-589.
26. STEWART, A. F., I. M. WILLIS & A. G. MACKINLAY. 1984. Nucleic Acids Res. **12:** 3895-3907.
27. DAYHOFF, M. O. 1976 *In* Atlas of Protein Sequence and Structure. M. O. Dayhoff, Ed. Vol. **5** (Suppl. 2). National Biomedical Research Foundation. Bethesda, MD.
28. JONES, W. K., L.-Y. YU-LEE, S. M. CLIFT, T. L. BROWN & J. M. ROSEN. 1985. J. Biol. Chem. **260:** 7042-7050.
29. YU-LEE, L.-Y., & J. M. ROSEN. 1983. J. Biol. Chem. **258:** 10794-10804.
30. YU-LEE, L.-Y., L. RICHTER-MANN, C. H. COUCH, A. F. STEWART, A. G. MACKINLAY & J. M. ROSEN. 1986. Nucleic Acids Res. **14:** 1883-1902.
31. MATYUKOV, V. S. & A. P. URNYSHEV. 1980. Genetika **16:** 884-886.
32. GUPTA, P., J. M. ROSEN, P. D'EUSTACHIO & F. H. RUDDLE. 1982. J. Cell Biol. **93:** 199-204.
33. ISHII, S., G. T. MERLINO & I. PASTAN. 1985. Science **230:** 1378-1381.
34. CONCINO, M., R. A. GOLDMAN, M. H. CARUTHERS & R. WEINMANN. 1983. J. Biol. Chem. **28:** 8493-8496.
35. CAMPBELL, S. M., J. M. ROSEN, L. G. HENNIGHAUSEN, U. STRECH-JURK & A. E. SIPPEL. 1984. Nucleic Acids Res. **12:** 8685-8697.
36. HAMADA, H., M. SEIDMAN, B. H. HOWARD & C. M. GORMAN. 1984. Mol. Cell. Biol. **4:** 2622-2630.
37. HENNIGHAUSEN, L. G. & A. E. SIPPEL. 1982. Nucleic Acids Res. **10:** 2677-2684.
38. GORMAN, C. M., L. F. MOFFAT & B. H. HOWARD. 1982. Mol. Cell. Biol. **2:** 1044-1051.
39. GRAHAM, F. L. & A. J. VAN DER EB. 1973. Virology **52:** 456-467.
40. LAW, M.-F., J. C. BYRNE & P. M. HOWLEY. 1983. Mol. Cell. Biol. **3:** 2110-2115.
41. SARVER, N., P. GRUSS, M.-F. LAW, G. KHOURY & P. M. HOWLEY. 1981. Mol. Cell. Biol. **1:** 486-496.
42. ROSEN, J. M., W. K. JONES, S. M. CAMPBELL, C. A. BISBEE & L.-Y. YU-LEE. 1985. *In* Membrane Receptors and Cellular Regulation. M. P. Czech & C. R. Kahn, Eds.: 385-396. Alan R. Liss. New York, NY.
43. SOUTHERN, P. J. & P. BERG. 1982. J. Mol. Appl. Genet. **1:** 327-341.

44. SCHÜTZ, G., W. SCHMID, M. JANTZEN, U. DANESCH, B. GLOSS, U. STRÄHLE, P. BECKER & M. BOSHART. 1986. Ann. N.Y. Acad. Sci. This volume.
45. GROSSCHEDL, R. & D. BALTIMORE. 1985. Cell **41:** 885-897.
46. DAVID-INOUYE, Y., C. H. COUCH & J. M. ROSEN. 1986. Ann. N.Y. Acad. Sci. This volume.
47. SARVER, N., R. MUSCHEL, J. C. BYRNE, G. KHOURY & P. M. HOWLEY. 1985. Mol. Cell. Biol. **5:** 3507-3516.

Regulation and Structure of Murine Malic Enzyme mRNA[a]

SRILATA BAGCHI, LEIGH S. WISE,
MARYANNE L. BROWN, HEI SOOK SUL,
DAVID BREGMAN, AND
CHARLES S. RUBIN[b]

Department of Molecular Pharmacology
Albert Einstein College of Medicine
New York, New York 10461

INTRODUCTION

Cytosolic malic enzyme is a homotetramer (M_r = 250 000) that catalyzes the oxidative decarboxylation of malate: malate + $NADP^+$ → pyruvate + CO_2 + NADPH + H^+.[1] NADPH produced by this reaction is a major source of the reducing equivalents necessary for the *de novo* biosynthesis of palmitate and other long-chain fatty acids.[1] Thus, malic enzyme is considered a member of a small family of "lipogenic" enzymes that includes ATP-citrate lyase, fatty acid synthetase, and acetyl-CoA carboxylase.[2]

The rates of synthesis of malic enzyme and other lipogenic enzymes are often coordinately regulated by hormones and dietary manipulations.[2-4] Alterations in specific mRNA levels account for parallel changes in the concentrations of malic enzyme, fatty acid synthetase, and ATP-citrate lyase observed in a starvation/carbohydrate feeding paradigm and in response to triiodothyronine,[5-9] thereby indicating that control is exerted at the pretranslational level. Goodridge and colleagues[10] have demonstrated that 3- to 5-fold increases in both the stability of malic enzyme mRNA and rate of transcription of the malic enzyme gene determine a net 20-fold elevation in hepatic malic enzyme mRNA content when starved neonatal ducks are fed a high-carbohydrate diet.

Malic enzyme is also regulated during differentiation and development in mammals. For example, malic enzyme is not expressed in fetal rat liver, but the enzyme rapidly accumulates in newborn animals.[11] We have shown that the activity, mass, and rate of synthesis of malic enzyme increase by approximately an order of magnitude during the differentiation of 3T3-L1 preadipocytes into adipocytes.[12] These changes correlated with an increase in translatable mRNA for malic enzyme. In considering further analyses of malic enzyme regulation in adipogenesis an additional level of complexity is introduced. Unlike avian tissues, which contain a unique 2.1-kb malic enzyme mRNA,[5] mouse and rat cells express two mRNAs of markedly different size (2.0 kb

[a] Supported by Grant AM 27165 from the National Institutes of Health.

[b] To whom correspondence should be addressed.

and 3.1 kb) that hybridize with cDNA probes for malic enzyme.[7,13] Both mRNAs are found on polysomes and appear to code for the same malic enzyme subunit polypeptide.[6,7,13] Numerous temporal, molecular, and structural aspects of malic enzyme regulation during adipocyte differentiation have not been previously studied.

We now report the characterization of the primary structure of murine malic enzyme mRNA and its derived amino acid sequence; an assessment of the relationships between the two malic enzyme mRNAs; temporal aspects of the regulation of malic enzyme mRNA during adipocyte differentiation; a comparison of the coordinate regulation of malic enzyme and ATP-citrate lyase mRNAs with the regulation of glycerol 3-phosphate dehydrogenase mRNA; and evidence that malic enzyme is controlled at a pretranslational level during differentiation, but is subject to translational control in preadipocytes.

EXPERIMENTAL PROCEDURES

Isolation of Malic Enzyme mRNA and Construction of Recombinant Plasmids

Murine malic enzyme mRNA from liver was enriched to a level of 20 to 25% purity by a polysome immunoadsorption procedure as previously described.[7,8] The purified mRNA (0.4 μg) was employed as a template for first-strand cDNA synthesis catalyzed by avian myeloblastosis virus reverse transcriptase, using the conditions described by Land *et al.*[14] The second strand was prepared by the Gubler and Hoffman modification[15] of the method of Okayama and Berg.[16] The tailing of cDNA, its insertion into pBR 322, and the transformation of *Escherichia coli* RR 1 were carried out as previously described.[8] Recombinant plasmids containing large inserts with 5′ end and middle sequences were detected by *in situ* hybridization[17] with a 5′ end-labeled, 80-bp *Pst*I-*Bam*HI fragment derived from a double digest of plasmid pME4 (FIG. 1).

The construction and characterization of another set of cDNA clones (including pME1, pME4, and pME6) that contain sequences corresponding to the middle and 3′ end sequences of malic enzyme mRNA were described in an earlier report.[7]

cDNA Sequencing

Fragments of cDNA inserts were generated by single and double digestions with the restriction endonucleases indicated in FIGURE 1 and subcloned into the replicative forms of M13 mp18 and M13 mp19.[18] The cDNA sequence was determined by the dideoxynucleotide-chain-termination procedure.[19,20]

Preparation of Synthetic Oligonucleotides

Two deoxyoligonucleotides, the 17-mer 5′-AAAGAGAACAAAGCTAA-3′ and the 26-mer 5′-ATAGGTAAATACATTCATTTACAACT-3′, were synthesized on an

Applied Biosystems 380B DNA synthesizer. After deblocking, the oligonucleotides were purified by electrophoresis in a 20% polyacrylamide gel containing 7 M urea.[21]

Electrophoresis and Northern Blot Hybridization of mRNA

Total RNA was isolated from 3T3-L1 cells by the guanidinium thiocyanate procedure of Chirgwin *et al.*[22] Subsequently, poly(A)$^+$ RNA was purified by affinity chromatography on oligo dT cellulose as described by Krystosek *et al.*[23] Samples of poly(A)$^+$ RNA (7 μg) were denatured and subjected to electrophoresis under denaturing conditions[24] in 0.7% agarose gels as previously indicated.[8] After the resolved RNAs were transferred to nitrocellulose, the Northern blots were probed with plasmid pME1 that was labeled with ^{32}P via nick-translation as previously described.[7,8] Radiolabeled bands corresponding to malic enzyme mRNAs on autoradiograms of Northern blots were quantified by scanning densitometry.

Cell Culture

3T3-L1 cells were grown and induced to differentiate as previously described.[12,25]

Relative Rates of Lipogenic Enzyme Synthesis

The relative rates of synthesis of malic enzyme and ATP-citrate lyase were determined by a combination of pulse-labeling with [^{35}S]Met, immunoprecipitation of the enzymes from cell cytosol, SDS-polyacrylamide gel electrophoresis, autoradiography, and scintillation spectrometry as described previously.[12]

Analysis of Malic Enzyme mRNA and Amino Acid Sequence Data

The cDNA sequence was analyzed for amino acid coding regions, restriction enzyme sites, internal symmetries and repeats, translational initiation and termination signals, and poly(A) addition sites using the programs of Staden.[26,27] Hydrophilic and hydrophobic regions of the polypeptide were identified by using the algorithm described by Kyte and Doolittle.[28] Secondary structure predictions are based on the Chou-Fasman rules[29] as applied in the program of George and Barker.[30] The protein sequence data base of the National Biomedical Research Foundation at Georgetown University was searched with the complete malic enzyme amino acid sequence by the methods of Lipman and Pearson[31] and Orcutt *et al.*[32] to reveal possible homologies with other proteins or domains of other proteins.

Materials

A plasmid (pGPD1) that contains a 350-bp sequence corresponding to murine glycerol 3-phosphate dehydrogenase mRNA[33] was a generous gift from Dr. Leslie Kozak, Jackson Laboratory, Bar Harbor, ME.

RESULTS AND DISCUSSION

Isolation of cDNA Clones Corresponding to Malic Enzyme mRNA Sequences

We have already described the construction and isolation of 25 recombinant plasmids containing malic enzyme cDNA sequences.[7] The largest insert (1327 bp) was found in plasmid pME1, and a poly(A) tail was detected in pME6 in preliminary sequencing analyses (FIG. 1). The identification of the 3′ end of malic enzyme mRNA

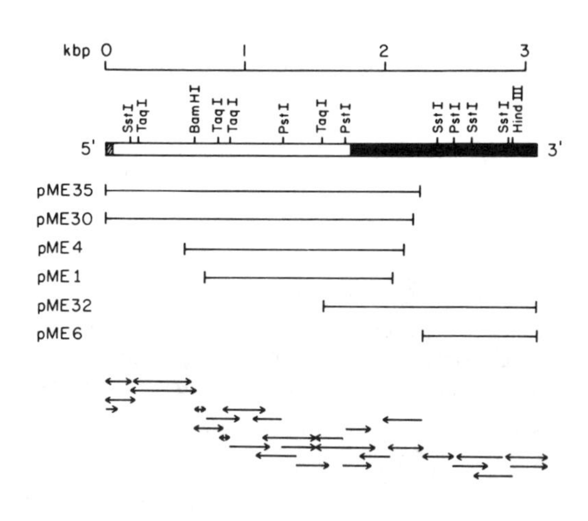

FIGURE 1. Restriction map and sequencing strategy for malic enzyme cDNA. The diagram is presented in the orientation corresponding to the 5′-to-3′ orientation of the mRNA. The positions of restriction sites used for sequencing are indicated. The open portion of the diagram is the protein-coding sequence; the hatched and solid portions are the 5′ and 3′ untranslated sequences, respectively. Lines immediately below the restriction map show the sizes and designations for cDNA inserts from recombinant plasmids that were used to establish the sequence. The 5′ and 3′ termini of all inserts are *Pst*I sites. Lines with arrowheads show the direction and extent of the sequences obtained from restriction fragments subcloned into M13 mp18 and M13 mp19.

and restriction endonuclease mapping of a series of overlapping recombinants revealed that only pME4 (FIG. 1) had a cDNA sequence that included nucleotides upstream from the 5′ end of the pME1 insert. Moreover, this series of 25 recombinants accounted for only approximately 70% of the 3.1-kb malic enzyme mRNA.

To obtain sequences corresponding to the 5′ end of the mRNA, a second library was constructed from an immunopurified malic enzyme mRNA template by the Gubler and Hoffman modification[15] of the procedure of Okayama and Berg.[16] This library was screened with an end-labeled *Bam*HI-*Pst*I fragment derived from pME4 (FIG. 1). Several recombinant plasmids containing inserts with substantial 5′ extensions of the mRNA sequence, including pME30 and pME35, were isolated and characterized (FIG. 1). The restriction endonuclease sites used for mapping, subcloning, and sequencing and the sequencing strategy are shown in FIGURE 1.

Sequence of Malic Enzyme mRNA

The primary sequence of the protein coding and 3′ untranslated portions of malic enzyme mRNA (3064 nucleotides) was established by dideoxynucleotide-chain-termination sequencing[19,20] and is presented in FIGURE 2. An open reading frame of 1716 bases is linked to 1348 nucleotides of the 3′ untranslated sequence. The open reading frame is followed by a series of 5 translation stop codons among the 12 succeeding codons (FIG. 2). Other reading frames in the 3.1-kb sequence are far too short to code for a subunit polypeptide with an M_r of approximately 60 000. The amino acid coding sequence accounts for 55% of the nucleotides in the larger malic enzyme mRNA and more than 85% of the primary structure of the 2-kb mRNA.

The first Met codon in the long open reading frame occurs at the nucleotide position designated "1" in FIGURE 2. Another AUG triplet is found at nucleotide 122, but utilization of this translation start site generates a reading frame with many stop codons interspersed with relatively short open reading frames. The sequence immediately surrounding the assigned initiator AUG exhibits two characteristics typical of translation start sites in eukaryotic mRNAs.[34] The AUG is preceded by a purine nucleotide three bases upstream and is immediately adjacent to dG in the 3′ direction (FIG. 2). In contrast, potential Met initiator codons beginning at nucleotides 122 and 226 are preceded by dT three nucleotides upstream. Utilization of the indicated AUG initiation codon also produces a derived sequence of 572 amino acids that is consistent with both the previously established subunit M_r[6,35,36] and the amino acid composition of malic enzyme (see below). Although all of the foregoing observations strongly support the conclusion that the initiator AUG is correctly assigned, final proof will require future sequence analyses of cDNA clones encompassing the complete 5′ untranslated region of malic enzyme mRNA and analysis of the amino acid sequence of the NH_2-terminal region of the enzyme subunit.

Genetic studies in mice[37] and Southern blot analyses of restriction digests of murine liver DNA (S. Bagchi, H. S. Sul, and C. S. Rubin, unpublished results) indicate that a single gene codes for malic enzyme. Thus it is probable that the co-equal expression of the 3.1-kb and 2-kb malic enzyme mRNAs in various murine tissues and cells (Sul *et al.*,[7] and FIGS. 6 & 8, in the last part of this section) results from a regulated, posttranscriptional processing step. Examination of the 3′ untranslated region of the sequence in FIGURE 2 suggests two possible explanations for the origin of the larger and smaller malic enzyme mRNAs. A sequence that exhibits a high degree of homology with a consensus "intervening sequence" (IVS) donor site for splicing[38] appears at nucleotide 1821 (FIG. 3A). Moreover, a second sequence that accurately matches the consensus acceptor site for splice junctions[38] is found at nucleotide 2961 (FIG. 3B). A possible interpretation is that approximately 50% of malic enzyme transcripts escape a nuclear processing step, and that a 1140-nucleotide IVS is retained in the 3′ untranslated portion of the mature cytoplasmic form of the 3.1-kb mRNA. Elimination of this sequence by splicing would produce a smaller mRNA species of 1924 nucleotides, plus poly(A) tail and 5′ untranslated sequence.

A second possibility is raised by the occurrence of a major consensus poly(A) addition signal at nucleotide 2028. Alternative utilization of this poly(A) addition site

```
                                    1                                                     30                                      60
                               C GGG GGG ATG GAG CCC CGA GCC CCC CGC CGC CGA CAC ACC CAC CAG CGC GGC TAC CTG CTG ACG CGG
                                         MET Glu Pro Arg Ala Pro Arg Arg Arg His Thr His Gln Arg Gly Tyr Leu Leu Thr Arg  20
                                    90                                         120                                     150
GAC CCG CAT CTC AAC AAG GAC TTG GCT TTT ACT CTG GAA GAG AGA CAG CAG TTG AAC ATT CAT GGA TTG TTG CCG CCC TGC ATC ATC AGC
Asp Pro His Leu Asn Lys Asp Leu Ala Phe Thr Leu Glu Glu Arg Gln Gln Leu Asn Ile His Gly Leu Leu Pro Pro Cys Ile Ile Ser  50
                                    180                                        210                                     240
CAG GAG CTC CAG GTC CTT AGA ATA ATT AAG AAT TTC GAA CGA CTG AAC TCT GAC TTC GAC AGG TAT CTC CTG TTA ATG GAC CTG CAA GAC
Gln Glu Leu Gln Val Leu Arg Ile Ile Lys Asn Phe Glu Arg Leu Asn Ser Asp Phe Asp Arg Tyr Leu Leu Leu Met Asp Leu Gln Asp  80
                                    270                                        300                                     330
AGA AAT GAG AAG CTC TTC TAC AGC GTG CTC ATG TCT GAT GTT GAA AAG TTC ATG CCT ATT GTT TAC ACC CCC ACC GTG GGC CTC GCA TGC
Arg Asn Glu Lys Leu Phe Tyr Ser Val Leu Met Ser Asp Val Glu Lys Phe Met Pro Ile Val Tyr Thr Pro Thr Val Gly Leu Ala Cys 110
                                    360                                        390                                     420
CAG CAG TAC AGT TTG GCA TTC CGG AAG CCA AGA GGC CTC TTT ATT AGT ATC CAT GAC AAA GGG CAC ATT GCT TCA GTT CTT AAT GCA TGG
Gln Gln Tyr Ser Leu Ala Phe Arg Lys Pro Arg Gly Leu Phe Ile Ser Ile His Asp Lys Gly His Ile Ala Ser Val Leu Asn Ala Trp 140
                                    450                                        480                                     510
CCA GAG GAT GTC GTC AAG GCT ATT GTG GTA ACT GAT GGA GAG CGC ATC CTT GGC TTG GGA GAC CTT GGC TGT AAT GGG ATG GGC ATC CCT
Pro Glu Asp Val Val Lys Ala Ile Val Val Thr Asp Gly Glu Arg Ile Leu Gly Leu Gly Asp Leu Gly Cys Asn Gly Met Gly Ile Pro 170
                                    540                                        570                                     600
GTG GGT AAA CTG GCC CTT TAC ACG GCA TGT GGA GGG GTG AAC CCA CAA CAG TGT CTA CCC ATC ACT TTG GAT GTG GGA ACA GAA AAT GAG
Val Gly Lys Leu Ala Leu Tyr Thr Ala Cys Gly Gly Val Asn Pro Gln Gln Cys Leu Pro Ile Thr Leu Asp Val Gly Thr Glu Asn Glu 200
                                    630                                        660                                     690
GAG TTA CTT AAG GAT CCA CTG TAC ATC GGG CTG CGG CAC CGG CGA GTC AGA GGC CCT GAG TAT GAC GCC TTC CTG GAT GAG TTC ATG GAG
Glu Leu Leu Lys Asp Pro Leu Tyr Ile Gly Leu Arg His Arg Arg Val Arg Gly Pro Glu Tyr Asp Ala Phe Leu Asp Glu Phe Met Glu 230
                                    720                                        750                                     780
GCA GCG TCT TCC AAA TAT GGC ATG AAT TGC CTT ATT CAG TTT GAA GAT TTT GCC AAT CGG AAT GCA TTT CGT CTC CTG AAC AAG TAT CGA
Ala Ala Ser Ser Lys Tyr Gly Met Asn Cys Leu Ile Gln Phe Glu Asp Phe Ala Asn Arg Asn Ala Phe Arg Leu Leu Asn Lys Tyr Arg 260
                                    810                                        840                                     870
AAC AAG TAT TGC ACA TTT AAC GAT GAT ATT CAA GGA ACA GCG TCT GTT GCG GTT GCG GGT CTC CTT GCA GCT CTT CGA ATA ACC AAG AAC
Asn Lys Tyr Cys Thr Phe Asn Asp Asp Ile Gln Gly Thr Ala Ser Val Ala Val Ala Gly Leu Leu Ala Ala Leu Arg Ile Thr Lys Asn 290
                                    900                                        930                                     960
AAG CTC TCT GAT CAG ACA GTG CTG TTC CAG GGA GCT GGA GAG GCT GCC TTG GGG ATT GCT CAC TTG GTT GTT ATG GCC ATG GAG AAA GAA
Lys Leu Ser Asp Gln Thr Val Leu Phe Gln Gly Ala Gly Glu Ala Ala Leu Gly Ile Ala His Leu Val Val Met Ala Met Glu Lys Glu 320
                                    990                                        1020                                    1050
GGT TTA TCA AAG GAG AAT GCT AGA AAG AAG ATA TGG TTG GTT GAC TCA AAA GGA CTA ATA GTT AAG GGT CGT GCA TCT CTC ACA GAA GAG
Gly Leu Ser Lys Glu Asn Ala Arg Lys Lys Ile Trp Leu Val Asp Ser Lys Gly Leu Ile Val Lys Gly Arg Ala Ser Leu Thr Glu Glu 350
                                    1080                                       1110                                    1140
AAA GAG GTG TTT GCC CAT GAA CAT GAA GAA ATG AAG AAT CTG GAA GCC ATT GTT CAA AAG ATA AAA CCA ACT GCC CTC ATA GGA GTT GCT
Lys Glu Val Phe Ala His Glu His Glu Glu Met Lys Asn Leu Glu Ala Ile Val Gln Lys Ile Lys Pro Thr Ala Leu Ile Gly Val Ala 380
                                    1170                                       1200                                    1230
GCA ATT GGT GGT GCT TTC ACT GAA CAA ATT CTC AAG GAT ATG GCT GCC TTC AAC GAG CGG CCC ATC ATC TTT GCT TTG AGT AGT CCG ACC
Ala Ile Gly Gly Ala Phe Thr Glu Gln Ile Leu Lys Asp Met Ala Ala Phe Asn Glu Arg Pro Ile Ile Phe Ala Leu Ser Ser Pro Thr 410
                                    1260                                       1290                                    1320
AGC AAA GCG GAG TGC TCT GCA GAC GAG TGC TAC AAG GTG ACC AAG GGA CGT GCA ATC TTT GCC AGC GGC AGT CCT TTT GAT CCA GTC ACT
Ser Lys Ala Glu Cys Ser Ala Asp Glu Cys Tyr Lys Val Thr Lys Gly Arg Ala Ile Phe Ala Ser Gly Ser Pro Phe Asp Pro Val Thr 440
                                    1350                                       1380                                    1410
CTC CCA GAT GGA CGG ACT CTG TTT CCT GGC CAA GGC AAC AAT TCC TAC GTG TTC CCT GGA GTT GCT CTT GGG GTG GTG GCC TGC GGA CTG
Leu Pro Asp Gly Arg Thr Leu Phe Pro Gly Gln Gly Asn Asn Ser Tyr Val Phe Pro Gly Val Ala Leu Gly Val Val Ala Cys Gly Leu 470
                                    1440                                       1470                                    1500
AGA CAC ATC GAT GAT AAG GTC TTC CTC ACC ACT CGT GAG GTC ATA TCT CAG CAA GTG TCA GAT AAA CAC CTG CAA GAA GGC CGG CTC TAT
Arg His Ile Asp Asp Lys Val Phe Leu Thr Thr Arg Glu Val Ile Ser Gln Gln Val Ser Asp Lys His Leu Gln Glu Gly Arg Leu Tyr 500
                                    1530                                       1560                                    1590
CCT CCT TTG AAT ACC ATT CGA GGC GTT TCG TTG AAA ATT GCA GTA AAG ATT GTG CAA GAT GCA TAC AAA GAA AAG ATG GCC ACT GTT TAT
Pro Pro Leu Asn Thr Ile Arg Gly Val Ser Leu Lys Ile Ala Val Lys Ile Val Gln Asp Ala Tyr Lys Glu Lys Met Ala Thr Val Tyr 530
                                    1620                                       1650                                    1680
CCT GAA CCC CAA AAC AAA GAA GAA TTT GTC TCC TCC CAG ATG TAC AGC ACT AAT TAT GAC CAG ATC CTA CCT GAT TGT TAT CCG TGG CCT
Pro Glu Pro Gln Asn Lys Glu Glu Phe Val Ser Ser Gln Met Tyr Ser Thr Asn Tyr Asp Gln Ile Leu Pro Asp Cys Tyr Pro Trp Pro 560
                                    1710                                       1740
GCA GAA GTC CAG AAA ATA CAG ACC AAA GTC AAC CAG TAA CGCAACAGCTAGGATTTTTAACTTTATTAGTAAAATCTTGAAGTTTTCATGATCTTTAAGGGTCAGAA 1787
Ala Glu Val Gln Lys Ile Gln Thr Lys Val Asn Gln  .              .        .       . .                                    572

TCTTTTATGATGATTCATAGAGAGCTTAGAATAAGGTGATTTTAGTTTAATAACAACTCATGGGAGTCTATTAGGATAAATTAGGATAAATTTCACACCAGACGGTTTTGTTTCACTTAC 1907
TGTGGATATTTATGTTTTCTCTGTTGATTATTCTCTTTATGAATTCTGTTTAAAAGCTACTGTACCTGCTGCTAGAAAGTCCTCACTGATATGTAGGAGGCTAATGGAAGACCCACTAGT 2027
AATAAATTAATATAGCATAACTTGATTATATTTAAGTGCCTACAGTTCTTTCTTGACTATTTTGCTAAAATCTCTAAACAGAAAAGATAAACACAAACTTGGGTATAGCTGAACTTTTAC 2147
TAAACAGAAGCACTACTTTGTTGCCTAGAGAAAAATCTTCTCAGGACTTTTATTCCAGGCCTCCGTTAGCTTTGTTCTCTTTGTACACCTGACTCAACACCTCTGAGAAAGCTCACTGCT 2267
GTTTACAGTACCCTTGCGTAGCCTTAGCTCATCAGCGTCTTCTGTCGTTGTTATGTTATATCCCATAGAGTAGAGCTCTCGAACCCAAACACTCCATAGGAAACACCCTTTCTCATCTCT 2387
GAGCAACCCTGGCCCTGTCGAGATATCGGGTGTTTTTGTTAGTGTAGCCTGGGACGTGAGAGGGCTGCAGGAGGGTCCTTGAGACGGGGCCCTGGAACCCACCTCTGAGACAAGGGAGTC 2507
AGATGCCAGACAGTGGTTCCCAGACAAGCTCAGGCCTCCATGAAGATCACCTGCTCTAATGTCCCTGTGCTTAGTTCGGAGGACTGAGAGCTCATGGCATGAGTAAATACATCTCTCAAT 2627
GCCTACCTTTCTATCAGATATTAAAATATGTTAATTACCAAAACCATTCTCTGAGAAAAAAAAAACCAAGCCTTTCCAGTGTATTAATTTACTGGACACGTTGATAATGGCATGACTAGA 2747
AACAGCCTTAACTCCTAAGTCAGGTTCAAGAACATTCTGTGTATCTAGAGACTCCTGACTTTGAAGTTGCTTTAAAGCCTGTGTGGTTTCCGGCGGGCAGGCTCTGTACAGTGAGCTCCT 2867
TGAAGCTTTCAAGGTGTGAGCTAAAACGGGTACAGACTTCCTAATGACAACATTGTGACTAACGGTTTCAACAGTGTAGTTATTTGAGAAAGCCATTTCAGAATTTCTATCTTTTCTTGT 2987
ATGTTTCCATGTTGTCAGGTAGTTGTAAATGAATGTATTTACCTATGCAAAAGATTTATTAAAGCCTAGAGAATATG 3064 + a 75 base poly A tail.
```

would generate a smaller malic enzyme mRNA of approximately 2045 bases, plus a poly(A) tail and 5′ untranslated sequence. In previous studies we reported that pME6 which contains an 830-bp cDNA insert corresponding to the 3′ terminus of the 3.1-kb mRNA (FIG. 1) hybridized only with the larger malic enzyme mRNA. This suggests that the smaller malic enzyme mRNA is produced by selection of an alternative poly(A) addition site and contains no sequences in common with the 3.1-kb mRNA beyond nucleotides 2045-2060. To test this idea, 26-mer and 17-mer oligonucleotides complementary to nucleotides 3008-3033 (probe 1) and 2213-2229 (probe 2), respectively, were synthesized and used for Northern gel analysis of mouse liver poly(A)$^+$ RNA (FIG. 4). Probe 1 will hybridize to a sequence beyond the splice acceptor site (position 2961, see FIG. 3B) and should appear in both mRNAs if alternate splicing pathways determine the generation of the two mRNAs. ^{32}P-labeled probe 1, however, hybridizes only with the 3.1-kb mRNA (FIG. 4). Probe 2 also failed to complex with the 2-kb mRNA, confirming the absence of common sequences only 175 bases, approximately, beyond the upstream poly(A) addition signal. Thus selection of alternative poly(A) addition sites appears to account for the sizes of the two malic enzyme mRNAs. The hexanucleotide signal at position 2028 is the most frequently used in eukaryotic mRNAs[39] whereas that directing poly(A) addition to the larger mRNA at nucleotide 3045 (AUUAAA) is only observed in approximately 10% of mRNAs.[40] Another AUUAAA sequence appears at nucleotide 2647, but this hexanucleotide apparently does not serve as a polyadenylation signal. Thus, other structural features in either the nuclear transcripts of the malic enzyme gene and/or the 3.1-kb mRNA itself must contribute to the maintenance of the 1:1 ratio of the two mRNA species (Sul *et al.*,[7] and FIGS. 6 & 8, in the last part of this section). It is possible to speculate that the striking difference between the 3′ untranslated regions of the two malic enzyme mRNAs could have effects on the efficiency of malic enzyme translation or mRNA stability, but the physiological significance is not known at present.

The utilization of several polyadenylation sites to generate multiple mRNAs that contain a single protein coding sequence is a characteristic shared with dihydrofolate reductase,[41] α-amylase,[42] and several other mRNAs. The example of dihydrofolate reductase is particularly striking because seven mRNAs, exhibiting 3′ untranslated sequences ranging from 80 to nearly 5000 nucleotides in length, are produced from a single gene.[41] In all instances the physiological and regulatory implications of size and sequence diversity in the 3′ noncoding regions of mRNAs remain to be elucidated.

Amino Acid Sequence of Malic Enzyme

The primary structure of murine malic enzyme was deduced from the 1716-nucleotide sequence corresponding to the long open reading frame in the mRNA (FIG. 2). Malic enzyme is a neutral polypeptide comprising 572 amino acid residues. Its calculated subunit M_r of 64 000 (TABLE 1) is in good agreement with the M_r of 268 000

FIGURE 2. Nucleotide sequence of malic enzyme mRNA and the deduced amino acid sequence of the malic enzyme subunit. The numbers on the far right correspond to the amino acid sequence. The polypeptide sequence contains 572 amino acids, beginning with the initiator Met at position 1. The peptide sequences (230-238 and 560-572) were confirmed by gas-phase microsequencing analysis. Five translation termination codons that follow the COOH-terminal Gln are indicated with dots. Hexanucleotide sequences constituting possible poly(A) addition signals are underlined.

determined for the native tetramer[35] and subunit M_r values of 58 000 to 63 000 estimated by SDS-polyacrylamide gel electrophoresis of denatured mouse and rat malic enzymes.[6,36] Because no partial amino acid sequence data were available for malic enzyme, the purified protein was cleaved with CNBr, and two fragments were purified to homogeneity by reverse-phase HPLC (R. Bradshaw and C. S. Rubin, unpublished results). Sequences corresponding to residues 230-238 and 560-572 (FIG. 2) were confirmed by gas-phase microsequencing (R. Bradshaw and C. S. Rubin, unpublished results). The latter fragment was derived from a pH-dependent cleavage of a Trp-Pro bond at the COOH-terminus during CNBr treatment in formic acid. The derived amino acid composition of murine malic enzyme closely matches the data obtained by direct amino acid analysis (TABLE 1).

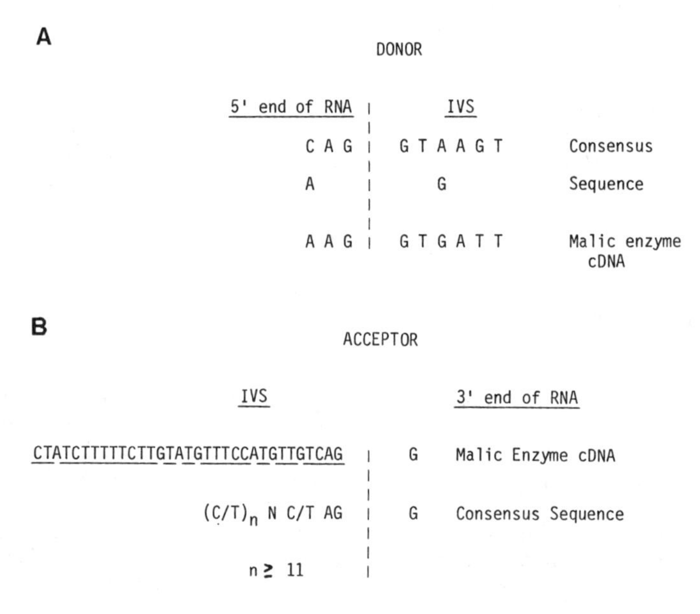

FIGURE 3. Comparison of consensus donor (**A**) and acceptor (**B**) splice sequences with the malic enzyme mRNA sequence at the boundaries of an apparent 1140-nucleotide "intervening sequence" that is not processed. The malic enzyme cDNA sequence shown in **A** includes nucleotides 1821-1829; that shown in **B** includes nucleotides 2974-3006. Consensus sequences were taken from Mount.[38]

A highly conserved feature of the AMP-binding portion of the dinucleotide fold[43] that accommodates NAD and NADP in several mammalian dehydrogenases is the octapeptide sequence X-Y-Gly-X-Gly-Y-X-Gly, where X is a hydrophobic residue; Y is variable; and the indicated spacing of the three Gly residues is retained. This sequence occurs at residues 156-163 of malic enzyme (FIG. 2), and its homology with components of the NAD(P) binding sites in alcohol dehydrogenase (ADH), glutamate dehydrogenase (GluDH), and three isozymes of lactate dehydrogenase (LDH) is illustrated in FIGURE 5A. Most of the differences between the malic enzyme octapeptide and the other sequences presented in FIGURE 5A could arise from single base changes

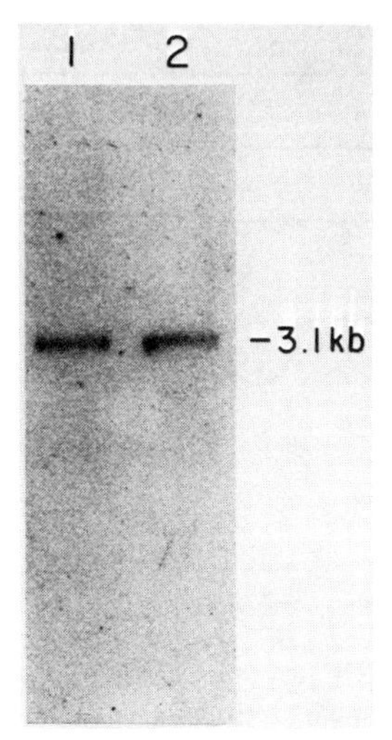

FIGURE 4. Hybridization of malic enzyme mRNAs with oligonucleotide probes complementary to portions of the 3′ untranslated sequence. Murine liver poly(A)$^+$ RNA (5μg/lane) was denatured, resolved by electrophoresis in 0.7% agarose, and transferred to nitrocellulose as described in the experimental procedures section. The Northern blot in lane 1 was probed with the ^{32}P-labeled 17-mer complementary to nucleotides 2013-2029 (probe 2, see text); that in lane 2 was probed with the ^{32}P-labeled 26-mer complementary to nucleotides 3008-3033 (probe 1, see text). Malic enzyme mRNA was visualized by autoradiography.

in codons. Another sequence, located more proximal to the NH_2-terminus (residues 101-109) of malic enzyme, can be aligned with a tryptic fragment of an NADP binding site in goose fatty acid synthetase[46] to yield six identical residues and two conservative substitutions that are potentially attributable to single base changes in codons (FIG. 5B). This high degree of homology suggests that the nonapeptide is a component of the coenzyme binding domain. A computer-assisted comparison of the derived primary sequence of malic enzyme with the data in protein sequence banks[31,32] failed to reveal more extensive homologies with the family of dehydrogenases that use NAD(P) as a coenzyme. In contrast, a segment of malic enzyme bounded by residues 402 and 458 (FIG. 2) aligns with a portion (residues 48-102) of the amino acid sequence of human, rabbit, and chicken triose phosphate isomerases (TPIs) to yield 28% identity (15 residues) and 44% conservative substitutions (24 residues) (FIG. 5C). Application of the Chou-Fasman rules[29] suggests that this region of malic enzyme has neither α-helical nor β-strand character. This prediction and the disposition of the TPI-like sequence between two hydrophobic regions (residues 374-406 and 456-473, FIG. 2) suggests that residues 402-458 comprise a hinge region connecting two functional domains. By analogy, this region might also be available for subunit-subunit interactions in the native malic enzyme tetramer because 21 of 27 side chains involved in intersubunit contacts in crystals of TPI are included in the homologous sequence (FIG. 5C).[47]

Relative Levels of Malic Enzyme mRNA in 3T3-L1 Preadipocytes and Adipocytes

We have already demonstrated that malic enzyme activity, mass, and relative rate of synthesis increase by an order of magnitude during the conversion of 3T3-L1

TABLE 1. Amino Acid Composition of Murine Malic Acid Enzyme

	Moles of Amino Acid per Mole Subunit	
Amino Acid	Measured[a]	Predicted
Ala	45	45
Arg	32	31
Asn	57[b]	25
Asp		30
Cys	11	11
Gln	70[c]	27
Glu		38
Gly	39	39
His	12	12
Ile	28	35
Leu	64	61
Lys	37	37
Met	12	13
Phe	24	24
Pro	29	29
Ser	25	27
Thr	25	26
Trp		3
Tyr	19	19
Val	37	40

NOTE: The predicted M_r of the malic enzyme subunit is 64 000.
[a]Amino acid analysis was carried out as previously described.[56]
[b]The sum of Asn plus Asp was determined.
[c]The sum of Gln plus Glu was determined.

preadipocytes into adipocytes.[12] To determine whether the increased expression of malic enzyme is controlled at a pretranslational level, poly(A)$^+$ RNA was prepared from cells at early (days 0 and 2), middle (day 4), and late (day 6) stages of adipocyte development (see Wise *et al.*[12] and Rubin *et al.*[25] for descriptions of morphological and biochemical changes that are characteristic of 3T3-L1 adipocyte differentiation) and was analyzed by a Northern gel procedure (see the experimental procedures section). After denaturation and electrophoresis on a 0.7% agarose gel, the resolved mRNA species were transferred to nitrocellulose and probed with plasmid pME1 that was labeled with ^{32}P by nick translation. These experiments were replicated four times and yielded similar results in each instance. Typical results are presented in FIGURES 6 & 7.

At the initial stages of adipocyte differentiation (FIG. 6B, lanes 2 and 3), the levels of both the 2-kb and 3.1-kb malic enzyme mRNAs are low. The relative concentrations of the two mRNAs increase markedly as adipogenesis ensues (FIG. 6B, lanes 4 and 5), and the levels reached by day 6 (FIG. 6B, lane 5) represent maximal values that are maintained for several succeeding days (data not shown). Although some slight variability in the relative proportions of the two malic enzyme mRNAs is occasionally observed during adipose conversion, approximately equal amounts of the 3.1-kb and 2-kb malic enzyme mRNAs are consistently found in the mature adipocytes (FIG. 6B, lane 5). Thus, the levels of both malic enzyme mRNAs appear to be regulated in parallel throughout the progression of the differentiation program. Moreover, the utilization of two alternative poly(A) addition signals (see FIG. 2 and the text above) is strictly controlled in 3T3-L1 cells to produce a near-stoichiometric balance between two mRNA species that differ by 1.1 kb in their 3′ untranslated regions, irrespective of the state of differentiation or total abundance of malic enzyme mRNA sequences (FIG. 6B).

Northern blot analyses performed with a cDNA probe for ATP-citrate lyase,[8] a second lipogenic enzyme, revealed that the kinetics of accumulation of ATP-citrate lyase and malic enzyme mRNAs are very similar (FIGS. 6A & 6B). These data were quantified by scanning densitometry, and the malic enzyme and ATP-citrate lyase mRNA levels were compared with the relative rates of synthesis of the two enzymes in the same batch of sells (FIG. 7). During adipocyte differentiation (days 0-6, FIGS. 6 & 7) there is a good correlation between mRNA content and the relative rates at which the two enzymes are synthesized. The same results are obtained when malic

A

Malic Enzyme	Ile (156)	Leu	<u>Gly</u>	Leu	<u>Gly</u>	Asp	Leu	<u>Gly</u> (163)
LDH-C	Val (26)	Val	<u>Gly</u>	Val	<u>Gly</u>	Asx	Val	<u>Gly</u> (33)
LDH-A	Val (26)	Val	<u>Gly</u>	Val	<u>Gly</u>	Ala	Val	<u>Gly</u> (33)
LDH-B	Val (26)	Val	<u>Gly</u>	Val	<u>Gly</u>	Gln	Val	<u>Gly</u> (33)
GLuDH	Val (249)	Gln	<u>Gly</u>	Phe	<u>Gly</u>	Asn	Val	<u>Gly</u> (256)
ADH	Val (197)	Phe	<u>Gly</u>	Leu	<u>Gly</u>	Gly	Val	<u>Gly</u> (204)

B

Malic Enzyme	<u>Val</u> (101)	Tyr	<u>Thr</u>	Pro	<u>Thr</u>	<u>Val</u>	<u>Gly</u>	Leu	Ala (108)
Fatty Acid Synthetase NADP-binding fragment	<u>Val</u>	Phe	<u>Thr</u>	—	<u>Thr</u>	<u>Val</u>	<u>Gly</u>	Ser	<u>Ala</u>

C

```
                402                                                     458
Malic Enzyme    IIFALSSPTSKAECSADECYKVTKGRAIFASGSPFDPVTLPDGRTLFPGQGNNSYVF
                : ::  . ..: . .:..:::::.:   :... . . ..   ..  . :......::
TPI             IDFARQKLDPKIAVAAQNCYKVTNG--AFTGEISPGMIKDCGATWVVLGHSERRHVF
                48                                                      102
```

: = identity

. = conservative substitution

FIGURE 5. Alignments of sequences showing homology of malic enzyme with (**A**) a portion of the NAD- and NADP-binding regions of several dehydrogenases, including glutamate dehydrogenase (GluDH), alcohol dehydrogenase (ADH), and three isozymes (A, B, and C) of lactate dehydrogenase (LDH); (**B**) a tryptic fragment of an NADP-binding domain in fatty acid synthetase; and (**C**) the subunit-subunit interaction site in triose phosphate isomerase (TPI). The sequences in **A** are from Rossman *et al.*,[43] Branden *et al.*,[44] and Smith *et al.*;[45] the sequence in **B** is from Poulouse and Kolattukudy;[46] and the sequence in **C** is taken from Banner *et al.*[47]

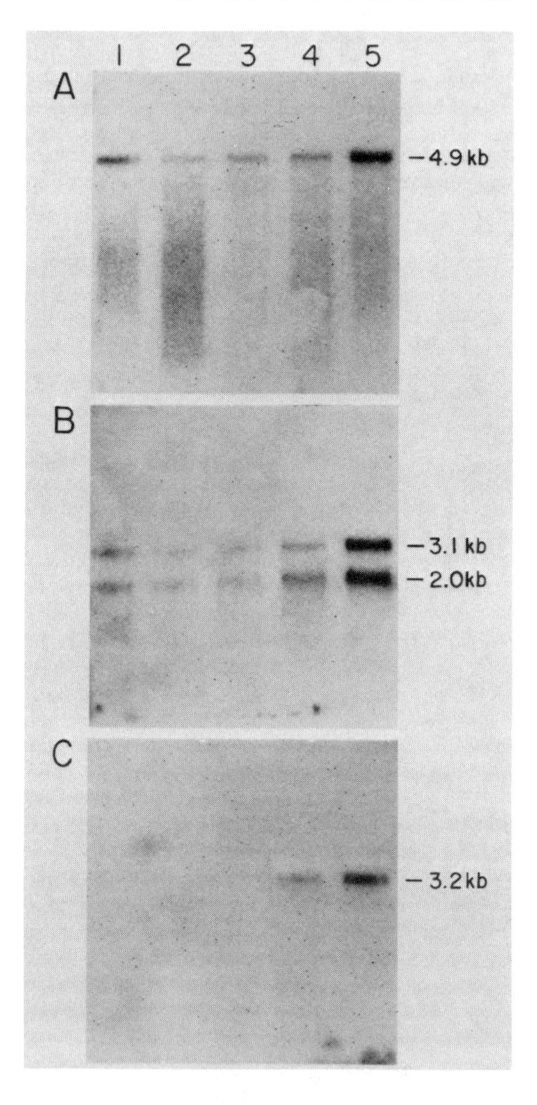

FIGURE 6. Expression of malic enzyme, ATP-citrate lyase, and glycerol 3-phosphate dehydrogenase mRNAs in undifferentiated and differentiating 3T3-L1 cells. Poly(A)$^+$ RNAs (7μg/lane) were denatured, resolved by electrophoresis in 0.7% agarose, and transferred to nitrocellulose as described in the experimental procedures section. Poly(A)$^+$ RNA applied to lane 1 was isolated from preconfluent preadipocytes (day -1). Lane 2 contained mRNA from confluent preadipocytes. Lanes 3, 4, and 5 received mRNA prepared from developing adipocytes 2, 4, and 6 days after the initiation of differentiation, respectively.[25] The Northern blot in **A** was probed for ATP-citrate lyase mRNA (4.9 kb) sequences with the nick-translated, ^{32}P-labeled plasmid pACL 2;[8] the blot in **B** was hybridized with ^{32}P-labeled plasmid pME1; and the blot in **C** was hybridized with ^{32}P-labeled plasmid pGPD1[33] to detect GPD mRNA (3.2 kb). Murine 28S and 18S mRNAs and *E. coli* 23S and 16S mRNAs were used as size markers. An autoradiogram is shown.

enzyme and ATP-citrate lyase mRNA levels are assessed by dot-blot analysis (data not shown). Therefore, the expression of the two lipogenic enzymes appears to be coordinately controlled at a pretranslational level during adipocyte development.

Another aspect of lipogenic enzyme regulation was discerned in 3T3-L1 preadipocytes. Substantial levels of both malic enzyme and ATP-citrate lyase mRNAs are detected in confluent, undifferentiated cells (day -1: FIG. 6A, lane 1, and FIG. 6B, lane 1). On day -2 and at earlier times of exponential cell growth the relative amounts of ATP-citrate lyase and malic enzyme mRNAs are the same as the levels found at day 0 of the differentiation protocol (FIG. 6A, lane 2, and FIG. 6B, lane 2). Despite the three- to fourfold elevation in lipogenic enzyme mRNA content, the rates of synthesis of the two enzymes are only enhanced by 50-100% in preadipocytes on day -1 (FIG. 7). Because both ATP-citrate lyase and malic enzyme mRNAs from preadipocytes

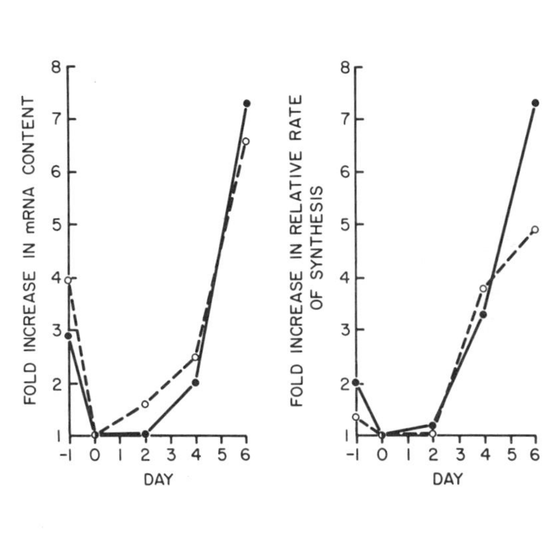

FIGURE 7. Relative mRNA abundance and rates of enzyme synthesis for malic enzyme and ATP-citrate lyase during adipocyte differentiation. The relative amounts of malic enzyme (●——●) and ATP-citrate lyase (○----○) mRNAs were estimated by quantitative densitometry (left panel) of the autoradiograms presented in FIGURES 6A & 6B. The relative amount of malic enzyme mRNA includes the sum of the integrated values for the 3.1-kb and 2-kb mRNAs. The densitometry data were normalized to the values obtained on day 0 and are presented as the ratio day n/day 0 (that is, as fold increase). The same results were obtained from dot-blot analyses of total RNA (not shown). The right panel depicts the relative rates of synthesis of malic enzyme and ATP-citrate lyase obtained from pulse-labeling with [^{35}S]Met (see Wise *et al.*[12] and the experimental procedures section).

(day -1, 0) and adipocytes are translated with similar efficiencies *in vitro*,[12] the data suggest that the expression of ATP-citrate lyase and malic enzyme is under translational control in preadipocytes. The discrepancy between the mRNA level and the rate of synthesis for malic enzyme is striking whereas a more modest effect is observed for ATP-citrate lyase. This form of regulation appears to be relieved during adipogenesis. The elevation in ATP-citrate lyase and malic enzyme mRNA content in confluent preadipocytes on day -1 could be a function of density-dependent or cell-cycle-specific factors or could arise from the depletion of a key nutrient in the medium.

Although the molecular mechanisms responsible for controlling lipogenic enzyme mRNA expression remain to be explored, it is relevant to distinguish between gene products that 1) are elevated during adipocyte differentiation, 2) are subject to tissue-specific hormonal control, and 3) are also constitutively expressed in a variety of other tissues from those that are highly abundant in fat cells (and perhaps a few other cells) but are rare in most tissues and organs. Malic enzyme typifies the first, more complex category, being constitutively expressed at substantial levels in a variety of mouse (FIG. 8) and rat[13] tissues, but subject to marked hormonal, dietary, and developmental regulation in adipocytes (FIG. 6), hepatocytes (FIG. 8), and mammary tissue.[1] The

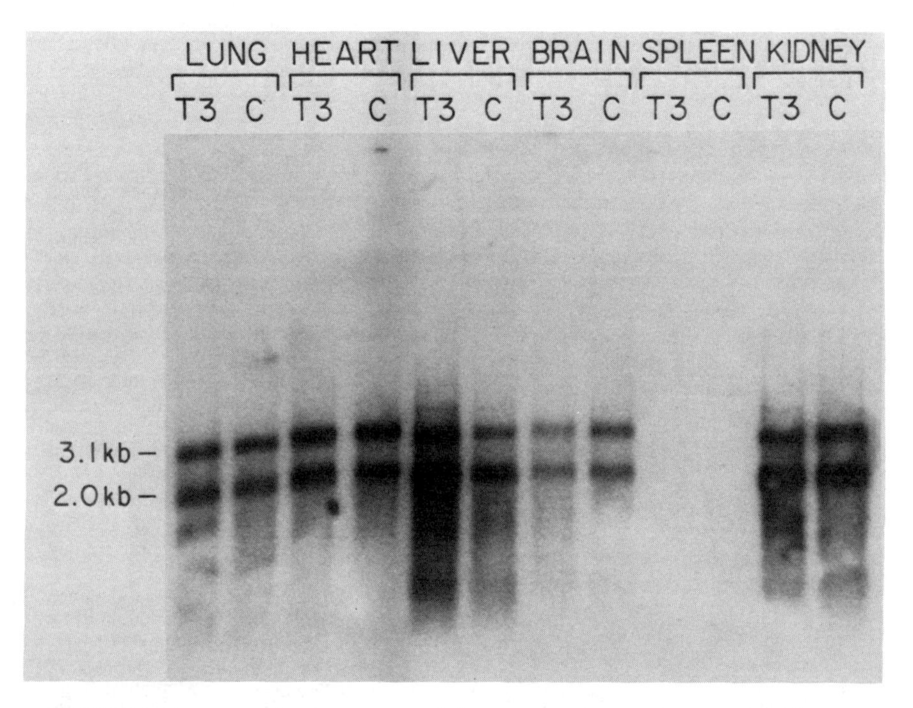

FIGURE 8. Expression of malic enzyme mRNAs in various mouse tissues. Poly(A)$^+$ mRNAs (5 μg/lane) isolated from the indicated tissues were denatured, resolved by electrophoresis in 0.7% agarose, and transferred to nitrocellulose as described in the experimental procedures section. The Northern blot was probed with nick-translated ^{32}P-labeled pME1, and the malic enzyme 3.1-kb and 2-kb mRNAs were visualized by autoradiography. The relative amounts of malic enzyme mRNAs were compared in tissues obtained either from animals injected with thyroid hormone (sufficient to saturate triiodothyronine receptors for 48 hr) 36 hr prior to RNA isolation or from saline-injected controls. Quantitative densitometry indicated that malic enzyme mRNA content rose 7-fold in liver and 2.3-fold in heart in response to the hormone. Triiodothyronine did not significantly alter malic enzyme mRNA content in the other tissues.

"adult" isozyme of glycerol 3-phosphate dehydrogenase (GPD), an anabolic enzyme that converts a glycolytic intermediary metabolite to a precursor of the glycerol backbone of triacylglycerides, is a prototype for the second category.[33,48] As originally reported by Spiegelman and Green and their colleagues,[49–51] GPD activity and mass and GPD mRNA are undetectable in 3T3-F442A preadipocytes, but appear and increase precipitously during adipose conversion. Similarly, GPD mRNA was absent from confluent 3T3-L1 preadipocytes, but became relatively abundant in adipocytes (FIG. 6C).

It is clear that the regulation of the two categories of mRNAs must be fundamentally different in adipocytes and preadipocytes (FIG. 8). Moreover, the mechanisms employed by 3T3-L1 adipocytes and other adipocyte cell lines to modulate the amount of previously expressed mRNAs and to switch on adipocyte-specific genes include increased transcription rates, altered mRNA stability, and a combination of these two processes.[52-55] This diversity is further complicated by the differential kinetics of accumulation of specific mRNAs, which may reflect the serial induction of regulatory proteins.[52] Further studies on the molecular basis of malic enzyme regulation in fibroblasts, preadipocytes, and adipocytes may provide novel insights on multiple factors involved in its expression as a constitutive "housekeeping" gene product in some cells and tissues, and as a developmentally and hormonally regulated enzyme in others.

REFERENCES

1. FRENKEL, R. 1975. Curr. Top. Cell. Regul. **9:** 157-181.
2. GIBSON, D. M., R. T. LYONS, D. F. SCOTT & Y. MUTO. 1972. Adv. Enzyme Regul. **10:** 187-204
3. FISCHER, P. W. F. & A. G. GOODRIDGE. 1978. Arch. Biochem. Biophys. **190:** 332-344.
4. MARIASH, C. N., F. E. KAISSER & J. H. OPPENHEIMER. 1980. Endocrinology **106:** 22-27.
5. WINBERRY, L. K., S. M. MORRIS, J. E. FISCH, M. J. GLYNIAS, R. A. JENIK & A. G. GOODRIDGE. 1983. J. Biol. Chem. **258:** 555-559.
6. MAGNUSON, M. & V. M. NIKODEM. 1983. J. Biol. Chem. **258:** 12712-12717.
7. SUL, H. S., L. S. WISE, M. L. BROWN & C. S. RUBIN. 1984. J. Biol. Chem. **259:** 555-559.
8. SUL, H. S., L. S. WISE, M. L. BROWN & C. S. RUBIN. 1984. J. Biol. Chem. **259:** 1201-1205.
9. MORRIS, S. M., J. H. NILSON, R. A. JENIK, L. K. WINBERRY, M. A. MCDEVITT & A. G. GOODRIDGE. 1982. J. Biol. Chem **257:** 3225-3229.
10. GOLDMAN, M. J., D. W. BACK & A. G. GOODRIDGE. 1985. J. Biol. Chem. **260:** 4404-4408.
11. BALLARD, F. J. & R. W. HANSON. 1967. Biochem. J. **102:** 952-958.
12. WISE, L. S., H. S. SUL & C. S. RUBIN. 1984. J. Biol. Chem **259:** 4827-4832.
13. DOZIN, B., M. A. MAGNUSON & V. M. NIKODEM. 1985. Biochemistry **24:** 5581-5586.
14. LAND, H., M. GREZ, H. HAUSER, W. LINDENMAIER & G. SCHUTZ. 1981. Nucleic Acids Res. **9:** 2251-1166.
15. GUBLER, U. & B. HOFFMAN. 1983. Gene **25:** 263-269.
16. OKAYAMA, H. & P. BERG. 1982. Mol. Cell. Biol. **2:** 161-170.
17. THAYER, R. E. 1979. Anal. Biochem. **98:** 60-63.
18. MESSING, J. 1983. Methods Enzymol. **101:** 20-78.
19. SANGER, F., S. NICKLEN & A. R. COULSON. 1977. Proc. Natl. Acad. Sci. USA **74:** 5463-5467.
20. SANGER, F., A. R. COULSON, B. G. BARRELL, A. J. H. SMITH & B. A. ROE. 1980. J. Mol. Biol. **143:** 161-178.
21. MAXAM, A. & W. GILBERT. 1980. Methods Enzymol. **65:** 499-560.
22. CHIRGWIN, J. W., A. E. PRZBYLA, R. J. MACDONALD & W. J. RUTTER. 1979. Biochemistry **18:** 5294-5299.
23. KRYSTOSEK, A., M. L. CAWTHON & D. KABAT. 1975. J. Biol. Chem. **250:** 6077-6084.
24. RAVE, R., R. CRKVENJAKOV & H. BOEDTKER. 1979. Nucleic Acids Res. **6:** 3559-3567.
25. RUBIN, C. S., A. H. HIRSCH, C. FUNG & O. M. ROSEN. 1978. J. Biol. Chem. **253:** 7570-7578.
26. STADEN, R. 1977. Nucleic Acids Res. **4:** 4037-4051.
27. STADEN, R. 1984. Nucleic Acids Res. **12:** 505-519.
28. KYTE, J. & R. F. DOOLITTLE. 1984. J. Mol. Biol. **157:** 105-132.
29. CHOU, P. Y. & G. D. FASMAN. 1978. Annu. Rev. Biochem. **47:** 251-276.

30. GEORGE, D. G. & W. C. BARKER. 1985. PRPLOT: PIR Report PRP-0185. National Biomedical Research Foundation. Washington, DC.
31. LIPMAN, D. J. & W. R. PEARSON. 1985. Science **227:** 1435-1441.
32. ORCUTT, B. C., M. O. DAYHOFF & W. C. BARKER. SEARCH: NBR Report 820503-08710. National Biomedical Research Foundation. Washington, DC.
33. KOZAK, L. P. & E. H. BIRKENMEIER. 1983. Proc. Natl. Acad. Sci. USA **80:** 3020-3024.
34. KOZAK, M. 1984. Nucleic Acids Res. **12:** 857-872.
35. LI, J. J., C. R. ROSS, H. M. TEPPERMAN & J. TEPPERMAN. 1975. J. Biol. Chem. **250:** 141-148.
36. WISE, L. S. & C. S. RUBIN. 1984. Anal. Biochem. **140:** 256-263.
37. SHOWS, T. B., V. M. CHAPMAN & F. H. RUDDLE. 1970. Biochem. Genet. **4:** 707-718.
38. MOUNT, S. M. 1982. Nucleic Acids Res. **10:** 459-472.
39. BROWNLEE, N. J. & N. PROUDFOOT. 1976. Nature **263:** 211-214.
40. WICKENS, M. & P. STEPHENSON. 1984. Science **226:** 1045-1051.
41. SETZER, D. R., M. MCGROGAN & R. T. SCHIMKE. 1982. J. Biol. Chem. **257:** 5143-5147.
42. TOSI, M., R. A. YOUNG, O. HAGENBUCHLE & U. SCHIBLER. 1981. Nucleic Acids Res. **9:** 2313-2323.
43. ROSSMANN, M. G., A. LILJAS, C. I. BRANDEN & L. J. BANASZAK. 1975. *In* The Enzymes. P. D. Boyer, Ed.: 61-102. Academic Press. New York, NY.
44. BRANDEN, C. I., H. JORNVALL, H. EKLUND & B. FURUGREN. 1975. *In* The Enzymes. P. D. Boyer, Ed.: 103-190. Academic Press. New York, NY.
45. SMITH, E. L., B. M. AUSTEN, K. M. BLUMENTHAL & J. F. NYC. 1975. *In* The Enzymes. P. D. Boyer, Ed.: 293-367. Academic Press. New York, NY.
46. POULOUSE, A. J. & P. E. KOLATTUKUDY. 1983. Arch. Biochem. Biophys. **220:** 652-656.
47. BANNER, D. W., A. C. BLOOMER, G. A. PETSKO, D. C. PHILLIPS, C. I. POGSON, I. A. WILSON, P. H. CORRAN, A. J. FURTH, J. D. MILMAN, R. E. OFFORD, J. D. PRIDDLE & S. G. WALEY. 1975. Nature **255:** 609-614.
48. RATNER, P. L., M. FISHER, D. BURKART, J. R. COOK & L. P. KOZAK. 1981. J. Biol. Chem. **256:** 3576-3579.
49. SPIEGELMAN, B. M., M. FRANK & H. GREEN. 1983. J. Biol. Chem. **258:** 10083-10089.
50. WISE, L. S. & H. GREEN. 1979. J. Biol. Chem. **254:** 273-275.
51. SPIEGELMAN, B. & H. GREEN. 1980. J. Biol. Chem. **255:** 8811-8818.
52. CHAPMAN, A. B., D. M. KNIGHT, B. S. DIECKMANN & G. M. RINGOLD. 1984. J. Biol. Chem. **259:** 15548-15555.
53. COOK, K. S., C. R. HUNT & B. M. SPIEGELMAN. J. Cell Biol. **100:** 514-520.
54. BERNLOHR, D. A., M. A. BOLANOWSKI, T. J. KELLY & M. D. LANE. 1985. J. Biol. Chem. **260:** 5563-5567.
55. CHAPMAN, A. B., D. M. KNIGHT & G. M. RINGOLD. 1985. J. Cell Biol. **101:** 1227-1235.
56. STEIN, J. C. & C. S. RUBIN. 1985. J. Biol. Chem. **260:** 10991-10995.

Molecular Basis for the Hormonal Regulation of the Tyrosine Aminotransferase and Tryptophan Oxygenase Genes

GÜNTHER SCHÜTZ, WOLFGANG SCHMID, MICHAEL JANTZEN, ULRICH DANESCH, BERND GLOSS, UWE STRÄHLE, PETER BECKER, AND MICHAEL BOSHART

Institute of Cell and Tumor Biology
German Cancer Research Center
D-6900 Heidelberg, Federal Republic of Germany

INTRODUCTION

A classical example for hormonal control of gene expression is the induction of the enzymes tyrosine aminotransferase (TAT) and tryptophan oxygenase (TO).[1-3] These two enzymes are expressed exclusively in the parenchymal cells of the liver.[4] The expression of both genes is affected by glucocorticoids.[5-10] In nuclear run-on assays it could be shown that hormonal induction involves changes in the relative rates of transcription of these two genes. Apart from glucocorticoid induction, the activity of the TAT gene is also regulated by cAMP.[11-14] The accumulation of TAT mRNA after cAMP stimulation is a consequence of transcriptional activation of the TAT gene.[15] Combined dexamethasone and cAMP treatment leads to higher TAT mRNA concentrations than treatment with either inducer alone, which implies that dexamethasone and cAMP act by distinct mechanisms.[15]

In addition to hormonal regulation, both TAT and TO genes are under developmental control. Whereas TAT enzyme activity appears around birth, TO enzyme activity can be detected only 2 weeks later.[16] The activity of the TAT gene is affected by two distinct genetic loci, which appear to operate in *trans* on the expression of the TAT gene. By genetic and biochemical analysis of several albino lethal mutants of the mouse, a control region required for expression and inducibility of several liver enzymes including TAT has been assigned to the region of the albino locus on chromosome 7 of the mouse.[17] The basal level of TAT mRNA is severely decreased, and the concentration of the mRNA is no longer inducible by glucocorticoids and cAMP, although the structural gene for TAT is still present.[18] Because the mouse TAT structural gene is located on chromosome 8,[19] the effect of this presumptive regulatory locus must operate in *trans*.

Another locus, the so-called tissue-specific extinguisher-1 (*Tse-1*) has been mapped to chromosome 11 of the mouse.[20] When chromosome 11 of a fibroblast cell is introduced by microcell fusion into a hepatoma cell expressing TAT, TAT gene activity is selectively extinguished, suggesting that a negatively acting factor encoded in chromosome 11 is responsible for this effect.

To study the complex control mechanism operating on these two liver-specific genes at the molecular level, we have isolated the TAT and TO genes from the rat as well as the TAT gene of the mouse.[7,9,10,19] The availability of these DNA clones have allowed us to determine the level at which hormonal regulation of expression of these two genes occurs and, furthermore, allowed us to identify those sequences that are required for hormonal induction in *cis*. We demonstrate here that these sites represent binding sites for the glucocorticoid receptor.

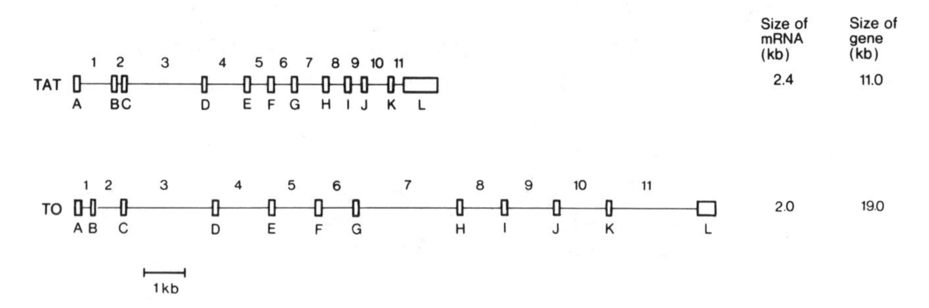

FIGURE 1. Exon-intron structure of the rat TAT and TO genes. Open boxes represent the exons; connecting lines represent the intervening sequences, and are drawn to scale. The size of the mRNA and the size of the gene are indicated in kb pairs.

RESULTS

To allow analysis at the molecular level of the sequences important for the hormonal regulation of expression of the TO and TAT genes, we have cloned cDNA sequences representing these two mRNAs[9,10] as well as genomic DNA containing these two genes.[7,19] The exon-intron organization and the start site of transcription of these two genes was determined by electron microscopic analysis of heteroduplexes of mRNA with the cloned genomic DNA, as well as by S1 nuclease mapping.[7,10] FIGURE 1 shows the exon-intron organization of these two genes. The coding portion of these genes is interrupted by multiple intervening sequences; thus the genes are much longer than the respective mRNAs. In order to allow studies of the molecular effects of the albino deletion mutations, we have isolated and characterized the mouse TAT gene and have shown that the rat and mouse TAT genes are very similar in exon-intron structure. Sequence comparison around the 5′ end of the rat and mouse genes shows extensive homology over the entire sequence (approximately 1 kb), indicating the importance of these sequences for the regulation of these two genes.[19]

In order to obtain clues to possible regulatory sequences, we have analyzed the chromatin structure in the vicinity of the 5′ end of the TO and TAT genes with regard to DNase I hypersensitivity. Hypersensitive sites frequently map near the 5′ end of actively transcribed genes. Their presence has been correlated in many systems with the state of expression. We have found DNase I hypersensitive sites at each of the two promoters in liver cells of the rat[22] and mouse (FIG. 2). DNase I hypersensitive

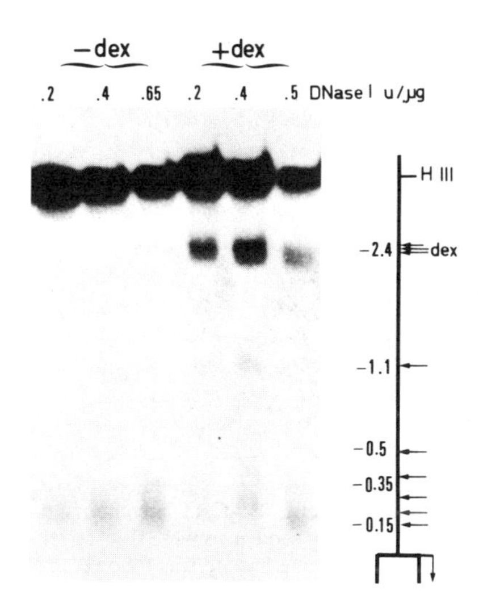

FIGURE 2. A glucocorticoid-inducible DNase I hypersensitive site far upstream of the mouse TAT gene. Nuclei from hormone-induced (+ dex) and uninduced (− dex) mouse livers were treated with increasing amounts of DNase I. After purification, the DNAs were cut with the restriction enzyme abutting with the probe used, separated on a 1% agarose gel, and blotted to nitrocellulose filters. The end-labeling technique was used for visualizing the preferential DNase I cleavage sites. On the right part of the figure these hypersensitive sites are aligned with the 5′ flanking region of the TAT gene.

sites of both genes are absent in kidney nuclei and therefore appear to be specific for the tissue expressing the genes.[22] Most remarkable is the alteration in chromatin structure following glucocorticoid induction (FIG. 2). A strong DNase I hypersensitive site is detected about 2.4 kb upstream of the start site of transcription of the mouse TAT gene. This dexamethasone-dependent hypersensitive site suggested to us the possibility that glucocorticoid response elements might be located far upstream of the start site of transcription.

To identify sequences important for the hormonal control of these two genes by glucocorticoids, fusion genes were constructed containing the 5′ flanking DNA of these genes upstream of a suitable indicator gene. We used either the bacterial gene for neomycin resistance (neo) or the gene for the enzyme chloramphenicol transacetylase (CAT). A typical hybrid gene is shown in FIGURE 3. Because chromatin studies have suggested that the control sequence might be remote from the cap-site[22] (FIG. 2), 2.9 kb of 5′ flanking DNA was included in the construction of this hybrid gene. In the case of the TO promoter, a CAT fusion gene containing 1.9 kb of 5′ flanking DNA was constructed. These fusion genes were introduced into hepatoma cells and fibroblasts, and the effect of steroid administration on expression of these genes was analyzed by CAT enzyme activity measurements or analysis of the RNA with RNA filter hybridization experiments and/or with S1 nuclease analysis. Both

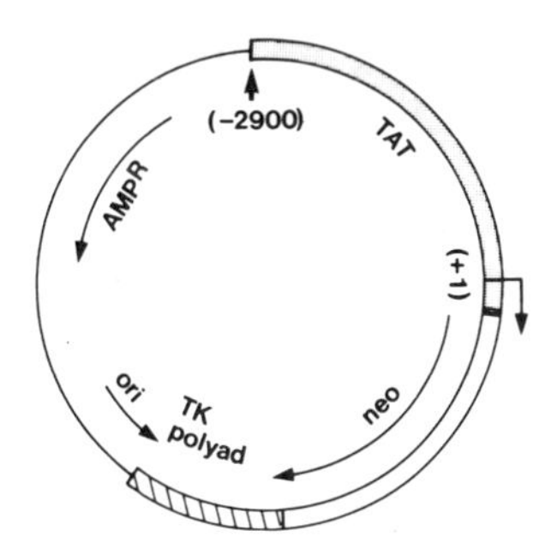

FIGURE 3. Structure of pTATneo, a fusion gene of the TAT 5′ flanking DNA and the bacterial neomycin resistance gene.

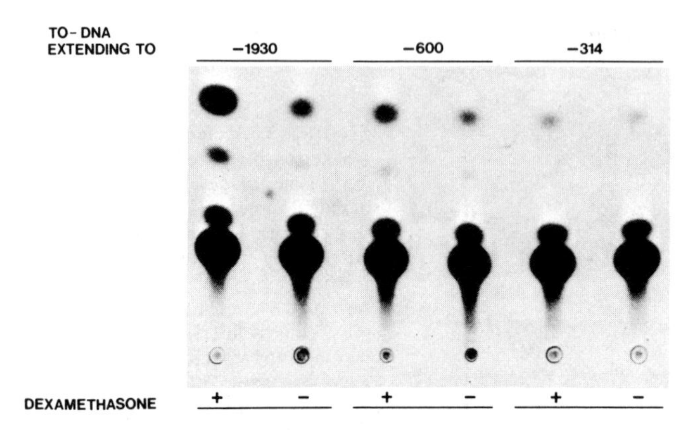

FIGURE 4. Expression of TO-CAT recombinants in mouse L cells. TO-CAT recombinants containing approximately 1930 bp, 600 bp, and 340 bp of 5′ flanking sequences were transferred into mouse L cells. Expression in the presence and absence of dexamethasone was followed by measuring CAT activity in extracts of the cells 48 hr after transfection.

genes, when introduced into heterologous or homologous cells, are expressed, and the expression is regulated by glucocorticoids (data not shown). In order to delineate the important regulatory elements in more detail, deletion mutants of these parental plasmids were constructed and tested. A typical experiment is shown in FIGURE 4 for the TO-CAT recombinants expressed in mouse L cells. It is seen that as long as 1930 nucleotides remain of 5′ flanking sequences of the TO fusion gene, strong in-

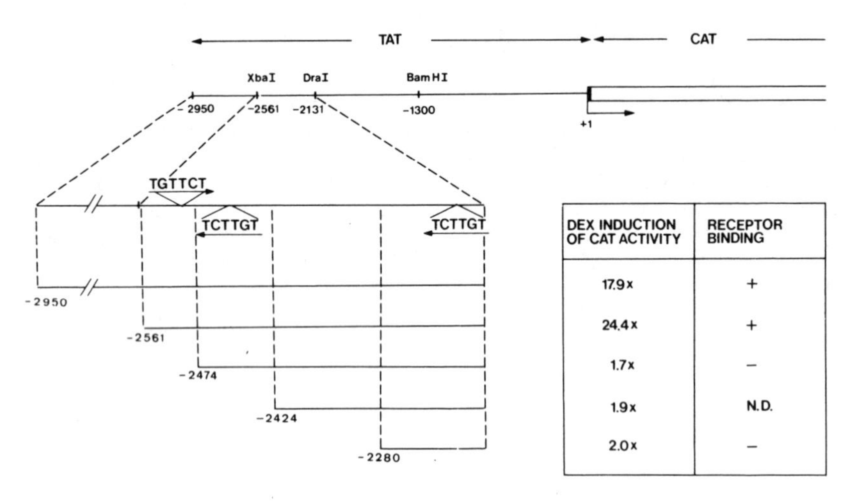

FIGURE 5. A remote DNA sequence confers glucocorticoid responsiveness and receptor binding. The upper part of the figure shows the 5′ flanking part of the TAT gene fused onto the CAT gene. Deletion derivatives as indicated by the deletion end points were tested for dexamethasone inducibility of expression of the CAT recombinants in L cells 12 hr after hormone treatment as well as for receptor binding. The approximate position of the hexanucleotide (TGTTCT) indicative of glucocorticoid receptor binding sites is shown.

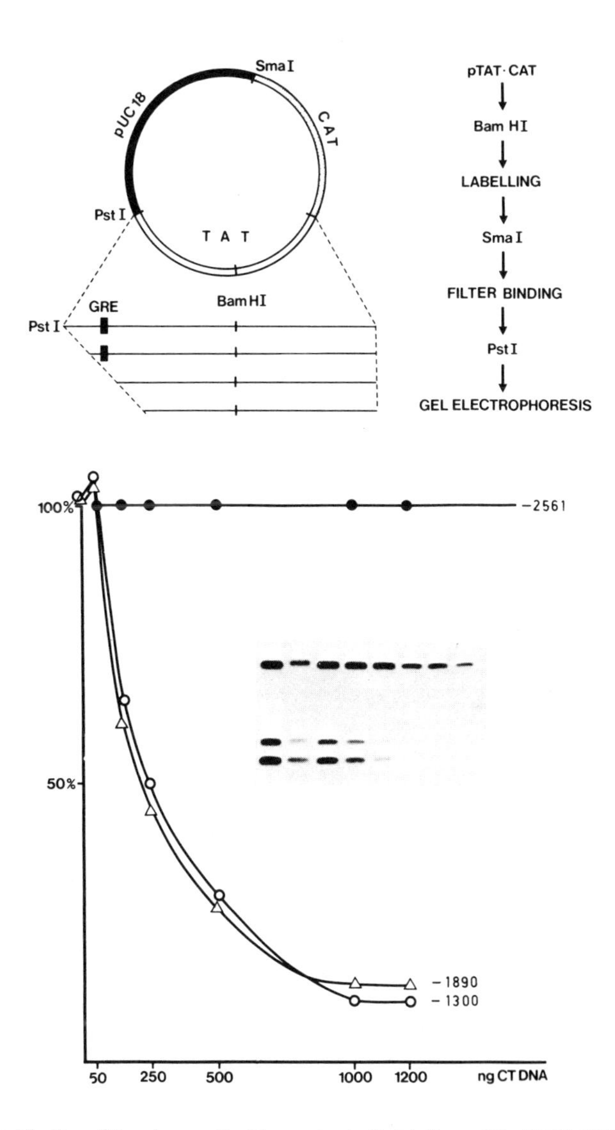

FIGURE 6. Binding of the glucocorticoid receptor to 5′ deletions of the TAT-CAT recombinant. The upper part of the figure shows the method employed to determine binding of the purified glucocorticoid receptor to 5′ deletion derivatives of the TAT-CAT plasmids. The lower part of the figure shows the results of a filter binding experiment with three plasmids containing 5′ flanking TAT DNA as indicated in the presence of increasing amounts of competing calf thymus DNA (CT DNA). The inset shows the autoradiogram of the fragments preferentially retained on nitrocellulose filters. The top set of fragments represents the deletion ending at -2561; the middle set, the deletion ending at -1890; and the lower set, the deletion ending at -1300. For quantitation the bands were cut out and counted. The data are plotted as percentages of binding of the deletion ending at -2561.

duction of expression by glucocorticoids can be achieved. Deletion of 5′ flanking sequences up to about -500 resulted in considerable loss of inducibility, even though some induction of the TO-CAT recombinant can still be observed. Further deletion to -314 abolished all inducibility. These findings suggest the presence of two hormone-responsive elements in the 5′ flanking region of the TO gene. These results were further confirmed by more finely spaced deletion mutants.

A similar series of experiments was conducted with the TAT fusion genes, which were introduced in rat hepatoma cells and mouse fibroblasts. In mouse L cells the TAT 5′ flanking sequence of 2.95 kb conferred strong inducibility of expression. Expression studies of 5′ deletion derivatives of the parental TAT-CAT recombinant

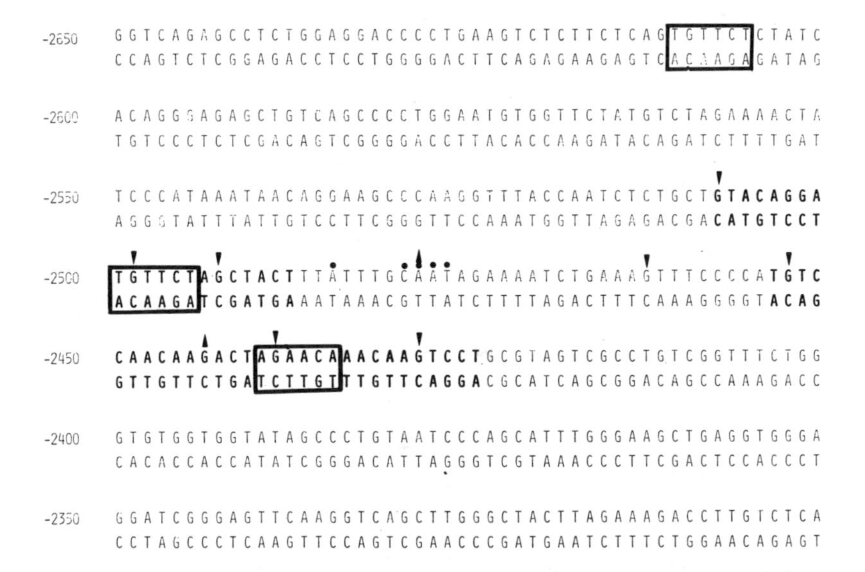

FIGURE 7. Sequence of the far 5′ flanking DNA of the TAT gene. The DNA sequence between -2650 and -2200 of the TAT gene is presented. The three hexanucleotides TGTTCT characteristic of receptor binding sites are shown in boxes. The regions protected from DNase I digestion after binding of the glucocorticoid receptor *in vitro* are shown in boldface type. DNase I hypersensitive regions are shown with dots. Protected and enhanced guanine residues as evidenced from dimethyl sulfate protection experiments are shown with upward- and downward-pointing arrows, respectively. The DNase I and dimethyl sulfate protection experiments have been performed with the upper strand only.

defined important control elements in the region that showed a local alteration in chromatin structure in the previous DNase I hypersensitivity studies (FIG. 5). Inspection of the DNA sequence within this region revealed sequences homologous to the glucocorticoid receptor binding sites of the mouse mammary tumor virus,[23,24] methallothionin,[25] and lysozyme[26,27] genes.

To determine whether this control element located so far upstream of the initiation site of transcription functions as a glucocorticoid receptor binding site, filter binding studies with the purified glucocorticoid receptor and the 5′ deletion mutants were performed. Binding of the glucocorticoid receptor to TAT 5′ deletion derivatives was investigated as shown in the upper panel of FIGURE 6. The lower panel shows that the purified receptor interacts preferentially with the deletion mutant that contains the control sequence identified *in vivo*, but not with derivatives ending at -1890 or -1300. The DNA filter binding studies were complemented by DNase I footprint analyses as well as by dimethyl sulfate protection experiments (data not shown). Two footprinting regions could be identified in the far upstream sequence of the TAT gene using a purified glucocorticoid receptor. These protected regions are indicated by the shaded sequence in FIGURE 7. Dimethyl sulfate protection experiments indicate contact sites of the receptor. Protections and enhancements indicated by the arrows in FIGURE 7 are seen in the sequences defined by the DNase I footprints. These nuclease protection experiments identify partially homologous receptor binding sites located in the region that confer glucocorticoid responsiveness *in vivo*.

CONCLUSIONS

Induction of enzyme activity for TO and TAT, two gluconeogenic enzymes, results from increased expression of the respective genes at the transcriptional level. The effect of the hormone on transcription is rapid and does not require new protein synthesis, suggesting a direct effect of the hormone receptor complex on the activity of the gene. In order to identify sequences important for regulation, fusion genes containing the presumptive control sequences were constructed and their expression studied under the influence of steroid hormones. This has allowed identification of two control sequences in the case of the TO gene located around -420 and around -1190 and control elements located at about -2500 bp upstream of the TAT initiation site. The responsible sequences show strong sequence homology to each other as well as to previously characterized glucocorticoid control elements.[23–27] What role these hormonal control elements play in the course of developmental activation of these genes will now be tested by analysis of expression of these genes in transgenic animals.

REFERENCES

1. FEIGELSON, P. & O. GREENGARD. 1962. J. Biol. Chem. **237:** 3714-3717.
2. SCHIMKE, K. T., E. A. SWEENEY & C. M. BERLIN. 1965. J. Biol. Chem. **240:** 322-331.
3. KENNEY, F. T. 1962. J. Biol. Chem. **237:** 3495-3498.
4. HARGROVE, J. L. & R. B. MACKIN. 1984. J. Biol. Chem. **259:** 386-393.
5. BAXTER, J. D. & G. G. ROUSSEAU, Eds. 1979. Glucocorticoid Hormone Action. Springer Verlag. Berlin.
6. SCHÜTZ, G., L. KILLEWICH, G. CHEN & P. FEIGELSON. 1975. Proc. Natl. Acad. Sci. USA **72:** 1017-1020.
7. SHINOMIYA, T., G. SCHERER, W. SCHMID, H.-W. ZENTGRAF & G. SCHÜTZ. 1984. Proc. Natl. Acad. Sci. USA **81:** 1346-1350.
8. DANESCH, U., S. HASHIMOTO, R. RENKAWITZ & G. SCHÜTZ. 1983. J. Biol. Chem. **258:** 4750-4753.
9. SCHERER, G., W. SCHMID, C. M. STRANGE, W. RÖWEKAMP & G. SCHÜTZ. 1982. Proc. Natl. Acad. Sci. USA **79:** 7205-7208.
10. SCHMID, W., G. SCHERER, U. DANESCH, H.-W. ZENTGRAF, P. MATTHIAS, C. M. STRANGE, W. RÖWEKAMP & G. SCHÜTZ. 1982. EMBO J. **1:** 1287-1293.
11. WICKS, W. D. 1974. Adv. Cyclic Nucleotide Res. **4:** 335-438.
12. ERNEST, M. J. & P. FEIGELSON. 1978. J. Biol. Chem. **253:** 319-322.

13. NOGUCHI, T., M. DIESTERHAFT & D. GRANNER. 1978. J. Biol. Chem. **253:** 1332-1335.
14. CULPEPPER, J. A. & A. Y.-C. LIU. 1983. J. Biol. Chem. **258:** 13812-13819.
15. HASHIMOTO, S., W. SCHMID & G. SCHÜTZ. 1984. Proc. Natl. Acad. Sci. USA **81:** 6637-6641.
16. GREENGARD, O. 1970. *In* Biochemical Actions of Hormones. G. Litwack, Ed. Vol. **1:** 53-87. Academic Press. New York, NY.
17. GLUECKSOHN-WAELSCH, S. 1979. Cell **16:** 225-237.
18. SCHMID, W., G. MÜLLER, G. SCHÜTZ & S. GLUECKSOHN-WAELSCH. 1985. Proc. Natl. Acad. Sci. USA **82:** 2866-2869.
19. MÜLLER, G., G. SCHERER, H.-W. ZENTGRAF, S. RUPPERT, B. HERRMANN, H. LEHRACH & G. SCHÜTZ. 1985. J. Mol. Biol. **184:** 367-373.
20. KILLARY, A. M. & R. E. K. FOURNIER. 1984. Cell **38:** 523-534.
21. ELGIN, S. C. R. 1981. Cell **27:** 413-415.
22. BECKER, P., R. RENKAWITZ & G. SCHÜTZ. 1984. EMBO J. **3:** 2015-2020.
23. PAYVAR, F., D. DEFRANCO, G. L. FIRESTONE, B. EDGAR, Ö. WRANGE, S. OKRET, J.-A. GUSTAFSSON & K. R. YAMAMOTO. 1983. Cell **35:** 381-392.
24. SCHEIDEREIT, E., S. GEISSE, H. M. WESTPHAL & M. BEATO. 1983. Nature **304:** 749-752.
25. KARIN, M., A. HASLINGER, H. HOLTGREVE, R. I. RICHARDS, P. KRAUTER, H. M. WESTPHAL & M. BEATO. 1984. Nature **308:** 513-519.
26. VAN DER AHE, D., S. JANICH, C. SCHEIDEREIT, R. RENKAWITZ, G. SCHÜTZ & M. BEATO. 1985. Nature **313:** 706-709.
27. RENKAWITZ, R., G. SCHÜTZ, D. VAN DER AHE & M. BEATO. 1984. Cell **37:** 503-520.

Developmental Genetics of Hepatic Gluconeogenic Enzymes[a]

SALOME GLUECKSOHN-WAELSCH

Department of Genetics
Albert Einstein College of Medicine
New York, New York 10461

One of the central problems in modern molecular biology and genetics as well as developmental biology, is that of the differentiation of cell specificity. The identity of the genetic endowment of each cell of the high eukaryote organism, in contrast to the heterogeneity of its cell and tissue types, raises questions concerning the mechanisms responsible for the cell-specific expression of certain genes in contrast to the repression of others. Progress in the analysis of such problems depends to a considerable degree on the selection of model systems best suited for an experimental approach. In the ensuing search, one would be hard pressed to find a system superior to that of the liver for an analysis of developmental regulation of cell-specific differentiation. This organ obviously includes a variety of cell types with quite different functions. But it is one particular cell, that is, the hepatocyte, that confers on the liver its characteristic distinction by expressing a cluster of specific functions shared by hardly any other organ. Within the group of liver-specific but heterogeneous gene products, a cluster of enzymes stands out that is concerned with various steps in sugar metabolism, in particular with gluconeogenesis. The metabolic regulation of the relevant liver enzymes and the mechanisms of their induction, for example, by hormones, have been the subject of many experimental approaches, some of which are included in the discussions in this volume. The questions that I would like to address here, however, do not concern the metabolic regulation of gluconeogenic enzymes expressed by the differentiated hepatocyte of the postnatal organism, but rather their developmental regulation. In particular, I propose to focus on the mechanisms of regulation that operate in the course of prenatal differentiation of both form and function of hepatocytes and that are responsible for the emergence of a cell type in which specific genes are activated and others repressed.

The particular system in high eukaryotes that has lent itself beautifully to an exploration of these questions is that of radiation-induced mutations in the mouse. In the history of developmental genetics, mutations have played a most significant role in offering experimental material for the analysis of gene action during development, and, in the words of a recent review,[1] "mutational studies provide the strongest causal link between a gene's product and a particular aspect of development." This holds true for organisms as different from each other as *Drosophila* and mice, both of them characterized by extensive knowledge of their genetics, thus facilitating their developmental analysis. The mutations in mice to which I refer were produced in studies of radiation-induced mutagenesis conducted in this country by the Russells at Oak

[a]This work was supported by Grant GM 27250 from the National Institutes of Health and Grant CD-38 from the American Cancer Society.

Ridge and in England by Searle at Harwell. Mutant homozygotes were recognized as albinos, and the mutations appeared to map at the albino locus in chromosome 7 of the mouse. It was the lethality of certain of these radiation-induced albino mutations in the homozygous state that attracted our attention, and we set out to find the reasons for the early postnatal death. The ensuing studies led eventually to the analysis, which I shall discuss here, of developmental regulation of the liver-specific enzymes.

Besides causing an absence of melanin formation, the radiation-induced lethal albino mutations were found to have a multitude of additional effects.[2] These concern specifically the liver, which shows defects of normal differentiation on ultrastructural as well as biochemical levels. Early in our studies we began to suspect that these mutations were actually chromosomal deletions, and cytogenetic evidence proved that they extended over several centimorgans, that is, crossover units.[2] Deletions of this magnitude would naturally be expected to include many gene sequences and therefore cause an array of independent deleterious effects. The particular defects on which I want to focus here, however, are not independent of each other but have a common denominator in the important liver function of gluconeogenesis. I am referring to a cluster of liver-specific enzymes that appear to be affected not singly but as a group by the deletion of particular nucleotide sequences in chromosome 7, and that fail to develop the normally expected perinatal rise in activity in mice homozygous for one of several albino deletions. Members of this cluster are glucose 6-phosphatase, phosphoenolpyruvate carboxykinase, and tyrosine aminotransferase, the first two enzymes involved directly in gluconeogenesis and the last one indirectly. The discovery of the glucose 6-phosphatase deficiency in the albino deletion homozygotes is due to Carl Cori,[3] without whom the present state of biochemical analysis of these mutants would never have been reached. It was the observation of low blood glucose levels in newborn mutant homozygotes unable to survive for more than a few hours after birth, that originally led us to contact Carl Cori in order to ask him if he might be interested in determining glucose 6-phosphatase levels in the livers of these newborn mice. This marked the beginning of a long and very happy collaborative association between Cori's laboratory and our laboratory, an association that also made possible the additional biochemical studies, for example, of tyrosine aminotransferase and other enzyme activities in the mutants.[4] During the later molecular phase of the work I enjoyed an active and most pleasant collaboration first with Schütz in Heidelberg and then with Hanson in Cleveland.

The overlapping chromosomal deletions with which we are concerned are shown in FIGURE 1. They map at and around the albino locus and extend over different stretches of chromosome 7. FIGURE 2 demonstrates the breeding scheme used in the maintenance of the lethal albino deletions and in the production of deletion homozygotes.

I shall skip the description of the various ultrastructural intracellular membrane abnormalities[5] as well as an account of the deficiencies of some other liver-specific enzymes and serum proteins, for example, albumin, transferrin, and α-fetoprotein, all of which are caused by the deletions as discussed in previous publications,[2] and I shall concern myself here exclusively with the three liver-specific enzymes involved in gluconeogenesis. Their activities, as shown in TABLE 1 for lethal homozygous newborns and their normal littermates, are reduced dramatically in deletion homozygotes (TABLE 1).

Because of the large size of the deletions, which may be roughly calculated to include a minimum of 25-50 active genes, the possibility had to be considered that the structural genes for all three enzymes were part of the deleted sequences. Actually, it would have been intriguing to identify such a cluster of neighboring genes with related functions. From the beginning, however, the presence of normal enzyme levels in heterozygotes had argued strongly against this possibility because mutations or deletions of structural genes in mammalian systems are expected to be subject to a dosage effect that would cause enzyme levels in heterozygotes to be intermediate between the two homozygous types.[6] Eventually, results of specifically designed somatic cell hybridization experiments were able to demonstrate the integrity of the three relevant structural genes in deletion homozygotes and assign their location to chromosomes other than chromosome 7.[7,8] These experiments made use of cell hybrids between mutant or normal newborn mouse hepatocytes and rat hepatoma cells. They succeeded in demonstrating not only that the gene for glucose 6-phosphatase[8] and that for tyrosine aminotransferase[7] mapped on chromosomes other than chromosome 7, but also that the two structural genes were intact and located on two separate chromosomes.

This chromosomal mapping information was obtained several years before the molecular cloning of the genes encoding the relevant enzymes was accomplished. Since that time, the phosphoenolpyruvate carboxykinase (PEPCK) and the tyrosine aminotransferase (TAT) gene have been cloned in the laboratories of Hanson[9] and Schütz,[10]

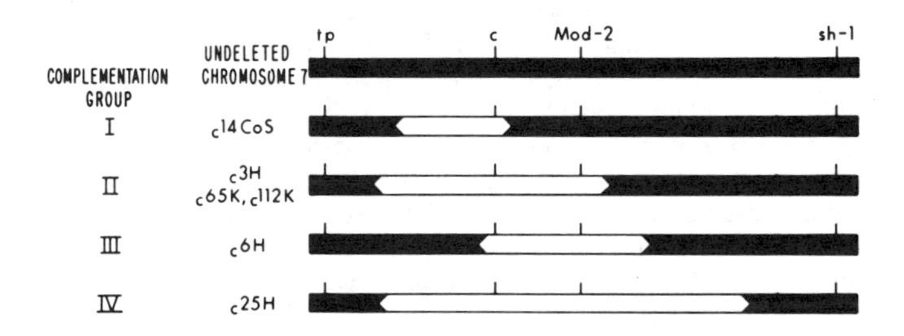

FIGURE 1. Complementation map of albino deletions.

respectively. Schütz and his collaborators mapped the TAT structural gene to chromosome 8 of the mouse by using restriction-fragment-length polymorphisms of genomic TAT fragments between different mouse species.[10] The structural gene for PEPCK has been assigned in the mouse to chromosome 2.[11] In a collaborative study with Schütz, the molecular integrity of the TAT structural gene in deletion homozygous mice was demonstrated on Southern blots, which showed identical restriction fragment patterns for TAT DNA from mutant and from normal mice.[12]

Having excluded possible chromosomal deletions of the respective structural genes as causes of the abnormalities, questions arose as to the nature of the gene sequences and their products whose absence is correlated with the specific enzyme deficiencies. Before I continue with the next phase of our analysis in detail, I want to eliminate one possibility that might account for the liver defects, and that is raised frequently. It has been argued that the described severe abnormalities of ultrastructure might have caused a dramatic reduction of hepatocyte numbers in the livers of deletion homozygotes. Such a quantitative reduction of hepatocytes should be reflected in a significant drop of hepatic protein synthesis in the deletion homozygotes. In collaboration with Carl Cori and his laboratory, experiments were performed that excluded a general defect of protein synthesis.[13] The actual reduction of hepatic protein synthesis in the deletion mutants was of an order of magnitude that could be accounted for easily by the previously demonstrated deficiencies of certain specific proteins, for example, the various perinatally developing enzymes and the plasma proteins. Re-

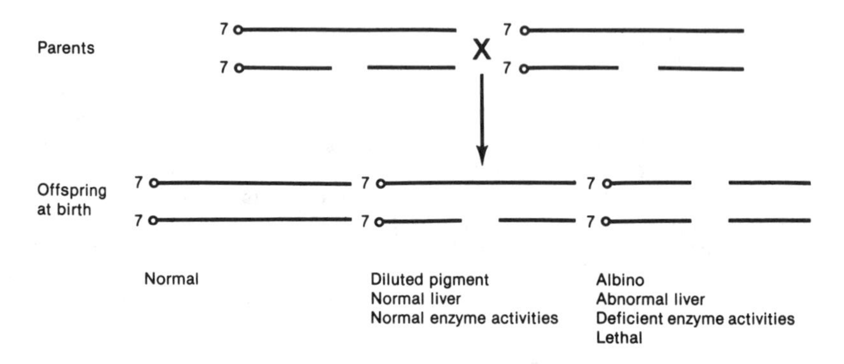

FIGURE 2. Breeding scheme of lethal albino deletions.

cently, a definitive answer to the question of hepatocyte numbers in livers of newborn deletion homozygotes was provided by experiments performed by Wolkoff and Stockert.[14] They found molecular weights and levels of two hepatocyte-specific, plasma-membrane-binding proteins, that is, the receptor for asialoglycoproteins and the organic-anion-binding protein probably concerned with organic anion transport, to be identical in normal newborn mice and their deletion homozygous littermates. These normal concentrations of binding proteins appear to reflect normal numbers of hepatocytes in the mutants, and a reduction of hepatocyte numbers can therefore not be the cause of the biochemical deficiencies in deletion homozygotes.

So far, all available evidence seems to suggest that the deletions include sequences of one or more regulatory genes that specifically control the expression of structural genes mapping elsewhere and encoding liver-cell-specific traits.[15] At this stage of the analysis, the big question concerns, of course, the level at which such regulation might take place. Approaches to these questions included the molecular analysis of TAT expression on the mRNA level made possible by the availability of Schütz's TAT DNA clone. As already mentioned, Schütz and his laboratory had provided the essential molecular evidence for the earlier conclusion obtained from somatic cell hybridization experiments that the structural gene for TAT was intact in the deletion homozygotes failing to express TAT enzyme activity. Examination of the accumulated steady state TAT mRNA in newborn deletion homozygotes, however, revealed a drastic reduction of its level to about 10% of normal, which, in contrast to normal littermates, could not be significantly modulated by glucocorticoids nor by cAMP.[12]

At this time it is not possible to decide whether the steady state mRNA deficiencies are the result of failure of normal transcription or of failure of normal processing and stabilization of the TAT message. Such distinction would require the determination of actual nuclear transcription rates.

Unfortunately, glucose 6-phosphatase, the first gluconeogenic enzyme discovered as being deficient in albino deletion homozygotes, has up to this time resisted all efforts of isolation, purification, etc., because of its tight association with the microsomal membranes. In the absence of possible antibodies to the enzyme or any amino acid sequence data, it is difficult to attempt its cloning. Thus it is not possible at this time to carry out molecular studies analogous to those on TAT in order to identify the level of regulation of expression of the glucose 6-phosphatase gene.

However, the other gluconeogenic enzyme, PEPCK, lent itself to a molecular analysis in collaboration with Hanson. As shown in TABLE 1, it was found (by Phyllis Shaw, working in our laboratory) that PEPCK enzyme activities were dramatically reduced in albino deletion homozygotes when compared to normal littermates. The deficiency of PEPCK activity could not be due to the possible inclusion of the PEPCK structural gene within the deleted sequences of chromosome 7 because the gene maps on chromosome 2. Restriction fragment patterns for PEPCK DNA were shown to be identical in homozygous deletion mutants and normal littermates by Southern blot analysis of the PEPCK gene as in the case of TAT. Furthermore, levels of PEPCK steady state mRNA also turned out to be severely reduced, and not inducible by cAMP.

The strong correlation between the reduced levels of enzyme activities as well as of steady state mRNA for TAT and PEPCK in deletion homozygotes as compared to normal littermates, adds emphasis to the question of actual transcription rates of the TAT and PEPCK genes in these mutants. Before these have been investigated it is not possible to decide if in the deletion homozygotes normal regulation fails at the transcriptional or the posttranscriptional level, for example, at the level of message processing or stabilization.

I would like to pause at this point in order to attempt a synthesis of results of the various experimental approaches employed in the analysis of the effects of the albino deletions in the mouse. To begin with, it is necessary to stress the remarkable specificity of effects on all levels: hepatocytes and restricted areas of kidney tubules only express the characteristic ultrastructural and biochemical defects of deletion homozygotes.[5,15] All other cell types remain completely normal, ultrastructurally as well as biochemically. The enzymes affected by the deletions appear to map on three separate chromosomes. The working hypothesis guiding the molecular approaches currently in progress assumes that a *trans*-acting factor encoded within the deleted sequences in chromosome 7 is responsible for the developmental regulation of a cluster of cell-type-specific functions. FIGURE 3 illustrates the steps by which such a *trans*-acting regulatory factor might discharge its responsibilities. Conceptually, this scheme owes much to the mating-type analysis by Herskowitz in yeast, where cell-specific expression of various genes was shown to be coordinately regulated by a *trans*-acting master regulatory locus.[16]

As already indicated, the *trans*-acting factor lacking in the mouse mutants and encoded within the deleted sequences of chromosome 7 is assumed to be a developmental regulatory factor responsible for rendering the relevant structural genes of the differentiated organism competent to receive and to respond to signals, such as those of hormones, instrumental in processes of metabolic regulation. The existence of such developmental regulation is in fact implied in the work of Kenney and collaborators,[17]

TABLE 1. Gluconeogenic Enzyme Activities in Livers of Lethal Albino Deletion Homozygotes and Their Normal Newborn Littermates

	Enzyme Activities		
	Glucose 6-Phosphatase[a]	Tyrosine Aminotransferase[a]	Phosphoenolpyruvate Carboxykinase[b]
Normal newborn littermates	3.78 ± 0.51	1.51 ± 0.08	185.7 ± 5.7
Deletion newborn homozygotes (c^{14CoS}/c^{14CoS})	0.63 ± 0.08	0.24 ± 0.01	26.4 ± 7.5

[a] μmol product/g liver/min.
[b] nmol/min/g liver.

who refer to a prenatal partial repression of the TAT gene and its derepression after birth which does not depend on inducing hormones. In the mechanism that removes the repression, that is, activates the gene, the proposed developmental regulatory factor might play the decisive role. It is only after this developmental step that the structural gene becomes competent to react to hormonal inducing stimuli. In fetal stages no induction has been achieved with glucocorticoids, and only a moderate and temporary induction is obtained with glucagon or cAMP. In this connection it is important to stress that in the deletion homozygotes the failure of enzyme inducibility by cAMP was found not to be due to abnormalities in the induction pathway. A study carried out some years ago in collaboration with Erickson and Siekewitz[18] failed to detect any differences in cAMP receptor activities between deletion homozygotes and their normal littermates. Subsequently, protein kinase activities were found to be normal

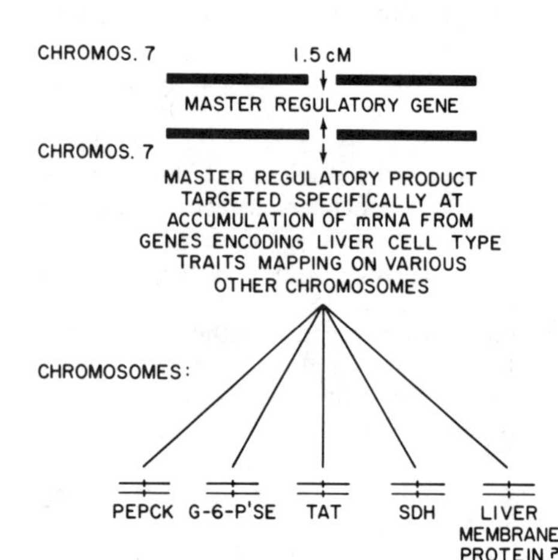

FIGURE 3. Regulation of gluconeogenic gene expression by a *trans*-acting factor.

in the mutants in preliminary studies (kindly carried out by Ora Rosen). Recently, these studies were extended and the earlier results confirmed with the help of Jack Erlichman and Charles Rubin. These latter two separated the two protein kinase isozymes in deletion homozygotes with specific antibodies and found them to be present in normal proportions, to have a normal ability to bind cAMP, and to catalyze the phosphotransferase reaction.

The restoration of normal mouse hepatocyte gene activities in somatic cell hybrids between rat hepatoma cells and mouse liver cells homozygous for the deletions may reflect the retention by the rat hepatoma cells of embryonic traits, such as prenatal expression of a gene encoding the factor responsible for rendering the relevant structural genes competent to react to inducing stimuli. The postulated *trans*-acting factor is assumed to be hepatocyte specific and fails to exert an effect on the same structural genes in other cell types, thus accounting for the absence of their expression in brain, skin, etc.

The TAT-extinguishing gene, tissue-specific extinguisher-1 (*Tse-1*),[19] may repress TAT expression preceding the derepression by the *trans*-acting factor in the course of normal hepatocyte differentiation. This is just another example of the multiplicity and complexity of positive and negative regulatory factors interacting during development.

Obviously not all hepatocyte-specific traits can be expected to be subject to the effects of identical regulatory factors: this follows from the work of Darnell and his collaborators[20] as well as from our own studies of the albino deletions that reveal coordinate developmental regulation of certain but not all liver-cell-type-specific genes. Several of the structural genes affected by the deletions participate in the same specific liver function, that is, gluconeogenesis, and encode the necessary enzymes. These appear to be coordinately regulated by the *trans*-acting factor encoded within the sequences in chromosome 7 and missing in the lethal albino deletions. In the course of differentiation, the regulatory factor is postulated to render the respective structural genes competent to receive and to respond to signals, such as those of hormones, perhaps by an effect on chromatin structure. These structural genes mapping on separate chromosomes and encoding a cluster of functionally related, that is, gluconeogenic enzymes, are coordinately controlled by the *trans*-acting factor. After these genes have been rendered competent to react, they are able to respond to hormonal and other metabolic signals.

Among our goals in the continuation of the analytical studies of the lethal albino deletions is of course that of obtaining molecular clones of the chromosomal region encoding the regulatory factor(s). Some progress in this direction has been made by Disteche, who succeeded in isolating a DNA fragment in the deletion region.[21] The fragment, however, did not include the sequences assumed to encode the regulatory factor. Recently, various molecular geneticists have expressed a strong interest in the possible molecular identification of this *trans*-acting regulatory gene and its product, encouraging the hope that in the not-too-distant future the discovery of a protein involved in developmental regulation as well as of its encoding gene may be accomplished.

REFERENCES

1. Beachy, P. 1985. *Drosophila* proteins pave the way to gene mutations. Nature **317:** 18-19.
2. Gluecksohn-Waelsch, S. 1979. Genetic control of morphogenetic and biochemical differentiation: Lethal albino deletions in the mouse. Cell **16:** 225-237.
3. Erickson, R. P., S. Gluecksohn-Waelsch & C. F. Cori. 1968. Glucose 6-phosphatase deficiency caused by radiation-induced alleles at the albino locus in the mouse. Proc. Natl. Acad. Sci. USA **59:** 437-444.
4. Thorndike, J., M. J. Trigg, R. Stockert, S. Gluecksohn-Waelsch & C. F. Cori. 1973. Multiple biochemical effects of a series of X-ray-induced mutations at the albino locus in the mouse. Biochem. Genet. **9:** 25-39.
5. Trigg, M. J. & S. Gluecksohn-Waelsch. 1973. Ultrastructural basis of biochemical effects in a series of lethal alleles in the mouse: Neonatal and developmental studies. J. Cell Biol. **58:** 549-563.
6. Gluecksohn-Waelsch, S. & C. F. Cori. 1970. Glucose 6-phosphatase deficiency: Mechanisms of genetic control and biochemistry. Biochemistry **4:** 195-201.
7. Cori, C. F., S. Gluecksohn-Waelsch, H. P. Klinger, L. Pick, S. L. Schlagman, L. Teicher & H. F. Wang-Chang. 1981. Complementation of gene deletions by cell hybridization. Proc. Natl. Acad. Sci. USA **78:** 479-483.
8. Cori, C. F., S. Gluecksohn-Waelsch, P. A. Shaw & C. Robinson. 1983. Correction of a genetically caused enzyme defect by somatic cell hybridization. Proc. Natl. Acad. Sci. USA **80:** 6611-6614.

9. YOO-WARREN, H., J. E. MONAHAN, J. SHORT, H. SHORT, A. BRUZEL, A. WYNSHAW-BORIS, W. M. MEISNER, D. SAMOLS & R. W. HANSON. 1983. Isolation and characterization of the gene coding for cytosolic phosphoenolpyruvate carboxykinase (GTP) from the rat. Proc. Natl. Acad. Sci. USA **80:** 3656-3660.
10. MÜLLER, G., G. SCHERER, H. ZENTGRAF, S. RUPPERT, B. HERRMAN, H. LEHRACH & G. SCHÜTZ. 1985. Isolation, characterization and chromosomal mapping of the mouse tyrosine aminotransferase gene. J. Mol. Biol. **184:** 367-373.
11. LEM, J. & R. E. K. FOURNIER. 1985. Assignment of the gene encoding cytosolic phosphoenolpyruvate carboxykinase (GTP) to *Mus musculus* chromosome 2. Somat. Cell Mol. Genet. **11:** 633-638.
12. SCHMID, W., G. MÜLLER, G. SCHÜTZ & S. GLUECKSOHN-WAELSCH. 1985. Deletions near the albino locus on chromosome 7 of the mouse affect the level of tyrosine aminotransferase mRNA. Proc. Natl. Acad. Sci. USA **82:** 2866-2869.
13. GARLAND, R. C., J. SASTRUSTEGUI, S. GLUECKSOHN-WAELSCH & C. F. CORI. 1976. Deficiency of plasma protein synthesis caused by X-ray-induced lethal albino alleles in mouse. Proc. Natl. Acad. Sci. USA **73:** 3376-3380.
14. WOLKOFF, A. W. & R. J. STOCKERT. 1986. Selective expression of hepatocellular membrane proteins in mice homozygous for a lethal chromosomal deletion. Proc. Soc. Exp. Biol. Med. **181:** 270-274.
15. GLUECKSOHN-WAELSCH, S. 1983. Genetic control of differentiation. *In* Teratocarcinoma Stem Cells. L. Silver, G. R. Martin & S. Strickland, Eds. Vol. **10:** 3-13. Cold Spring Harbor Laboratory. Cold Spring Harbor, NY.
16. HERSKOWITZ, I. 1985. Master regulatory loci in yeast and lambda. *In* Cold Spring Harbor Symp. Quant. Biol. **50:** 565-574.
17. PERRY, S. T., R. ROTHROCK, K. R. ISHAM, K. L. LEE & F. T. KENNEY. 1983. Development of tyrosine aminotransferase in perinatal rat liver: Changes in functional messenger RNA and the role of inducing hormones. J. Cell. Biochem. **21:** 47-61.
18. ERICKSON, R. P., P. SIEKEVITZ, K. JACOBS & S. GLUECKSOHN-WAELSCH. 1974. Chemical and immunological studies of liver microsomes from mouse mutants with ultrastructurally abnormal hepatic endoplasmic reticulum. Biochem. Genet. **12:** 81-95.
19. KILLARY, A. M. & R. E. K. FOURNIER. 1984. A genetic analysis of extinction: *Trans*-dominant loci regulate expression of liver-specific traits in hepatoma hybrid cells. Cell **38:** 523-534.
20. POWELL, D. J., J. M. FRIEDMAN, A. J. OULETTE, K. S. KRAUTER & J. E. DARNELL, JR. 1984. Transcriptional and post-transcriptional control of specific messenger RNAs in adult and embryonic liver. J. Mol. Biol. **179:** 21-35.
21. DISTECHE, C. M. & D. ALDER. 1984. Localization of cloned mouse chromosome 7-specific DNA to lethal albino deletions. Somat. Cell Mol. Genet. **10:** 211-215.

Hormonal Control of Adipogenesis[a]

GORDON M. RINGOLD,[b] ALGER B. CHAPMAN, DAVID M. KNIGHT, AND FRANK M. TORTI

Department of Pharmacology
Stanford University School of Medicine
Stanford, California 94305

INTRODUCTION

During development, cells that were pluripotent become increasingly restricted to specific differentiation pathways. This process, which culminates with differentiation into adult tissues, undoubtedly involves an ordered sequence of changes in gene expression reflected by synthesis of increasingly specialized proteins. Although little is known about how these changes are regulated, many of them involve interaction with specific inducers or hormones. Examples include thyroxine's induction of amphibian metamorphosis,[1] estrogen and progesterone's stimulation of oviduct development,[2,3] and ecdysone's triggering of insect molting.[4] Such hormone effects may provide useful insights into the mechanisms underlying the developmental control of gene expression. Analysis of such mechanisms, however, has been hampered by the complexity of the systems being studied and the difficulty in obtaining pure populations of precursor cells that undergo differentiation under controlled conditions. We recently described the isolation and characterization of a stable adipogenic cell line, TA1, derived from 5-azacytidine-treated 10T1/2 mouse embryo fibroblast.[5] The cells of this line are preadipocytes that resemble the 10T1/2 fibroblasts during growth; however, once growth is arrested by allowing cells to grow to confluence, they exhibit (over a period of 1-2 weeks) the morphology characteristic of adipocytes and accumulate lipid droplets. This morphologic change is accompanied by widespread alterations in the pattern of protein and RNA synthesis, and, as has also been found in other adipogenic cell lines, the appearance of enzymatic activities involved in fatty acid and triglyceride synthesis. We have isolated several cDNA clones that correspond to mRNAs that are induced during adipogenesis[5] and have used these clones to begin characterizing the mechanisms controlling gene induction during the developmental conversion of these cells.

The role(s) that specific hormones play in controlling the expression of the adipogenic state are slowly coming into focus. For example, serum factors that can be functionally replaced by growth hormone appear to be necessary for differentiation of adipocytes in culture.[6] Many other hormones and pharmacologic agents have been reported to accelerate the differentiation of such cells.[7,8] The assays used to evaluate differentiation, however, have often measured expression of the adipogenic phenotype

[a]This work was supported by grants from the National Institutes of Health, the Veterans Administration, the March of Dimes, and the Cetus Corporation.

[b]An Established Investigator of the American Heart Association.

(that is, lipid accumulation) rather than the proximal activation of adipose-specific genes. In this regard, we have recently documented that glucocorticoids accelerate the differentiation of TA1 preadipocytes by precociously stimulating the expression of such genes.[5,9]

A particularly intriguing metabolic abnormality associated with adipogenesis (or lack thereof) is that of cachexia. Many cancer patients as well as parasitically infected animals (and humans) undergo a severe and chronic loss of weight and appetite reflected most dramatically by the loss of adipose tissue. This catabolic wasting syndrome appears to be, at least in part, hormonally induced. The hormone produced by activated macrophages has been called cachectin.[10,11] Recent evidence indicates that cachectin and tumor necrosis factor (TNF) are one and the same[12,13] and that the TA1 adipogenic cell line serves as a useful model to study the mechanism by which TNF exerts its catabolic activity.

In this manuscript we analyze the basis for the acceleration of TA1 adipocyte differentiation by dexamethasone and indomethacin, a nonsteroidal anti-inflammatory drug, and suggest that both control the expression of one or more regulatory factors required for triggering the expression of differentiation-dependent genes. We also examine the effects of TNF on the differentiation of TA1 adipocytes and provide preliminary evidence that this hormone may reverse the "terminal" differentiation step(s) associated with activation of adipogenic genes.

EFFECTS OF DEXAMETHASONE AND INDOMETHACIN ON AN ADIPOSE DEVELOPMENTAL TRIGGER

We have previously shown that treatment of differentiating TA1 cells with certain hormones, notably the synthetic glucocorticoid dexamethasone and insulin, leads to an acceleration of the phenotypic changes associated with adipogenesis and to the precocious accumulation of specific mRNAs.[5] We show here that dexamethasone

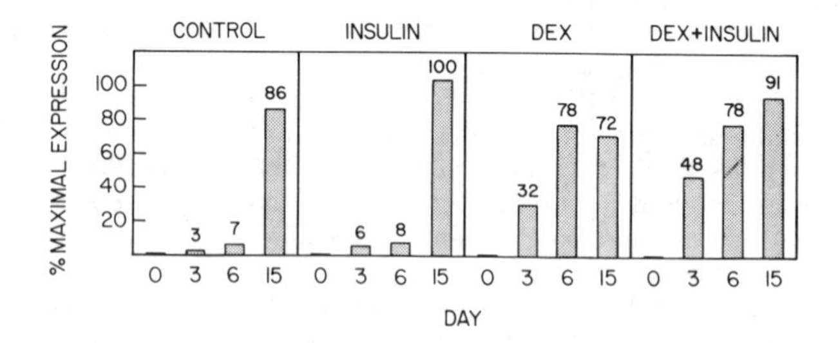

FIGURE 1. Hormonal effects on clone 1 RNA accumulation during adipocyte differentiation. TA1 preadipocytes were treated at confluence with either 1 μM dexamethasone, 5 μg/ml insulin, or a combination of both. RNA was isolated from cells 1 day prior to confluence (PC), at confluence (day 0), and every 3 days after confluence was reached, with continual hormone treatment.[3–15] Ten micrograms of total RNA were subjected to electrophoresis in an agarose formaldehyde gel and transferred to a nitrocellulose filter. This blot was then hybridized to nick-translated clone 1 DNA. Autoradiographic signals were quantitated and plotted as a percentage of maximal clone 1 RNA accumulation (normalized to levels obtained upon reprobing with a β-actin DNA).

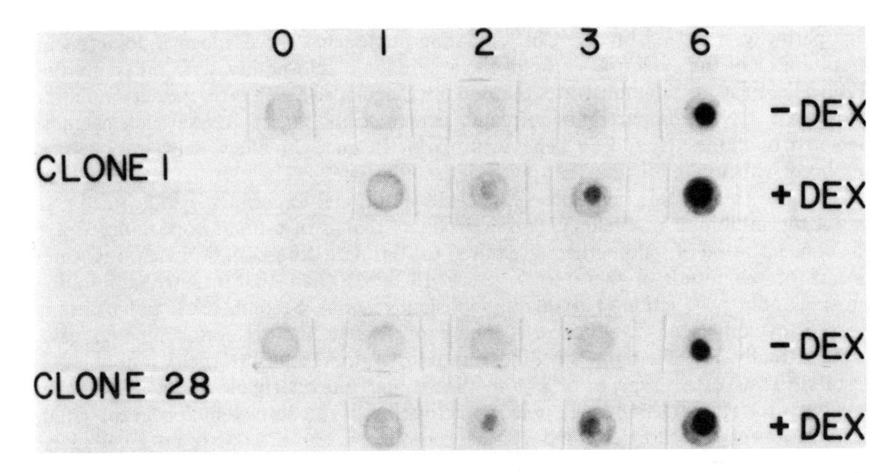

FIGURE 2. Acceleration of transcriptional activation of adipose-inducible genes by dexamethasone. TA1 cells were grown to confluence (day 0) and allowed to differentiate for the indicated number of days before harvesting for nuclear transcription assays. Parallel sets of cells were differentiated in the absence (−dex) or presence (+dex) of 1 μM dexamethasone. Cells harvested on day 0 received no hormone. Nuclei were isolated and transcription assays performed as described previously.[9] Labeled RNA was purified and hybridized to filter-immobilized plasmid containing the indicated DNAs. Filters were washed and subjected to autoradiography. Positive signals are visualized as darker dots within the circular areas defining filter background.

alone is capable of producing this effect (FIG. 1). For most of the genes examined, dexamethasone affects the timing of gene expression without altering the eventual steady state mRNA level.

In order to further delineate the role of glucocorticoids in temporal regulation of the developmental program, we determined transcription rates for clones 1 and 28 during the course of differentiation in the presence or absence of dexamethasone using a nuclear transcription assay.[9] The data in FIGURE 2 show that in the absence of dexamethasone, the transcription of clone 1 mRNA is first detectable at day 3 and is increased by day 6, and that the transcription of clone 28 mRNA is not detectable until day 6. When dexamethasone is included in the growth medium, however, transcription of both mRNAs is activated earlier. Clone 1 transcription is now detected by day 2, and clone 28 transcription can be seen as early as day 1. Thus the earlier accumulation of clone 1 and clone 28 mRNAs in response to dexamethasone results from earlier transcriptional activation of the corresponding genes. Other experiments indicate that the maximal transcriptional rates for clones 1 and 28 are similar with or without dexamethasone (data not shown). This result is unlike the higher maximal transcription rates seen in typical glucocorticoid-induced genes, suggesting that the hormone affects the triggering mechanisms that orchestrate the transcriptional induction of adipose genes rather than superimposing a second mode of regulation upon the genes directly.

The nonsteroidal anti-inflammatory drug, indomethacin, has been reported to accelerate differentiation of 3T3-L1 cells.[14] One of its major actions is to inhibit production of prostaglandins and other arachidonic acid products by inhibiting the enzyme cyclooxygenase.[15] Dexamethasone is known to limit production of arachidonic acid metabolites such as prostaglandins by inducing an inhibitor of phospholipase A_2, a key enzyme of that pathway (for review, see Schleimer[16]). To investigate whether this pathway is indeed involved in regulating adipogenesis, we added indomethacin, an inhibitor of the cyclooxygenase in the arachidonic acid pathway, to differentiating TA1 cells. FIGURE 3 demonstrates that indomethacin also accelerates the accumulation of clone 1 RNA compared to untreated control cells. Other RNAs such as those detected by clones 10 and 28 behave similarly. In fact, the effect is even more pronounced with indomethacin than with dexamethasone.

Despite this tantalizingly pleasing result, additional experiments indicate that it is not the inhibition of cyclooxygenase products that is of critical importance. First, the concentration of indomethacin required to stimulate differentiation is at least three orders of magnitude higher (10^{-4} versus 10^{-7} M) than that required to inhibit prostaglandin production (Knight, unpublished results). Second, other potent inhibitors of cyclooxygenase, which also inhibit prostaglandin production in TA1 cells, have virtually no effect on the differentiation of these cells (Knight, unpublished results). Thus other aspects of indomethacin and glucocorticoid function must be involved in triggering the developmental activation of the adipogenic program. Characterization of glucocorticoid-induced gene products present shortly after cells have reached confluence and identification of the precise metabolic pathway through which indomethacin exerts its effect on TA1 cells (perhaps unrelated to the arachidonic acid cascade or at least to a noncyclooxygenase-dependent pathway) might provide insight into the nature of this developmental switch.

Although we do not yet know the basis for indomethacin's ability to stimulate the differentiation of TA1 cells, this drug has been of great utility in our attempts to understand various aspects of the differentiation process itself. The major advantage

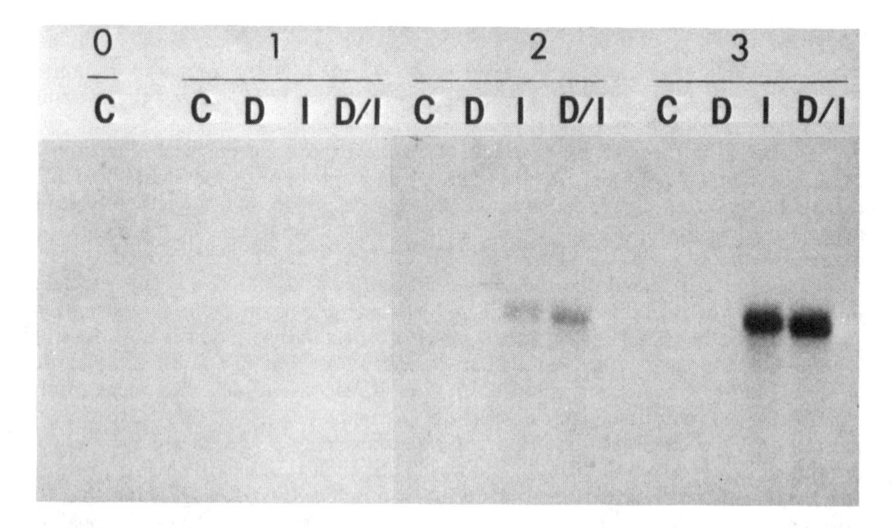

FIGURE 3. Acceleration of clone 1 RNA accumulation by indomethacin. TA1 cells were grown to confluence (day 0) and treated with 1 μM dexamethasone (D), 125 μM indomethacin (I), or both (D/I). Untreated control cells (C) were also included in this experiment. Cells were harvested at 0, 1, 2, or 3 days after confluence, and total RNA was extracted and subjected to electrophoresis on an agarose-formaldehyde gel. The RNA was transferred to nitrocellulose, and the filter was probed with nick-translated clone 1 cDNA and analyzed by autoradiography. On longer exposure of the autoradiogram, clone 1 RNA is detectable in the day 3 sample from dexamethasone- or dexamethasone/indomethacin-treated cells, but not in the day 0, 1, or 2 samples.

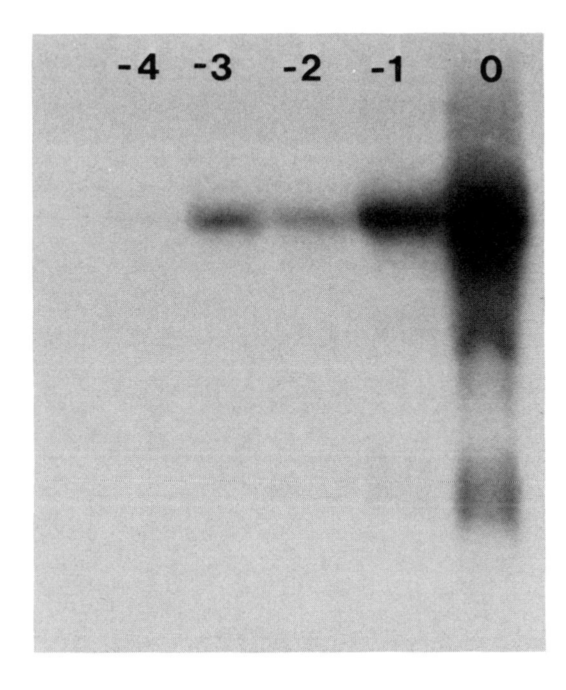

FIGURE 4. Increased cell density is required for efficient differentiation of TA1 cells. Indomethacin (125 μM) was added to TA1 cells at varying times prior to confluence, and RNAs were harvested from cells 48 hr later. The indicated times of addition represent the number of days prior to the cells reaching confluence. For example, day -4 cells were approximately 10% confluent at the time of exposure to indomethacin and were approximately 40% confluent at the time of harvesting (day -2). Expression of clone 1 RNA was analyzed as described in FIGURE 1.

of using indomethacin is that we can routinely obtain greater than 90% of the cells differentiating over a 2-3-day period. This represents a substantial improvement in both the synchrony and extent of differentiation compared to controls or glucocorticoid-treated cells.

To test whether cell contact (confluence) is required for differentiation, TA1 cells were plated at low density (approximately 10% of confluence) and treated with indomethacin every day until 100% confluence was reached (4-5 days later). RNAs were collected from cells 2 days after treatment and analyzed for expression of adipose-specific markers. FIGURE 4 shows that there is very little expression of clone 1 RNA until the cells approach confluence (days -2, -1, and 0) and that the highest levels of expression are obtained when the cells are fully confluent. The low levels of expression seen prior to that time may reflect differentiation of cells that were locally concentrated on the dish. Thus either cell-cell contact, elaboration of an extracellular matrix, or arrest in some specific portion of the cell cycle, or some combination of these factors, seems to be required for differentiation to proceed efficiently.

Clearly, the nature of the biochemical events involved in the activation of adipose-specific gene transcription remains mysterious. It is of particular interest to know whether continued exposure to indomethacin (or dexamethasone) is required for effective differentiation of TA1 cells. Confluent cells were incubated with indomethacin for intervals ranging from 12 hr to 6 days. RNAs were collected from all cells 6 days after confluence was reached, and were analyzed for expression of clone 1 RNA. The results shown in FIGURE 5 indicate that substantial activation is observed even during the 12-hr exposure to indomethacin and that nearly maximal differentiation is observed after a 1-day treatment. Thus, whatever indomethacin is initiating, it appears to be required only transiently. Similar results have been obtained with dexamethasone, although even shorter times of exposure may be sufficient to precociously activate the

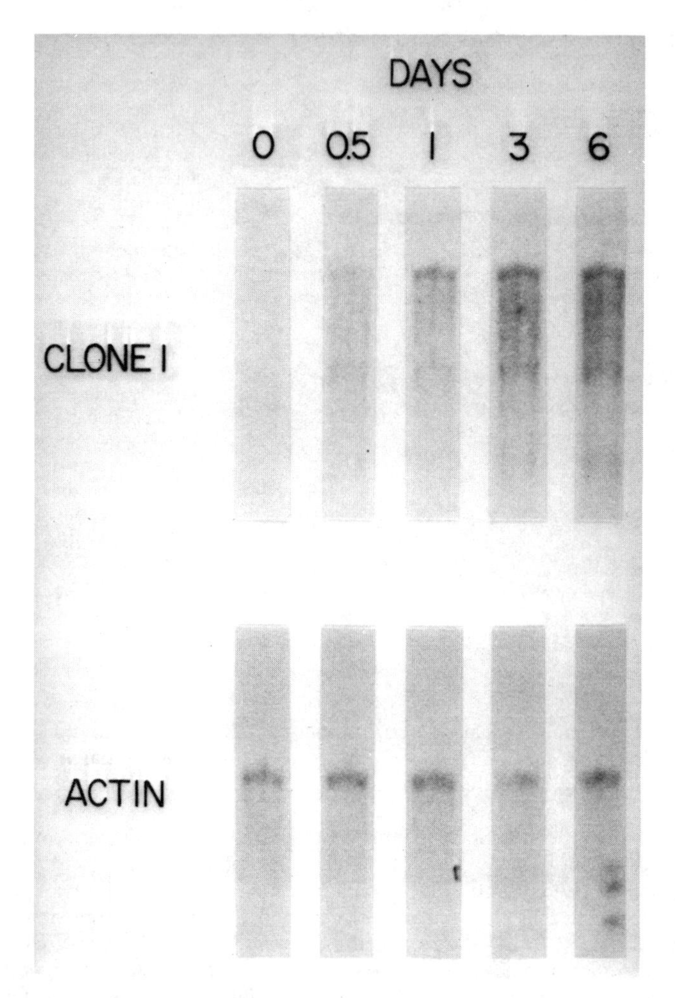

FIGURE 5. Indomethacin triggers differentiation of TA1 cells. TA1 cells were treated at confluence (day 0) with indomethacin (125 μM). At the indicated times the drug was washed out and the cells refed in the absence of drug. All cells were also refed (with or without drug, accordingly) at day 3. Six days after confluence was reached, RNAs were collected from all experimental groups and analyzed for expression of clone 1 and β-actin RNAs as previously described.[5,9]

developmental program (Chapman, unpublished results). Thus it appears that a critical regulatory event that triggers the activation of the developmental program can be manipulated by glucocorticoids and indomethacin. One might imagine that in the absence of these agents, an activating substance begins to accumulate towards a threshold level when cells reach confluence. Either the threshold itself could be lowered by glucocorticoids or indomethacin, or the rate at which the activating factor(s) accumulates could be increased by these agents. The net result in either case would be the precocious activation of the entire developmental program. It will of course be a challenge to identify such factors and perhaps even more difficult to understand the molecular basis for the existence of a threshold for triggering differentiation.

A GENERAL ROLE FOR GLUCOCORTICOIDS IN DEVELOPMENT

Neither glucocorticoids nor indomethacin appear to be absolutely required for the triggering of adipocyte differentiation: untreated cells still undergo adipose conversion, albeit at a slower rate. We have recently performed experiments using serum from which endogenous glucocorticoids have been removed by treatment with charcoal-dextran and have found that the rate and extent of TA1 cell differentiation is unaltered. Thus it is unlikely that low levels of hydrocortisone in serum simply parallel the dexamethasone effect. Rather, there may be two or more independent but functionally analogous pathways for triggering differentiation, one of which is glucocorticoid regulated. It is conceivable that glucocorticoids do not initiate the determination process itself, but rather play a role in accelerating the maturation process once differentiation has begun. It is striking that during most of fetal development in mammals cortisol levels remain low despite the presence of glucocorticoid receptors in most tissues. Late in gestation there is a rise in circulating cortisol levels coincident with the maturation of many tissues. Glucocorticoids have been implicated in stimulating maturation of the small intestine and lung[17,18] as well as in the induction of enzyme

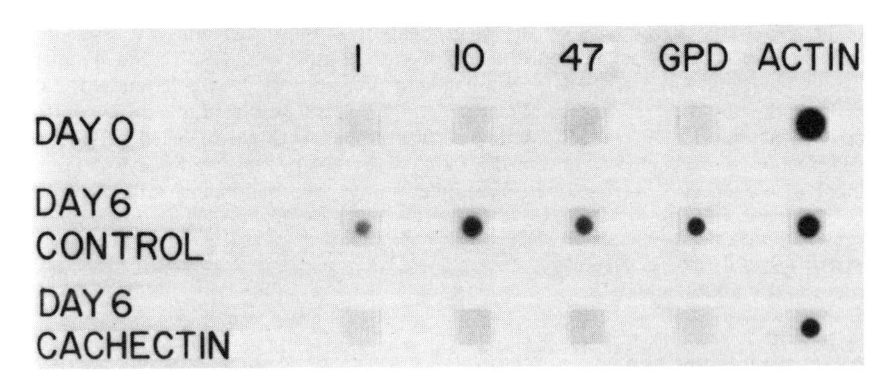

FIGURE 6. TA1 cells were grown in Eagle's basal medium supplemented with 10% heat-inactivated fetal calf serum. Dexamethasone (1 μM) was present in media for the first 3 days after confluence, and bovine insulin (5 μg/ml) for the first 6 days after confluence. Conditioned media from the macrophage cell line RAW 264 treated with endotoxin (24 hr at 10 μg/ml in serum-free medium) (cachectin) was first added to preadipocyte cultures 2 days prior to confluence at a concentration of 10 μl/ml, which inhibits 90% of lipoprotein lipase activity in cultured adipocytes. Cell cultures were resupplemented with hormones at day 0 (confluence) and day 3. Cells were harvested at day 6. Total RNA was applied to nitrocellulose in a dot blot apparatus. Nick-translated cDNA clones of genes, whose expression is seen only in differentiated TA1 adipocytes (clones 1, 10, and 47 and glycerophosphate dehydrogenase), as well as a β-actin cDNA clone were used to probe these filters under hybridization conditions as previously described.[5]

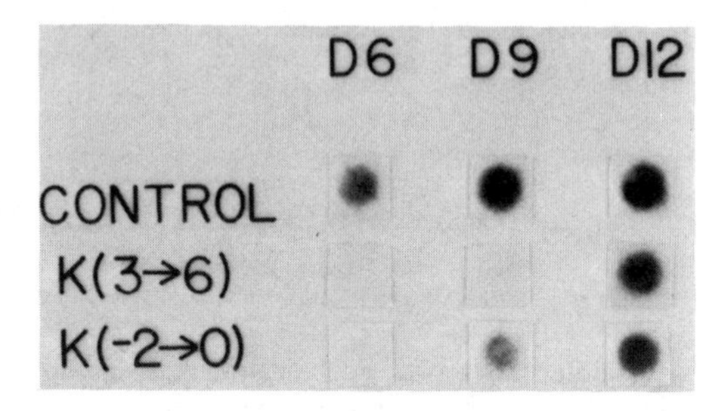

FIGURE 7. Dot blots were performed as described in FIGURE 6. Cachectin (K) was present in the media at the times indicated (3→6 represents days 3 to 6 after confluence; -2→0 represents the two days immediately prior to confluence). RNAs were analyzed with clone 1 cDNA at 6, 9, and 12 days (D) after confluence.

systems in various differentiating tissues.[19–21] Thus, a major role of glucocorticoids in development may be in accelerating prenatal maturation and differentiation of tissues and enzyme systems that facilitate extrauterine survival.[22]

NEGATIVE REGULATION OF ADIPOGENESIS BY CACHECTIN

The development of a chronic catabolic state is a hallmark of certain infections and malignancies. The weight loss that accompanies the development of such a state is termed cachexia, and is associated with the mobilization of triglycerides from adipose tissue, a process that often persists in spite of adequate caloric intake. In this study, we used the stable adipogenic cell line TA1 to examine the mechanism of inhibition of lipogenic enzyme activity by the macrophage hormone cachectin. To assess the influence of cachectin on the coordinate induction of adipose genes, we added supernatant from endotoxin-treated macrophages to preconfluent TA1 cells and to TA1 cells on the day they reached confluence. Total RNA isolated from these and control cells 6 days after they reached confluence was probed with radiolabeled cDNAs of genes whose expression was observed in adipocytes, but not in preadipocytes. Treatment with cachectin prevented the accumulation of adipose-inducible mRNAs (FIG.

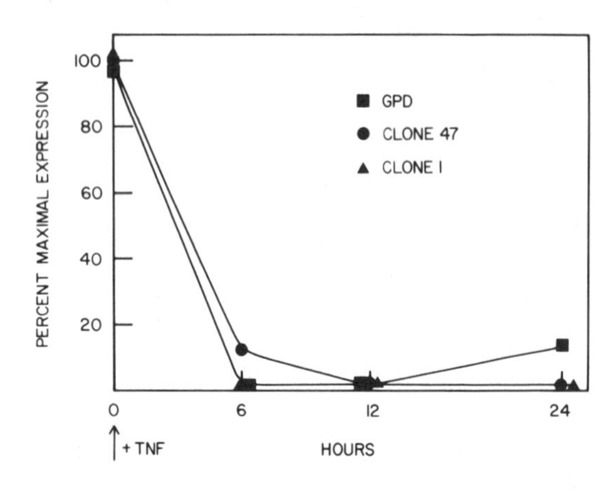

FIGURE 8. TNF (10 ng/ml) was added to day 6 TA1 cultures differentiated as described in FIGURE 6. Total RNA was isolated from cells at the indicated times after exposure, subjected to electrophoresis in a 1% agarose-formaldehyde gel, transferred to nitrocellulose, and probed with the indicated cDNAs as previously described.[5,13] Autoradiograms of the filters were scanned using a Hoeffer GS300 densitometer. The data were normalized for differences in amount of applied RNA as detected with a cDNA probe made to total cellular RNA and a β-actin cDNA.

6). Lipid accumulation was also completely inhibited by cachectin: cultures of TA1 cells treated with cachectin were maintained for as long as 23 days without the appearance of neutral lipid, as detected by staining with oil red O. On removal of cachectin from the media, however, adipocyte morphology returned, as did the expression of adipose-inducible genes (for example, see clone 1 in FIG. 7). It is noteworthy that under these conditions cachectin is not toxic to TA1 cells.

These effects of a crude cachectin preparation can be reproduced with great fidelity by pure TNF. Application of human TNF (produced by recombinant DNA technology) to TA1 cells prevents the morphological alteration and lipid accumulation characteristic of adipocytes. Moreover, the negative effect of TNF is dominant over

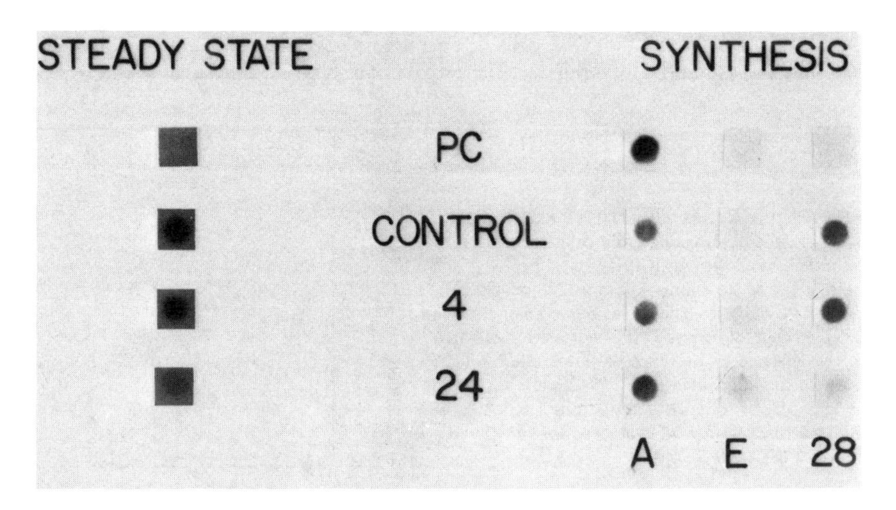

FIGURE 9. TA1 cells were stimulated to differentiate by addition of indomethacin (125 μM) for the first 3 days after confluence. Indomethacin was washed out, and 24 hr later TNF (10 ng/ml) was added to the cultures. At the indicated times (PC: preconfluent control cells; control: no TNF added after indomethacin removal; 4 and 24: 4 and 24 hr after TNF addition), steady state levels and relative synthetic rates of clone 28 RNA were determined as previously described.[9,13] Filters containing actin (A) and the plasmid EMBL (E) were included as controls for the nuclear transcription reactions in which clone 28 transcription was measured.

the inducing actions of glucocorticoids and indomethacin (Torti and Ringold, unpublished results). Analysis of several RNAs (detected with clones 1, 10, 28, and 47 and glycerophosphate dehydrogenase) reveals that TNF blocks, as does cachectin, the transcriptional activation of these genes during adipogenesis (data not shown).

In adult mammals, adipocytes undergo little or no proliferation. Thus the effect of cachectin on preconfluent TA1 cells in culture, although useful in investigating the coordinate regulation of adipose genes in development, probably does not provide a realistic model of mammalian cachexia. To model the *in vivo* situation more closely, we used mature adipocyte cultures to which we added cachectin. After 4 to 6 days of exposure to crude cachectin or pure TNF, most cells lost their neutral lipid. In typical experiments, 70 to 80% of cells would become laden with large lipid droplets when cachectin was first added. Then, 6 days later, approximately 10% of cells would have identifiable triglycerides when stained with oil red O. Alterations in adipose-specific RNAs occurred more rapidly than lipid mobilization (FIG. 8). Within 24 hr after the addition of cachectin to mature TA1 adipocytes, there was a greater than 90% decrease in the levels of such RNAs. We also found that the decrease in the glycerophosphate dehydrogenase mRNA level after cachectin addition paralleled decreases in glycerophosphate dehydrogenase enzyme activity.

That this decrease is due to an inhibition of gene transcription is evident in FIGURE 9. Within 24 hr after administration of TNF to fully differentiated TA1 cells, transcription of the clone 28 gene has been reduced to the level seen in preadipocytes; similar results are observed with other adipose genes. Thus TNF (cachectin) has the selective and rapid ability to inactivate a set of developmentally regulated adipose genes. As described below, this effect on fully differentiated TA1 cells is also reversible.

The results of the experiments just described suggest that TNF either causes "terminally" differentiated TA1 cells to revert to a less differentiated preadipocyte-like state or, alternatively, suppresses the activity of a subset of genes that are expressed

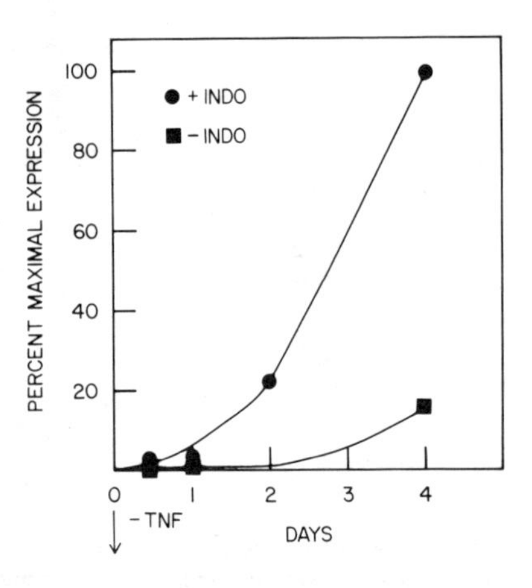

FIGURE 10. Reversibility of TNF inactivation of adipose gene expression. Fully differentiated TA1 cells were incubated with TNF for 24 hr, by which time little or no clone 1, clone 28, or glycerophosphate dehydrogenase was detectable (FIGS. 8 & 9). After removal of TNF, cells were incubated in the absence or presence of indomethacin (125 μM) for the indicated times. RNAs were harvested and analyzed for expression of clone 1 RNA (normalized to β-actin RNA) as described in the legend to FIGURE 8.

in adipocytes but does not alter the developmental state of the cells. In the latter case one could view TNF-treated TA1 cells as catabolic adipocytes as opposed to preadipocytes. Although we do not have a definitive answer to this question, the experiment shown in FIGURE 10 suggests that TNF may indeed cause TA1 cells to revert to a preadipocyte state. Fully differentiated TA1 cells were treated with TNF for 24 hr, TNF was then removed, and cells were allowed to reexpress adipose-specific RNAs in the presence or absence of indomethacin. As seen here, the accumulation of clone 1 RNA (as a representative example) occurred precociously in the indomethacin-treated cells. Thus, in addition to the fact that TNF's ability to suppress clone 1 RNA production is reversible, the reappearance of this RNA is sensitive to indomethacin in a fashion analogous to that seen in the triggering of preadipocytes described earlier. Although additional studies will be required to confirm this notion, it appears that TNF is indeed capable of acting to modulate the developmental state of these cells.

SUMMARY

The concept that hormonal substances can alter the expression of entire developmental programs is in itself not particularly new. The ability to define conditions under which a specific hormone can precociously activate the differentiation of a well-defined population of cells and under which another hormone can both block and reverse such a developmental progression, however, provides a major step forward toward unraveling the biochemical events that define the transition from a committed precursor to a fully differentiated cell. Further analysis of the molecular events initiated by glucocorticoids and TNF should provide insights into the control of adipogenesis and may generate a foundation for understanding the mechanisms by which other cells enter a particular differentiative lineage. In a more applied sense, such knowledge may also provide a rational approach to controlling metabolic disease syndromes related to adipogenesis gone awry such as obesity-associated diabetes and cachexia.

ACKNOWLEDGMENTS

We thank Drs. B. Beutler and A. Cerami for the preparations of cachectin and Drs. L. Lin and J. Larrick of the Cetus Corporation for the preparations of recombinant TNF.

REFERENCES

1. FRIEDEN, E. & J. J. JUST. 1970. *In* Biochemical Action of Hormones. G. Litwack, Ed. Vol. **1:** 1-52. Academic Press. New York, NY.
2. ROSEN, J. & B. W. O'MALLEY. 1975. *In* Biochemical Action of Hormones. G. Litwack, Ed. Vol. **3:** 271-315. Academic Press. New York, NY.
3. SCHIMKE, R. T., G. S. MCKNIGHT & D. J. SHAPIRO. 1975. *In* Biochemical Action of Hormones. G. Litwack, Ed. Vol. **3:** 245-269. Academic Press. New York, NY.
4. WYATT, G. R. 1972. *In* Biochemical Action of Hormones. G. Litwack, Ed. Vol. **2:** 386-490. Academic Press. New York, NY.
5. CHAPMAN, A. B., D. M. KNIGHT, B. S. DIECKMANN & G. M. RINGOLD. 1984. J. Biol. Chem. **259:** 15548-15555.
6. MORIKAWA, M., T. NIXON & H. GREEN. 1982. Cell **29:** 782-789.
7. RUBIN, C. S., A. HIRSCH, C. FUNG & O. M. ROSEN. 1978. J. Biol. Chem. **253:** 7570-7578.
8. RUSSEL, T. R. & R. HO. 1976. Proc. Natl. Acad. Sci. USA **73:** 4516-4520.
9. CHAPMAN, A. B., D. M. KNIGHT & G. M. RINGOLD. 1985. J. Cell Biol. **101:** 1227-1235.
10. KAWAKAMI, M., P. PEKALA, M. D. LANG & A. CERAMI. 1982. Proc. Natl. Acad. Sci. USA **79:** 912-916.
11. PEKALA, P., M. KAWAKAMI, C. W. ANGUS, M. D. LANG & A. CERAMI. 1983. Proc. Natl. Acad. Sci. USA **80:** 2743-2747.
12. BEUTLER, B., D. GREENWALD, J. D. HULMES, M. CHANG, Y.-C. E. PAN, J. MATHISON, R. ULTEVITCH & A. CERAMI. 1985. Nature **316:** 554-557.
13. TORTI, F. M., B. DIECKMANN, B. BEUTLER, A. CERAMI & G. M. RINGOLD. 1985. Science **229:** 867-869.
14. WILLIAMS, I. H. & S. E. POLAKIS. 1977. Biochem. Biophys. Res. Commun. **77:** 175-186.
15. VANE, J. R. 1971. Nature (London) New Biol. **231:** 232-235.
16. SCHLEIMER, R. P. 1985. Annu. Rev. Pharmacol. Toxicol. **25:** 381-412.
17. AVERY, M. E. 1975. Br. Med. Bull. **31:** 13-17.
18. MAHNOWSKA, K. W., R. N. HARDY & P. W. NATHANIELSZ. 1972. J. Endocrinol. **55:** 397-404.
19. JOST, A. & L. PICON. 1970. Adv. Metab. Disord. **4:** 123-184.
20. MARGOLIS, F. L., J. ROFFI & A. JOST. 1966. Science (Washington, DC) **154:** 275-276.
21. PADDINGTON, R. & A. A. MOSCONA. 1967. Biochim. Biophys. Acta **141:** 429-432.
22. LIGGINS, G. C. 1976. Am. J. Obstet. Gynecol. **126:** 931-939.

Tissue-specific Expression of Hepatic Functions

Genetic Aspects[a]

A. C. CHIN AND R. E. K. FOURNIER[b]

Department of Microbiology
and
Comprehensive Cancer Center
University of Southern California School of Medicine
Los Angeles, California 90033

INTRODUCTION

The regulation of cellular metabolism is a complex process in which transcriptional controls play a major role. Two conceptually distinct mechanisms for regulating gene activity seem to be involved. First, developmental controls restrict the array of tissue-specific genes that can be expressed in a given cell lineage. That is, particular gene sequences are identified for immediate or subsequent expression whereas others are rendered permanently silent. Second, the transcriptional activities of genes that are expressed can be modulated in response to environmental signals. Thus, developmental controls primarily determine the unique capabilities of a particular cell type whereas mechanisms for modulating gene activity tailor these capabilities to the metabolic needs of a particular moment. The studies described in this report concern the former aspect of gene control.

The developmental mechanisms that regulate gene activity have yet to be defined. At least part of the difficulty in studying these processes stems from our lack of information concerning the final product of cellular development: the fully differentiated adult cell type. With these considerations in mind, we have investigated the mechanisms that underlie cell-type-specific gene expression in terminally differentiated mammalian cells.

The model system we have employed has been to study the regulation of liver-specific gene expression in highly differentiated rat hepatoma cells growing *in vitro* and in various genetically altered lines derived from them.[1,2] This system has several experimental advantages. First, the biochemistry of liver has been studied intensively for many years. Thus, the metabolic specialization of this tissue is well defined, and numerous liver-specific markers and genes have been characterized. Second, a wide

[a]These studies were supported by Grant GM26449 from the National Institute of General Medical Sciences.

[b]Recipient of an American Cancer Society Faculty Research Award.

variety of hepatoma cell lines have been established *in vitro*. In particular, the family of lines derived from H4IIEC3 are noteworthy for their stable expression of a highly differentiated hepatic phenotype.[3] Thus, homogeneous cell populations with unlimited proliferative capacity and a stable hepatic phenotype can be obtained. Third, parasexual genetic crosses between hepatoma cells and other cell types have shown that liver-specific gene expression can be altered experimentally.[4–6] This observation forms the basis of a genetic strategy for investigating the control of tissue-specific gene expression in mammalian cells. Finally, liver is a relatively simple organ composed of few cell types, and it is easily accessible and available in quantity. Thus, the regulation of liver-specific gene expression *in vitro* can be related to the normal tissue functioning *in vivo*.

The expression of tissue-specific products is a phenotype that can be investigated at the cellular level. Thus, parasexual genetic crosses between different cell types provide an approach for investigating the control of tissue-specific gene expression.[7,8] In fact, genetic crosses of this type have been performed for nearly twenty years,[4] and a large literature has accumulated. These studies have clearly shown that systematic alterations in the expression of tissue-specific products occur in such intertypic hybrid cells. The genetic basis of these phenotypic alterations had not previously been defined.

The changes in expression of tissue-specific traits that occur in intertypic hybrid cells can be summarized as follows. First, intertypic hybrids are generally characterized by a lack of expression of the tissue-specific products expressed by either parental cell, a phenomenon termed extinction.[4] Extinction affects expression of tissue-specific (but not constitutive) markers, operates across species boundaries, and has been observed in many different intertypic hybrid crosses.[1,5,9,10] Second, extinguished traits can be reexpressed in hybrid clones that have segregated chromosomes of one of the parental cells.[6,10] For example, reexpression of liver-specific traits can be observed in hepatoma x fibroblast hybrids that have segregated fibroblast chromosomes.[1,10] The reexpression of differentiated traits is independent; that is, particular hepatic traits are reexpressed whereas others remain extinguished, and the pattern of reexpression differs from clone to clone. Third, some reexpressing segregants express both parental forms of particular tissue-specific markers. This finding indicates that activation of previously silent, differentiation-specific genes can occur in the segregants.[9,11]

We recently provided evidence that the extinction of tissue-specific gene expression in intertypic hybrid cells has a specific genetic basis.[1] First, we demonstrated that the extinction of five liver-specific traits in hepatoma x fibroblast hybrids was strictly correlated with the retention of a particular group of fibroblast chromosomes. Second, we constructed hepatoma x fibroblast hybrids retaining single fibroblast chromosomes, and correlated the extinction of particular hepatic markers with the presence of genetic loci located on specific fibroblast chromosomes. For example, hepatoma x fibroblast hybrids retaining only fibroblast chromosome 11 were extinguished for hepatic tyrosine aminotransferase expression, and lacked tyrosine aminotransferase-specific mRNA. Removal of chromosome 11 from the cells resulted in tyrosine aminotransferase reexpression to full parental levels. These data define a genetic locus, *t*issue-*s*pecific *e*xtinguisher-1 *(Tse-1),* that negatively regulates hepatic tyrosine aminotransferase expression in *trans.*[1] Other *Tse* loci that affect expression of other liver-specific genes have been identified in a similar manner (unpublished observations).

The experiments summarized in this report were designed to provide information concerning the generality and complexity of extinction. This was accomplished by screening previously characterized hepatoma x fibroblast hybrids for expression of 15 different liver-specific mRNAs, and correlating their patterns of expression with the retention of specific fibroblast chromosomes. Our results indicate that extinction affects the vast majority of tissue-specific traits, that steady state levels of specific mRNAs are affected, and that *Tse* loci affecting hepatic functions are localized on relatively few murine chromosomes.

METHODS

Cell Lines and Culture Conditions

The rat hepatoma cell lines FAO-1, FTO-2B, and FT-1 have been described previously.[1,12] Primary cultures of mouse embryo fibroblasts were prepared from C57BL/6J mouse embryos at 12-14 days of gestation using standard methods.[13] The karyotypic and phenotypic properties of the hepatoma x fibroblast hybrid clones FF5-1 and FF1-9 are described in detail elsewhere,[1] as are those of the microcell hybrid clones F(11)J and F(11)U.

Microcell hybrid clones F(3)A and F(3)B were prepared by isolating microcells from the ouabain-resistant (Oua^r) mouse line C3H10T1/2 Oua^r Cl 2[14] and fusing them with FT-1 recipients, and by selecting microcell hybrids in medium containing 3 mM ouabain according to published procedures.[1] The cells were cultured in a 1:1 mixture of Ham's F12:Dulbecco's minimal essential medium + 10% fetal bovine serum. FF and F(11) hybrids were cultured in medium containing hypoxanthine, aminopterin, and thymidine (HAT). Antibiotics were not employed. All cell lines were free of mycoplasma as judged by Hoechst 33258 staining[15] and by direct culturing.[16]

Northern Blotting Analyses

Cytoplasmic RNA (5 μg per lane) from various cell lines was extracted by a modification of previously described methods, run through formaldehyde-agarose gels, and transferred to Zetabind sheets (AMF Cuno, Meriden, CT) according to standard techniques.[17] Blots were prehybridized for 14-24 hr at 42° C in 50% formamide/5X SSC (1X = 0.15 M NaCl/0.015 M sodium citrate)/10X Denhardt's solution (1X = 0.02% polyvinylpyrrolidone/0.02% Ficoll 400/0.02% bovine serum albumin)/ 50 mM sodium phosphate buffer, pH 6.7/500 μg sonicated salmon sperm DNA per ml/5% dextran sulfate. Hybridizations were carried out for 17-24 hr at 42-55° C in 50% formamide/5X SSC/1X Denhardt's solution/20 mM sodium phosphate buffer, pH 6.7/100 μg sonicated salmon sperm per ml/10% dextran sulfate/1×10^6 cpm ^{32}P-labeled nick-translated cDNA probe (specific activity $\geq 1 \times 10^8$ cpm/μg) per ml. Filters were sequentially washed with agitation in two changes of 2X SSC/0.1% sodium dodecyl sulfate (SDS) at room temperature, one change of 0.2X SSC/0.1% SDS at room temperature, two changes of 0.2X SSC/0.1% SDS at 55° C. Autoradiography was for 1-3 days using Kodak XRP-1 film with a single intensifying screen. Blots were reused after stripping by boiling in water for 2 min and prehybridizing as above.

DNA Probes

Recombinant cDNA clones encoding 15 hepatic functions were used as markers in these studies. The clones encoded four liver-specific enzymes (tyrosine aminotransferase, phosphoenolpyruvate carboxykinase, alcohol dehydrogenase, and aldolase B), six serum proteins produced in liver (serum albumin, transferrin, ferritin, α-antitrypsin, retinol-binding protein, and complement component C3), and five anonymous liver-specific traits (pliv2, -8, -9, -10, and -11). An α-tubulin probe served as a constitutive function control. Plasmid designations and original references describing the isolation of these cDNA clones are compiled in TABLE 1.

RESULTS

Properties of the Cell Lines

FAO-1, FTO-2B, and FT-1 are highly differentiated rat hepatoma cell lines derived from H4IIEC3. These lines stably express numerous hepatic traits *in vitro*, including tyrosine aminotransferase, alanine aminotransferase, alcohol dehydrogenase, aldehyde dehydrogenase, and aldolase B activities.[1]

The FF series hybrids were generated by fusing FAO-1 rat hepatoma cells with diploid mouse embryo fibroblasts. Clone FF5-1 retained a complete complement of hepatoma chromosomes (mode = 53) plus 30 chromosomes derived from the fibroblast parent. Detailed karyotyping demonstrated that every mouse chromosome (autosomes 1-19 plus the X chromosome) was retained at high frequency in this hybrid population. Thus, FF5-1 was a karyotypically complete hybrid retaining both parental genomes. FF5-1 was completely extinguished for hepatic tyrosine aminotransferase, alanine aminotransferase, alcohol dehydrogenase, aldehyde dehydrogenase, and aldolase B expression.[1]

Hybrid clone FF1-9 also contained a complete hepatoma genome, but this clone retained only 9-10 fibroblast chromosomes. At passage 6, the particular murine chromosomes retained by this line were autosomes 1, 2, 3, 8, 9, 10, 11, 13, and 16, and each of these chromosomes was present at high frequency in the hybrid population. In spite of its retention of relatively few fibroblast chromosomes, FF1-9_6 was extinguished for expression of the liver-specific markers tyrosine aminotransferase, alanine aminotransferase, alcohol dehydrogenase, aldehyde dehydrogenase, and aldolase B.[1]

In contrast to clone FF5-1, which displayed a very stable hybrid karyotype, FF1-9 segregated fibroblast chromosomes when propagated *in vitro*. Between passages 6 and 14, the modal number of fibroblast chromosomes fell from 10 to 5. Concomitant with chromosome loss, a dramatic shift in the cellular morphology from fibroblastic to hepatic was observed. As described in detail elsewhere,[1] tyrosine aminotransferase, alanine aminotransferase, alcohol dehydrogenase, aldehyde dehydrogenase, and aldolase B activities were reexpressed in FF1-9 populations by passage 14.

The fibroblast chromosomes retained by FF1-9 at passage 6 but segregated at later passage potentially carried loci involved in extinction. To test this possibility, monochromosomal hybrids retaining individual fibroblast chromosomes were constructed by microcell fusion.[18,19] Clones F(11)J and F(11)U were prepared by fusing mouse

TABLE 1. Expression of Liver-specific mRNAs in Hepatoma x Fibroblast Hybrids

Liver-specific Marker	cDNA Probe	Probe Reference	mRNA Expression[a]				
			FAO-1	MEF	FF5-1	$FF1\text{-}9_6$	$FF1\text{-}9_{14}$
Tyrosine aminotransferase	pcTAT-3	20	+	−	−	−	+
Phosphoenolpyruvate carboxykinase	pPCK-10	21	+	−	−	−	+
Alcohol dehydrogenase	pZK6-6	22	+	−	−	−	+
Aldolase B	pHL413	23	+	−	−	−	+
Serum albumin	pRSA57	24	+	−	−	−	+
Transferrin	pliv6	25	+	−	−	−	+
Ferritin	pliv7	25	+	−	−	−	+
α_1-Antitrypsin	pliv3	25	+	−	−	−	+
Retinol-binding protein	CP2-6	26	+	−	−	−	+
pliv2	pliv2	25	+	−	−	−	+
pliv9	pliv9	25	+	−	−	−	+
pliv10	pliv10	25	+	−	−	−	+
pliv11	pliv11	25	+	−	−	+	+
C3	pMLC3-4	27	+	−	+	+	+
pliv8	pliv8	25	+	−	+	+	+
α-Tubulin	pKα-1	28	+	+	+	+	+

[a]A minus sign indicates a steady state mRNA level $< 10\%$ of that of FAO-1 hepatoma cells was obtained.

embryo fibroblasts with thymidine kinase-deficient FTO-2B hepatoma recipients and selecting hybrids expressing thymidine kinase activity in HAT medium. Because the murine thymidine kinase gene resides on chromosome 11, the monochromosomal microcell hybrids produced in this genetic cross were rat hepatoma cells selectively retaining this single fibroblast chromosome. As reported previously,[1] the F(11) hybrids were extinguished for hepatic tyrosine aminotransferase expression, but expressed full parental levels of alanine aminotransferase, alcohol dehydrogenase, aldehyde dehydrogenase, and aldolase B. These data defined the genetic locus *Tse-1,* a chromosome 11-linked locus that extinguishes hepatic tyrosine aminotransferase expression in *trans.*

F(3)A and F(3)B were monochromosomal microcell hybrids that retained fibroblast chromosome 3, another chromosome potentially involved in extinction. These lines were constructed as described in the preceding section.

Thus, the following informative hybrid clones were used in this study: FF5-1, a karyotypically complete hepatoma x fibroblast hybrid retaining all fibroblast chromosomes; FF1-9_{6}, an extinguished hybrid retaining 10 fibroblast chromosomes; FF1-9_{14}, a hybrid segregant that reexpressed hepatic markers; F(11)J and F(11)U, monochromosomal clones extinguished for tyrosine aminotransferase expression; and F(3)A and F(3)B, monochromosomal hybrids retaining another implicated chromosome.

Expression of Liver-specific mRNAs

We carried out a series of Northern blotting experiments to examine the expression of 15 distinct liver-specific mRNAs in the hybrid cell lines described in the preceding subsection.[20–27] Results of this analysis are summarized in TABLE 1. In these studies, an α-tubulin probe[28] served as a constitutive marker control: this is a housekeeping function expressed in both hepatic and nonhepatic cells. Hybridization of our Northern blots with the α-tubulin probe demonstrated the uniform presence of a 1.7-kb mRNA in all the cell lines tested (FIG. 1A). Thus, α-tubulin expression was not altered in any of the hepatoma x fibroblast hybrids.

The liver-specific cDNA clones corresponded to four hepatic enzymes,[20–23] six liver-derived serum proteins,[24–27] and five anonymous markers of hepatic differentiation. FAO-1 rat hepatoma cells expressed mRNAs corresponding to all of these hepatic markers (TABLE 1). In each case, a discreet mRNA species of the appropriate molecular size and approximate relative abundance was detected in FAO-1 cytoplasmic RNA (FIGS. 1B & 1C). In contrast, none of the 15 messages were detected in RNA samples prepared from fibroblasts (TABLE 1 and FIGS. 1B & 1C). These results are in accord with the established tissue-specific nature of these cDNA clones.

Expression of 13 of the 15 liver-specific mRNAs was extinguished in the karyotypically complete hybrid FF5-1. The two exceptional markers displayed a tissue-specific pattern of expression in the FAO-1 versus mouse embryo fibroblast comparison, but continued to be expressed in FF5-1. These markers were complement component C3 and pliv8, an anonymous liver-specific mRNA. Thus, the extinction mechanism seems to affect expression of the vast majority of tissue-specific traits, and is generally mediated through controls that affect steady state levels of specific mRNAs.

Twelve of the 13 traits extinguished in FF5-1 were also extinguished in hybrid clone FF1-9 at passage 6. Thus, the majority of extinguisher loci affecting expression of the markers assayed seems to reside on the fibroblast chromosomes retained by

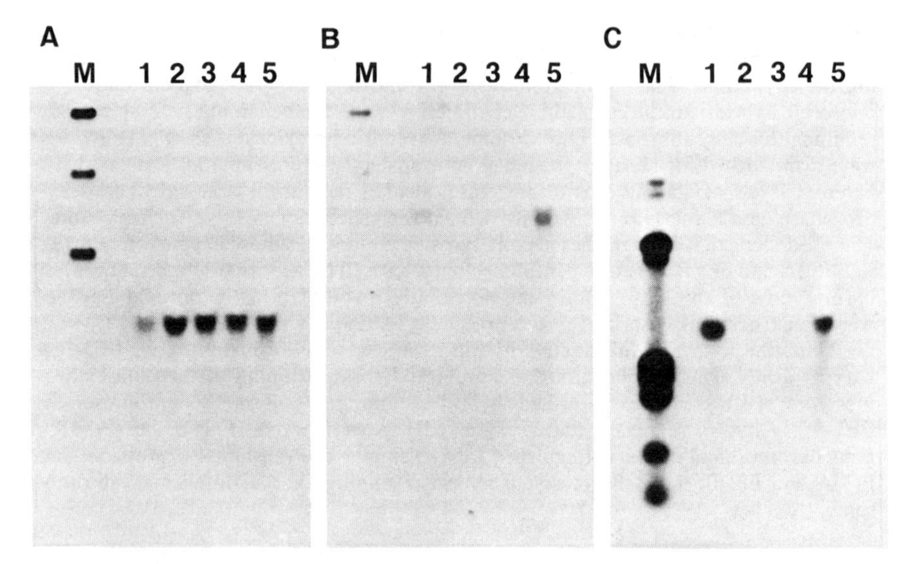

FIGURE 1. Expression of liver-specific mRNAs in hepatoma x fibroblast hybrids. Northern blots were prepared as described in the text and hybridized with ^{32}P-labeled cDNA probes pKα-1 (**A**), pPCK-10 (**B**), and pliv10 (**C**). Lane 1: FAO-1 (hepatoma parent); lane 2: mouse embryo fibroblast (fibroblast parent); lane 3: FF5-1 (karyotypically complete hybrid); lane 4: FF1-9_6 (hybrid segregant at early passage); lane 5: FF1-9_{14} (hybrid segregant at late passage). M: size markers.

FF1-9_6. Because only 9-10 fibroblast chromosomes were retained, the total number of extinguisher loci affecting hepatic gene expression would seem to be limited.

Finally, each of the liver-specific mRNAs extinguished in FF1-9_6 was reexpressed in this clone at later passage. This observation establishes unequivocally that the markers were extinguished in *trans* by fibroblast genetic loci that were segregating in the hybrid cell population.

Genetic Basis of Albumin Extinction

We previously demonstrated that tyrosine aminotransferase extinction is mediated by a specific *trans*-acting locus located on fibroblast chromosome 11. This locus has been termed *Tse-1.* Similarly, the *Tse* loci that affect phosphoenolpyruvate carboxykinase and alcohol dehydrogenase expression in hybrids have also been localized to single murine chromosomes (unpublished observations). These findings, together with results described in the preceding subsection, prompted us to determine whether any of the chromosomes retained by FF1-9_6 were individually capable of extinguishing expression of particular hepatic markers.

Monochromosomal hybrids F(3)A and F(3)B selectively retained mouse fibroblast chromosome 3. Both clones expressed most hepatic markers assayed, including tyrosine aminotransferase mRNA (FIG. 2B). These hybrids, however, seemed extinguished for albumin mRNA expression (FIG. 2A). Monochromosomal hybrids retaining other fibroblast chromosomes did not display this phenotype; for example, F(11)J and F(11)U, though extinguished for tyrosine aminotransferase mRNA expression (FIG. 2B), expressed high steady state levels of albumin mRNA (FIG. 2A). Thus, extinction of serum albumin expression may involve genetic loci located on mouse chromosome 3. These observations are in accord with the recent results of Pillot and Weiss, who have documented albumin extinction in hepatoma microcell hybrids retaining an X:3 translocation derived from the mouse cell line C11D (personal communication).

DISCUSSION

The results described in this report allow us to draw several general conclusions concerning the extinction phenomenon. First, the extinction mechanism seems to affect the vast majority of tissue-specific traits expressed by a given cell type. In this study, 13 of 15 liver-specific mRNAs were extinguished in intertypic hepatoma x fibroblast hybrids. Extinction of five other hepatic markers had previously been documented in the same hybrid clones.[1] Thus, 18 of 20 liver-specific traits assayed have responded to extinction, and all have been reexpressed in hybrid segregants that had eliminated fibroblast chromosomes. The sort of *trans* regulation evidenced in hybrids seems to be a common form of tissue-specific gene control.

Second, the expression of seven liver-specific genes has been assayed at both the protein and mRNA levels. In every case, phenotypic extinction of the protein product was correlated with a lack of the corresponding mRNA. These observations, together

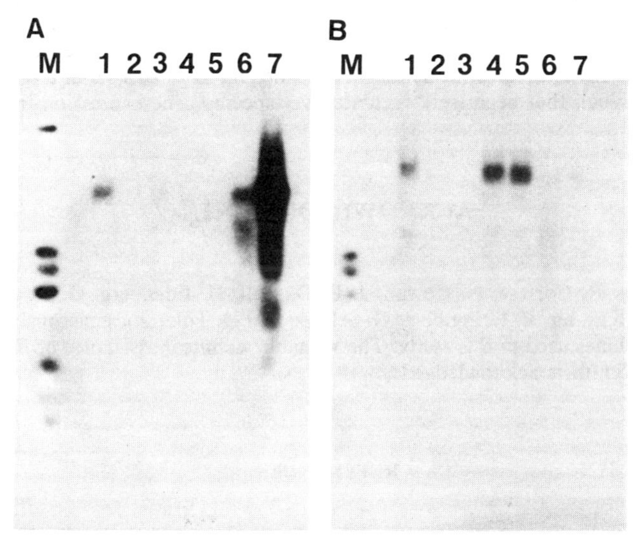

FIGURE 2. Expression of serum albumin and tyrosine aminotransferase mRNAs in hybrid and microcell hybrid clones. Northern blots were hybridized with ^{32}P-labeled pRSA57 (**A**) and pcTAT-3 (**B**). Lane 1: FT-1 (rat hepatoma parent); lane 2: mouse embryo fibroblast (fibroblast parent); lane 3: FF5-1; lane 4: F(3)A; lane 5: F(3)B; lane 6: F(11)J; lane 7: F(11)U. M: size markers.

with the results reported here, clearly establish that extinction is generally mediated through controls that affect steady state levels of tissue-specific mRNAs. Detailed studies of the transcription rates of the tyrosine aminotransferase, phosphoenolpyruvate carboxykinase, and serum albumin genes in various hybrid clones indicate that extinction involves transcriptional controls (unpublished observations).

Finally, extinction of 12 (of 13) liver-specific mRNAs was observed in a hybrid clone that retained only nine fibroblast chromosomes. This hybrid was previously shown to be extinguished for expression of five (of five) other hepatic markers.[1] Thus, the extinguisher loci affecting hepatic gene expression localize to relatively few fibroblast chromosomes, and the total number of such *Tse* loci might be limited. This would necessarily imply that particular *Tse* loci affected expression of multiple liver-specific structural genes. Genotypically simple hybrids retaining single fibroblast chromosomes or specific subchromosomal regions should allow this possibility to be explored.

SUMMARY

Intertypic hybrids formed by fusing dissimilar cell types generally fail to express the tissue-specific products of either parent, a phenomenon termed extinction. To explore the genetic basis and phenotypic complexity of this phenomenon, 15 liver-specific cDNA clones have been used as probes to assay expression of the corresponding mRNAs in a series of hepatoma x fibroblast hybrids retaining different fibroblast chromosomes. Thirteen of the 15 liver-specific mRNAs were extinguished in karyotypically complete hepatoma x fibroblast hybrids, but were reexpressed in hybrid segregants that had eliminated fibroblast chromosomes. In three cases analyzed in detail, extinction of a particular hepatic marker required the presence of a single, specific fibroblast chromosome. These data define and localize a set of discrete genetic loci, the *Tse* loci, that negatively regulate liver-specific gene expression in *trans*.

ACKNOWLEDGMENTS

We thank R. Cortese, N. Cowan, J. E. Darnell, H. Edenberg, G. H. Fey, R. W. Hanson, K. Krauter, E. E. Penhoet, G. Schütz, and D. Tolan for generously providing the cDNA clones used in this study. The valuable technical assistance of F. R. Parker and M. M. Smith is acknowledged.

REFERENCES

1. Killary, A. M. & R. E. K. Fournier. 1984. A genetic analysis of extinction: *trans*-Dominant loci regulate expression of liver-specific traits in hepatoma hybrid cells. Cell **38:** 523-534.
2. Peterson, T. C., A. M. Killary & R. E. K. Fournier. 1985. Chromosomal assignment and *trans* regulation of the tyrosine aminotransferase structural gene in hepatoma hybrid cells. Mol. Cell. Biol. **5:** 2491-2494.

3. PITOT, H. C., C. PERAINO, P. A. MORSE & V. R. PATTER. 1964. Hepatoma in tissue culture compared with adapting liver *in vivo*. Natl. Cancer Inst. Monogr. **13:** 229-242.
4. DAVIDSON, R. L., B. EPHRUSSI & K. YAMAMOTO. 1966. Regulation of pigment synthesis in mammalian cells as studied by somatic hybridization. Proc. Natl. Acad. Sci. USA **56:** 1437-1440.
5. SCHNEIDER, J. A. & M. C. WEISS. 1971. Expression of differentiated functions in hepatoma cell hybrids. I. Tyrosine aminotransferase in hepatoma-fibroblast hybrids. Proc. Natl. Acad. Sci. USA **68:** 127-131.
6. WEISS, M. C. & M. CHAPLAIN. 1971. Expression of differentiated functions in hepatoma cell hybrids: Reappearance of tyrosine aminotransferase inducibility after the loss of chromosomes. Proc. Natl. Acad. Sci. USA **68:** 3026-3030.
7. DAVIS, F. M. & E. A. ADELBERG. 1973. Use of somatic cell hybrids for analysis of the differentiated state. Bacteriol. Rev. **37:** 197-214.
8. DAVIDSON, R. L. 1974. Gene expression in somatic cell hybrids. Annu. Rev. Genet. **8:** 195-218.
9. BROWN, J. E. & M. C. WEISS. 1975. Activation of production of mouse liver enzymes in rat hepatoma-mouse lymphoid cell hybrids. Cell **6:** 481-494.
10. WEISS, M. C., R. S. SPARKES & R. BERTOLOTTI. 1975. Expression differentiated functions in hepatoma cell hybrids. IX. Extinction and reexpression of liver-specific enzymes in rat hepatoma-Chinese hamster fibroblast hybrids. Somat. Cell Genet. **1:** 27-40.
11. PETERSON, J. A. & M. C. WEISS. 1972. Expression of differentiated functions in hepatoma cell hybrids: Induction of mouse albumin production in rat hepatoma-mouse fibroblast hybrids. Proc. Natl. Acad. Sci. USA **69:** 571-575.
12. KILLARY, A. M., T. G. LUGO & R. E. K. FOURNIER. 1984. Isolation of thymidine kinase-deficient rat hepatoma cells by selection with bromodoexyuridine, Hoechst 33258, and visible light. Biochem. Genet. **22:** 201-213.
13. KOZAK, C. A., E. A. NICHOLS & F. H. RUDDLE. 1975. Gene linkage analysis in the mouse by somatic cell hybridization: Assignment of adenine phosphoribosyltransferase to chromosome 8 and α-galactosidase to the X chromosome. Somat. Cell Genet. **1:** 371-382.
14. LANDOLPH, J. R. & R. E. K. FOURNIER. 1983. Microcell-mediated transfer of carcinogen-induced ouabain resistance from C3H/10T1/2 Cl 8 mouse fibroblasts to human cells. Mutat. Res. **107:** 447-463.
15. CHEN, T. R. 1977. *In situ* detection of mycoplasma contamination in cell cultures by fluorescent Hoechst 33258 stain. Exp. Cell Res. **104:** 255-262.
16. BARILE, M. F., G. P. BODEY, J. SNYDER, D. B. RIGGS & M. W. GRABOWSKI. 1966. Isolation of mycoplasma orale from leukemic bone marrow and blood by direct culture. J. Natl. Cancer Inst. **36:** 155-168.
17. MANIATIS, T., E. F. FITSCH & J. SANBROOK, EDS. 1982. Molecular Cloning: A Laboratory Manual. Cold Spring Harbor Laboratory. Cold Spring Harbor, NY.
18. FOURNIER, R. E. K. 1981. A general high-efficiency procedure for production of microcell hybrids. Proc. Natl. Acad. USA **78:** 6349-6353.
19. FOURNIER, R. E. K. & J. A. FRELINGER. 1982. Construction of microcell hybrid clones containing specific mouse chromosomes: Application to autosomes 8 and 17. Mol. Cell. Biol. **2:** 526-534.
20. SCHERER, G., W. SCHMID, C. M. STRANGE, W. RÖWEKAMP & G. SCHÜTZ. 1982. Isolation of cDNA clones coding for rat tyrosine aminotransferase. Proc. Natl. Acad. Sci. USA **79:** 7205-7208.
21. YOO-WARREN, H., J. E. MONAHAN, J. SHORT, H. SHORT, A. BRUZEL, A. WYNSHAW-BORIS, H. M. MEISNER, D. SANIOLS & R. W. HANSON. 1983. Isolation and characterization of the gene coding for cytosolic phosphoenolpyruvate carboxykinase (GTP) from the rat. Proc. Natl. Acad. Sci. USA **80:** 3656-3660.
22. EDENBERG, H. J., K. ZHANG, K. FONG, W. F. BOSRON & T. K. LI. 1985. Cloning and sequencing of cDNA encoding the complete mouse liver alcohol dehyrogenase. Proc. Natl. Acad. Sci. USA **82:** 2262-2266.
23. ROTTMANN, W. H., D. R. TOLAN & E. E. PENHOET. 1984. Complete amino acid sequence for human aldolase B derived from cDNA and genomic clones. Proc. Natl. Acad. Sci. USA **81:** 2738-2742.
24. SARGENT, T. D., J. R. WU, J. M. SALA-TREPAT, R. B. WALLACE, A. A. REYES & J.

BONNER. 1979. The rat serum albumin gene: Analysis of cloned sequences. Proc. Natl. Acad. Sci. USA **76:** 3256-3260.

25. DERMAN, E., K. KRAUTER, L. WALLING, C. WEINBERGER, M. RAY & J. E. DARNELL. 1981. Transcriptional control in production of liver-specific mRNAs. Cell **23:** 731-739.
26. COLANTUONI, V., V. ROMANO, G. BENSI, C. SANTORO, F. CONSTANZO, G. RAUGEI & R. CORTESE. 1983. Cloning and sequencing of a full-length cDNA for human retinol-binding protein. Nucleic Acids Res. **11:** 7769-7776.
27. DOMDEY, H., K. WIEBAUER, M. KAZMAIER, V. MÜLLER, K. ODINK & G. FEY. 1982. Characterization of the mRNA and cloned cDNA specifying the third component of mouse complement. Proc. Natl. Acad. Sci. USA **79:** 7619-7623.
28. COWAN, N. J., P. R. DOBNER, E. V. FUCHS & D. W. CLEVELAND. 1983. Expression of human α-tubulin genes: Interspecies conservation of 3′ untranslated regions. Mol. Cell. Biol. **3:** 1738-1745.

Tissue-specific Expression of Pancreatic Genes in Transgenic Mice[a]

RAYMOND J. MACDONALD,[b] ROBERT E. HAMMER,[c]
GALVIN H. SWIFT,[b] DAVID M. ORNITZ,[d]
BRIAN P. DAVIS,[b] RICHARD D. PALMITER,[d]
AND RALPH L. BRINSTER[c]

[b] *Department of Biochemistry*
Division of Molecular Biology
University of Texas Health Science Center at Dallas
Dallas, Texas 75235

[c] *Laboratory of Reproductive Physiology*
School of Veterinary Medicine
University of Pennsylvania
Philadelphia, Pennsylvania 19103

[d] *Department of Biochemistry*
Howard Hughes Medical Institute Laboratory
University of Washington
Seattle, Washington 98195

The differential expression of genes that determine cellular phenotype is largely regulated by controlling the rate of initiation of transcription of these genes. To understand this aspect of cell differentiation it is necessary to define the genetic and molecular mechanisms that determine the timing, the extent, and the tissue-specific nature of transcription of developmentally regulated genes. We have chosen to study the expression of a family of pancreas-specific genes—the pancreatic serine proteases—because of the advantages that accrue from the comparative analysis of similarly regulated genes. At least nine distinct serine protease genes (three chymotrypsin genes, three trypsin genes, two elastase genes, and one kallikrein gene) are expressed in the exocrine pancreas of the rat. The gene family has evolved from a common ancestral serine protease gene through a series of duplications.[1] The family members encode enzymes that are structurally and functionally homologous. Each gene is expressed to high levels and selectively in the acinar cells of the pancreas as part of the differentiative phenotype of this cell type. With the exception of kallikrein,[2] the presence of enzymes with similar activities in other tissues appears due to expression of the protein products of related but distinct genes. Moreover, expression in the acinar cells of the pancreas is very high: about 20% of the total protein synthesis of the gland is for these few serine proteases.[3]

[a] This research was supported by grants from the National Institutes of Health.

mRNAs OF THE EXOCRINE PANCREAS

As would be expected for a tissue that synthesizes a few prominent protein products, the polyadenylated RNA population of the pancreas contains a few dominant mRNAs. FIGURE 1 shows the profile of polyadenylated RNA isolated from total pancreatic RNA and resolved by electrophoresis in an agarose gel containing the denaturant methylmercury hydroxide. Major mRNAs for the secretory enzymes lipase; amylase; carboxypeptidases A, A1, and B; and proelastase I have been identified.[7] The mRNAs for proelastase II,[8] chymotrypsinogens,[7] trypsinogens,[9] prokallikrein,[10] and RNase[11] constitute a group of prominent RNAs ranging in length from 900 to 1100 nucleotides. In addition to a protein-coding domain, each serine protease mRNA contains short 5′ and 3′ untranslated regions. The 3′ untranslated region of the elastase I mRNA is relatively long (280 nucleotides), and accounts for the greater overall length of elastase I mRNA (1250 nucleotides) compared to the mRNAs of the other serine proteases (FIG. 1).

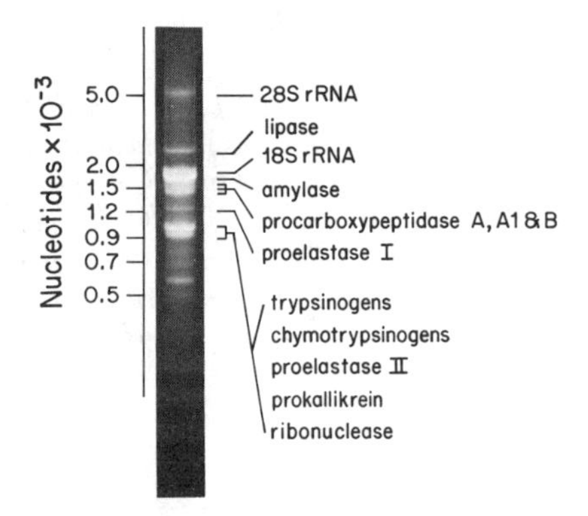

FIGURE 1. Prominent mRNAs for the secretory enzymes of the rat pancreas. Total RNA isolated from rat pancreas by the guanidine thiocyanate procedure of Chirgwin *et al.*[4] Polyadenylated RNA was selected by affinity chromatography on oligo-(dT) cellulose[5] and resolved by electrophoresis in a 1.5% agarose gel containing the denaturant methylmercury hydroxide.[6] The gel was stained with ethidium bromide and photographed. Small amounts of 18S and 28S ribosomal RNAs remained in this polyadenylated RNA preparation.

TISSUE-SPECIFIC EXPRESSION OF PANCREATIC SERINE PROTEASE mRNAs

The level of elastase I mRNA in several rat tissues (TABLE 1) illustrates the tissue-specific expression of a representative member of the pancreatic serine protease gene family. In the pancreas, elastase I mRNA constitutes about 1% of the total mRNA, or about 10 000 mRNAs per cell. Levels are at least 1000-fold lower in other tissues such as intestine, kidney, and liver. The physiological significance of this low, but detectable, expression in these tissues is unclear, but may represent expression at higher levels in a few specialized cells such as tissue mast cells.[8] In yet other tissues such as the parotid and submaxillary glands, elastase I mRNA is undetectable; expression is at least 100 000-fold greater in the pancreas than in these tissues. The elastase I gene appears maximally *on* in the pancreas, and *off* in tissues such as the parotid and submaxillary glands.

At least one level of regulation of the elastase I gene is transcriptional. Nuclear run-on experiments detect elastase I gene transcription in nuclei isolated from pancreas but not from liver (FIG. 2). Elastase I gene transcription in isolated pancreatic nuclei

TABLE 1. Elastase I mRNA Levels in the Rat

	Number of mRNAs per Cell[a]
Pancreas	10 000
Liver	10
Kidney	4
Intestine	3
Spleen	0.1
Submaxillary gland	<0.5
Parotid gland	<0.1
Testes	<0.1

[a]The mRNA levels were quantified by solution hybridization[21] using a single-stranded rat elastase I cDNA probe. The numbers of elastase I mRNA molecules per cell were calculated as described by Swift *et al.*[12]

accounts for about 0.2% of the total whereas transcription of the liver-specific transcript albumin is undetectable (less than 1 ppm). Conversely, elastase I gene transcription is undetectable in liver nuclei in which albumin gene transcription is prevalent. The limitations of the *in vitro* run-on transcription assay prevent confident measurements of transcription below 1 ppm, so the maximum level of differential transcription is only 150-fold, and is probably greater. Therefore, differential rates of transcription alone may account for the 1000-fold difference in elastase I mRNA levels of the pancreas and liver.

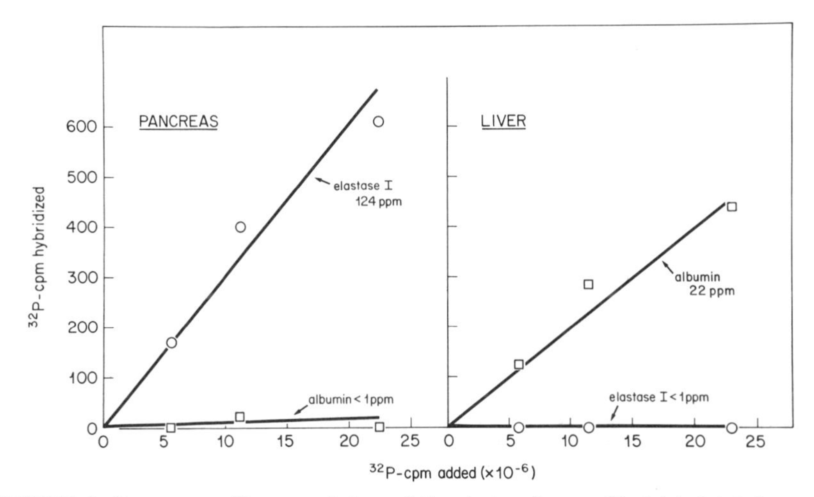

FIGURE 2. Pancreas-specific transcription of the elastase I gene. Nuclei isolated from rat pancreas and liver were incubated with [^{32}P]uridine 5′-triphosphate to extend nascent RNA transcripts. The *in vitro* transcribed RNA was isolated and hybridized to 4 μg of rat elastase I and rat albumin cDNA plasmids bound to nitrocellulose. The background was defined as the amount of hybridization to the cloning vector pBR322. The relative transcription rates (ppm) are expressed as the amount of hybridizing cpm minus background, divided by the number of input cpm and the length of the cloned cDNA probe in kb. All procedures were performed as described by McKnight and Palmiter.[13]

SEQUENCE ORGANIZATION OF THE ELASTASE I GENE

As is the case for most nuclear genes of higher eukaryotes, the rat pancreatic elastase I gene is split (FIG. 3). The elastase I mRNA coding sequence is divided into eight exons that span 12 kb of genomic DNA. The entire rat gene has been cloned in two recombinant lambda phages that contain genomic sequences that abut in the rat genome.[14] In addition to the exon and intron sequences, the cloned gene region contains 7.2 kb of 5′ flanking and 5 kb of 3′ flanking sequences.

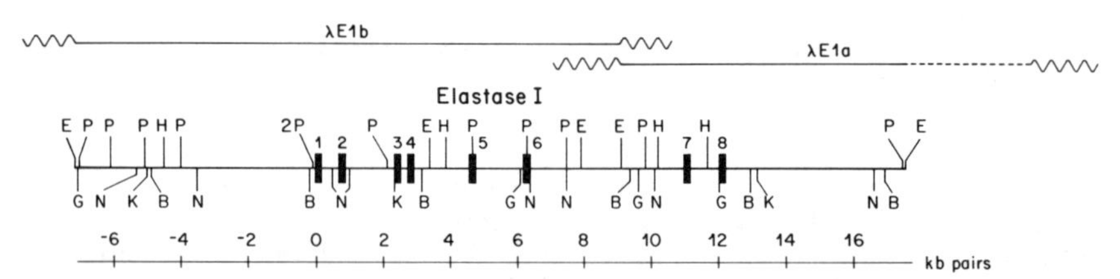

FIGURE 3. The sequence organization of the rat elastase I gene. The positions of the eight elastase I exons within 25 kb of rat nuclear DNA are shown relative to a restriction map of the region (E. *Eco*RI; G: *Bgl*II; P: *Pvu*II; N: *Nco*I; K: *Kpn*I; B: *Bam*HI; H: *Hind*III). The cloned gene region was obtained on two recombinant lambda phages (shown at the top), which contain inserts that abut in genomic DNA. For microinjection, the gene was reassembled by cloning into pBR322 and spanned the 24 kb shown, with the exception of the 1.5-kb *Eco*R1 fragment within intron 6. The gene was prepared for microinjection by cleavage within the pBR322 region with *Cla*I and *Nru*I to create a linear DNA fragment of 27 kb with dissimilar ends. This large fragment had 25 bp of pBR322 sequences at the 5′ end and 3.5 kb at the 3′ end of the genomic sequences shown.

INTRODUCTION OF THE CLONED RAT ELASTASE I GENE INTO MICE

Two methods are currently available to test for cell-specific expression of cloned genes. One approach is to introduce a purified gene into cells in culture by transfection. Cell-specific expression is measured by the differential expression of a gene in an appropriate differentiated cell line (such as the chymotrypsin gene in a pancreatic acinar cell carcinoma line) compared to an inappropriate cell line (such as a fibroblast or kidney cell line).[15] Major advantages of this method are the relative ease with which multiple genes or altered gene constructs can be tested and the speed of the assay. It is impractical, however, to screen a large number of differentiated cell lines representing many different cell types to measure the extended specificity of expression in a broad range of cell types. Moreover, cells in culture tend to express all transfected eukaryotic genes regardless of cell specificity at a low basal level, even though the endogenous gene is not expressed.[16] Expression in the appropriate differentiated cell type is generally only one to two orders of magnitude above this basal level whereas the normal differential expression in animals is much greater. Thus, parts of the regulatory mechanism responsible for rigorous cell-specific expression may be overlooked.

A second approach is to introduce cloned genes into all cells of an animal by microinjection into fertilized eggs to create transgenic animals.[17,18] A major advantage of this method is the ability to assay for expression in many tissues of an animal. We have introduced the rat elastase I gene into mice to determine whether expression of a transgene can occur in the proper tissue-specific manner.

In order to introduce the elastase I gene into mice, the gene was reassembled by cloning into pBR322.[12] Because the location of pancreas-specific regulatory sequences was unknown, extensive genomic regions were included. Thus, in addition to the 12 kb of intron and exon sequences, the reassembled elastase I gene contained the 7.2 kb of 5′ flanking and 5 kb of 3′ flanking sequences shown in FIGURE 3.

Approximately 250 rat elastase I gene copies were injected into the male pronucleus of fertilized mouse eggs, and the eggs implanted into oviducts of foster mother mice. To identify pups that have acquired the rat gene, DNA was isolated from a piece of the tail of each pup, spotted onto nitrocellulose, and hybridized with a DNA probe derived from the 3′ untranslated region of the rat elastase I mRNA. Under stringent hybridization conditions, this hybridization probe does not cross-hybridize with mouse elastase I sequences, and therefore can be used to selectively quantify the number of rat genes (or mRNA) even in the presence of endogenous mouse elastase I genes (or mRNA). In four microinjection experiments, 7 mouse pups from a total of 37 born had acquired the rat elastase I gene.

Several important parameters concerning the integration of the microinjected gene into the mouse genome cannot be controlled. 1) The number of integrated genes varies between transgenic mice, from a fragment of a single gene to several hundred copies per cell. The integration of multiple gene copies generally occurs at a single genomic location (occasionally animals with two independent integration sites are found) with the copies linked tandemly in a head-to-tail array. 2) The site of insertion, if not random, can occur at any of a large number of places on any chromosome. Integration does not occur preferentially at the position of the endogenous gene. 3) The timing of the integration event may vary. If integration occurs before the first cleavage, all cells of the animal carry the genes; integration at later times may limit the introduced gene(s) to only some cells of the animal (mosaicism).

The number of integrated rat elastase I genes, quantified by DNA tail blots, varied between 2 and 120 copies among the original transgenic mice. The number of independent integration loci for each founding transgenic mouse can be determined by following the segregation of the introduced genes in the progeny derived from mating with normal mice. Fifty percent of the progeny of a single locus transgenic mouse inherit the parental number of gene copies. Greater than 50% of the progeny inherit genes from a transgenic parent with multiple, unlinked loci of introduced genes; the number of genes inherited are characteristic of an individual locus or combination of loci. Fewer than 50% of the progeny inherit the genes if the founding mouse is mosaic; progeny that inherit the genes have a higher copy number than the apparent copy number of the mosaic parent. Mating experiments with the first five transgenic mice bearing the rat elastase I gene demonstrated the presence of each of these three kinds of integration classes. Transgenic mice 13-4, 6-3, and 9-4 had two, seven, and nine gene copies, respectively, integrated at a single locus. Mouse 13-1 had 100 gene copies divided into an 80-copy locus and a 20-copy locus. Mouse 22-5 with an apparent copy number of 120 was a mosaic: fewer than 50% of the progeny inherited the rat genes; those that did inherited twice the parental number of gene copies (FIG. 4).

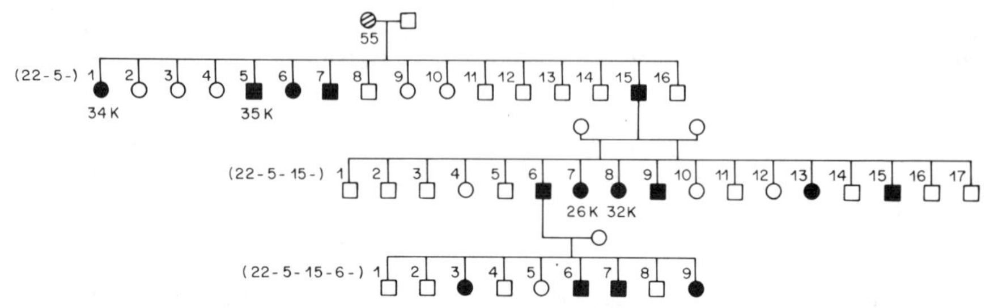

FIGURE 4. Pedigree of transgenic mouse line 22-5. Progeny that inherited rat elastase I genes were identified by hybridization of the rat-specific elastase I probe to DNA blots of tail DNA. The striped symbol indicates 120 copies of the transgene per cell; solid symbols indicate 250 copies of the transgene per cell; open symbols indicate the absence of transgenes. The levels of rat elastase I mRNA per cell (K = thousands of mRNA molecules) in the pancreas of five mice are shown.

EXPRESSION OF THE RAT ELASTASE I GENE IN TRANSGENIC MICE

To determine whether the introduced rat genes were expressed, and if so whether expression was pancreas specific, we assayed rat elastase I mRNA levels in eight transgenic mouse tissues including the pancreas.[12] In four of the first five animals tested, expression was indeed specific to the pancreas (FIG. 5). Mice 13-1, 9-4, 6-3, and 13-4 had pancreatic levels of the rat elastase I mRNA equal to or greater than the normal rat level. Expression in other tissues was not detectable by this analysis, with the exception of significant, but low, mRNA levels in the spleens of mice 13-1, 9-4, and 6-3, and in the liver of mouse 6-3. The level of this inappropriate expression, however, was at least two orders of magnitude below corresponding pancreatic levels.

Expression in mouse 22-5 was below the level of detection by Northern blot analysis for all tissues examined. More sensitive solution hybridization analysis revealed rat elastase I mRNA in the pancreas at about 0.5% of the normal rat level, and levels in nonpancreatic tissues were several times lower. Thus, some pancreas-specific expression was observed, but at an exceedingly low level. Because this transgenic mouse was mosaic (FIG. 4), one likely explanation for the lack of pancreatic expression is an extreme mosaicism of the pancreas such that rat elastase I genes were present in very few pancreatic cells. Consistent with this explanation, transgenic progeny derived from this founding mouse had high levels of rat elastase I mRNA in the pancreas (between 26 000 and 36 000 mRNAs per cell) (FIG. 4). Moreover, the expression was limited to the pancreas (see TABLE 2, discussed below). The appearance of high, pancreas-specific expression in the progeny of mouse 22-5 indicates that the chromosomal site of integration was not responsible for the lack of expression in the founding mouse.

The size of the rat elastase I mRNA in the pancreas of transgenic mice was indistinguishable from that of authentic rat elastase I mRNA (FIG. 5). This identical size indicates that initiation of transcription of the introduced genes and polyadenylation and splicing of the transcripts were correct. Moreover, accurate synthesis of the rat elastase I isozyme by transgenic mouse pancreas indicates that the entire amino acid coding region of the mRNA is intact and is also correctly spliced (see FIG. 7, discussed below).

TRANSGENIC PROGENY EXPRESS THE INTRODUCED GENES

Progeny of transgenic mice that inherit the introduced elastase I genes generally express them in the pancreas-specific manner of their parent. All transgenic progeny maintain high pancreatic (FIG. 6) and low-to-undetectable nonpancreatic levels of rat elastase I mRNA. Because the data are not age and sex matched, variations may be due to the differences between animals. Transgenic mouse 13-4 and its progeny have 10 000 to 18 000 rat elastase I mRNAs per pancreatic cell. Mice with greater than seven elastase I transgenes generally have 40 000 to 60 000 mRNAs per pancreatic cell, as if a maximum level is reached, independent of gene dosage above seven gene copies. When independent lines were derived carrying either the 20- or 80-copy locus of the two-loci mouse 13-1, mRNA levels were about at the same, apparently maximum

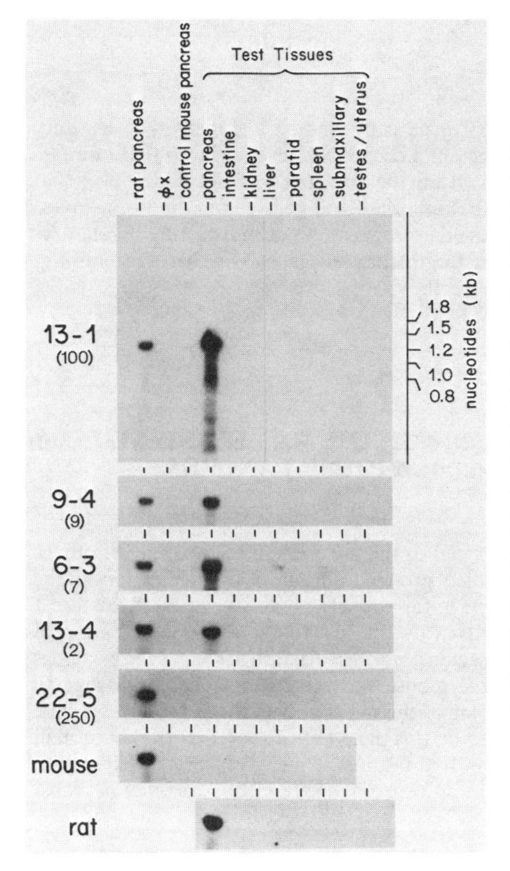

FIGURE 5. Northern blot analysis of rat elastase I mRNA expression in transgenic mice. Total RNA was isolated from the tissues listed at the top by the guanidine thiocyanate procedure of Chirgwin *et al.*[4] Ten micrograms of total RNA from each tissue was resolved by electrophoresis in 1.5% agarose gels containing methyl mercury hydroxide,[6] transferred,[19] covalently bound to activated filter paper,[20] and hybridized with a rat-specific single-stranded DNA probe derived from the 3′ untranslated region of the rat elastase mRNA.[12] The numbers in parentheses are the gene copy numbers per cell for each transgenic animal. Each blot also contained 10 μg of normal rat pancreatic RNA (first line) to show the size of authentic rat elastase I mRNA and the relative amount of elastase mRNA in normal rat pancreas. The absence of hybridization to mouse elastase mRNA present within the mouse pancreatic RNA sample on each blot (third lane) demonstrated the specificity of the probe.

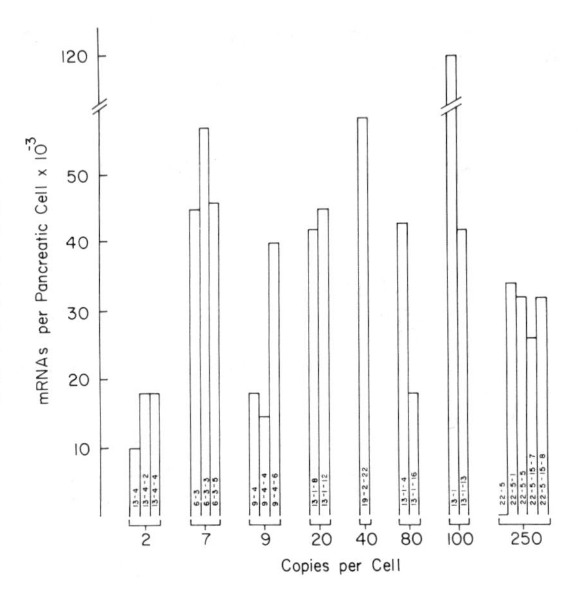

FIGURE 6. Pancreatic expression of the introduced rat elastase I genes is inherited in transgenic progeny. Rat elastase I mRNA levels were quantified by solution hybridization[21] using the rat-specific elastase I hybridization probe.[12] The number of integrated rat gene copies is given at the bottom for each transgenic line.

level, even with the fourfold difference in gene copy number. Moreover, transgenic progeny 13-1-13, which inherited both loci, had an mRNA level equal to that of mice with either locus, so that the expression from the two loci was not additive in the progeny. Even progeny of the 22-5 line that inherited 250 gene copies did not express levels above the apparent maximum. These data suggest that a *trans*-acting regulatory factor required for optimal expression of the introduced genes may become limiting at about 10 rat elastase I gene copies per cell.

SYNTHESIS AND SECRETION OF THE RAT ELASTASE I PROTEIN IN TRANSGENIC MOUSE PANCREAS

FIGURE 7 shows the proteins synthesized in the pancreas of the progeny of mouse 13-1. The presence of a few highly labeled proteins illustrates the dedication of the pancreas to the synthesis of a few enzymes for export. Three isozymes of elastase I were synthesized. Their identity was proven by the selective immunoprecipitation of these three proteins by antisera prepared against purified elastase I. Two of the isozymes (labeled mEI) are from two alleles of the mouse elastase I gene. The third isozyme (rEI), synthesized at a level similar to that of the mouse isozymes, is the rat enzyme. A similar analysis of the proteins secreted by this pancreas showed that the rat protein is also secreted at a high level and in a normal fashion. Thus, all aspects of expression of the introduced rat genes—transcription, processing of the mRNA precursor, translation, and secretion—appear normal.

SELECTIVENESS OF THE EXPRESSION OF THE RAT ELASTASE I GENES IN MICE

We assayed the pancreatic levels of the rat elastase I mRNA in eight transgenic mouse lines by solution hybridization (TABLE 2) to more accurately compare them with normal rat levels. The number of rat elastase I gene copies in these lines ranged from 2 to 250. Mice 13-1-12, 13-1-4, 19-2-22, and 22-5-5 were the first-generation progeny of four original transgenic mice. Mice 13-1-12 and 13-1-4 have inherited the 20-copy and 80-copy loci, respectively, of the two-loci 13-1 parent. Mouse 22-5-5 is an example of the appearance of high, pancreas-specific expression in the progeny of the very low expressor 22-5 mouse.

Each transgenic line listed (with the exception of mouse 13-1, which has two integration sites) represents an independent integration event of rat elastase I genes in the mouse genome. Each of these eight integration sites supports high, pancreas-specific expression of the foreign rat genes; pancreatic expression appears to be essentially independent of chromosomal position. Mouse 13-4, with two rat gene copies per cell, has an mRNA level equivalent to normal rat levels, as if each copy of the

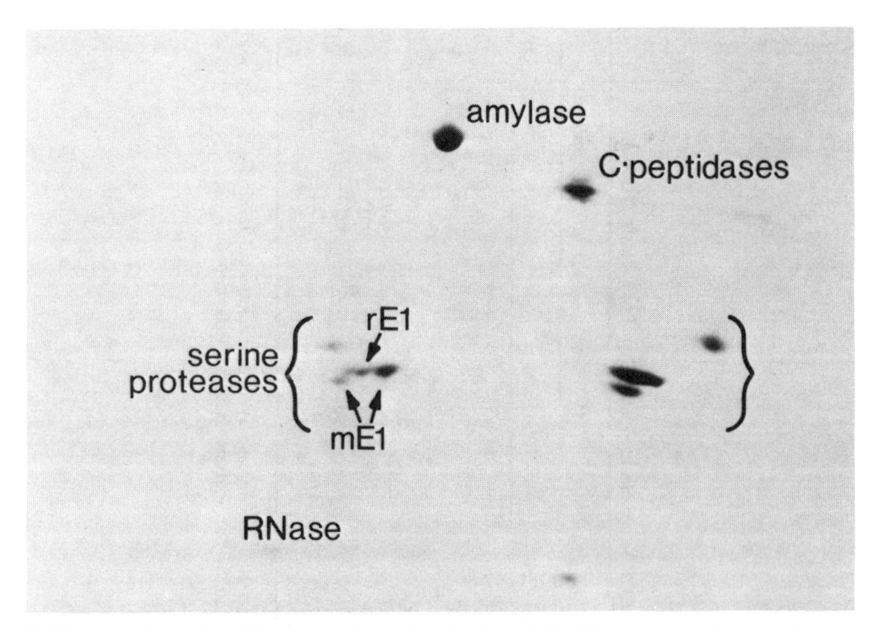

FIGURE 7. Synthesis of the elastase I protein at high levels by the pancreas of transgenic mice. Pancreatic lobules were dissected[22] from the pancreas of transgenic mouse 13-1-10 (a first-generation progeny of founder mouse 13-1) and cultured[3] for 6 hr in the presence of [^{3}H]leucine to radiolabel newly synthesized proteins. This transgenic mouse had approximately 100 copies of the rat gene per cell and 25 000 rat elastase I mRNA molecules per pancreatic cell. The labeled proteins were resolved by two-dimensional nonequilibrium pH gradient electrophoresis[23] and autoradiographed. The brackets delimit the 25-32 000 molecular weight region containing multiple serine proteases. Mouse and rat elastase I isozymes were identified by immunoprecipitation with monospecific antibody against purified rat elastase I (a gift from C. Largman).

TABLE 2. Rat Elastase I mRNA Levels in Tissues of Transgenic Mice

Animal	Genes per Cell	Rat Elastase I mRNAs per Cell[a]			Pancreas:Kidney	Pancreas:Liver
		Pancreas	Kidney	Liver		
Rat	2	10 000	4	10	2500	1000
13-4	2	10 000	<0.3	<1	>33 000	>10 000
6-3	7	45 000	<1	300	>45 000	150
9-4	9	18 000	<0.1	<0.6	>180 000	>30 000
13-1-12	20	45 000	<0.1	6	>450 000	>7500
19-2-22	40	48 000	ND	<1		>48 000
13-1-4	80	43 000	<0.1	ND	>430 000	
13-1	100	120 000	<0.2	3	>600 000	40 000
22-5-5	250	32 000	<1	<2	>32 000	>16 000

[a] A "less than" sign indicates that no increase over the background level of detection was observed. The detection level for a given tissue depended upon the amount of RNA available for analysis. ND: not determined.

rat gene in this transgenic mouse, present in a foreign genome at a novel location, is expressed as efficiently as a normal elastase I gene in the rat genome at its normal location. Thus, the establishment of extended chromatin domains by multiple, tandemly linked transgenes is not necessary for pancreas-specific expression. The most prodigious producer of rat elastase mRNA is mouse 13-1, with 120 000 mRNAs per average cell. The high pancreatic levels of rat elastase I mRNA in progeny 13-1-12 and 13-1-4 demonstrate that both loci are active.

The high pancreatic level of rat elastase I mRNA in transgenic mice permits a rigorous test of tissue-specific expression of the introduced genes. Messenger RNA levels in the pancreas and two representative nonpancreatic tissues (kidney and liver) of the eight transgenic mouse lines were compared to determine whether the differential expression of the transgenes was of the same magnitude as the normal endogenous genes (TABLE 2). Indeed, expression in nonpancreatic tissues of transgenic animals was often undetectable (fewer than two mRNAs per cell). The highest nonpancreatic expression was in the liver of mouse 6-3, with 300 copies per cell, and this level was still 150-fold lower than that for the expression in the pancreas of this animal. The level of expression in the kidney was 600 000-fold lower than the level in the pancreas of mouse 13-1. Therefore, even as many as 100 copies of the rat gene integrated at two different sites in the foreign mouse genome were controlled rigorously.

THE LOCATION OF PANCREAS-SPECIFIC REGULATORY SEQUENCES

The recurrent, successful pancreatic expression of the rat elastase I transgenes in mice suggested that it was feasible to identify tissue-specific regulatory sequences by trimming down the gene before introducing it into mice and assaying for pancreatic expression. Furthermore, results from even a few transgenic animals were expected to be significant. FIGURE 8 and TABLE 3 summarize the results of gene trimming experiments.[24] Removing 7 kb of upstream (5′) flanking sequences to leave only 205 bp upstream of the elastase I structural gene does not eliminate pancreas-specific expression. All four mice bearing this trimmed gene expressed rat elastase I mRNAs in the pancreas. It seems that removal of this extensive upstream region does not delete negative regulatory elements because expression in seven nonpancreatic tissues remained undetectable.

In a series of fusion gene experiments (FIG. 8) progressively shorter lengths of elastase I 5′ flanking sequences were linked to the structural gene for human growth hormone (hGH).[24] The fusion gene with 4.2 kb of flanking sequence linked to the hGH gene gave high, pancreas-specific expression in six of eight transgenic mice (TABLE 3). This result, coupled with the result from the trimmed gene experiment, indicated the presence of either 1) at least two independent pancreatic regulatory elements, one in the upstream region and one in the body of the elastase I gene or downstream flanking region, or 2) a single regulatory element within the only region common to the two gene constructs—the elastase gene region between −205 and +8. To test these two possibilities directly, only the elastase I region −205 to +8 was joined to the hGH gene. In four of six transgenic animals bearing this fusion gene, hGH mRNA was found at high levels in the pancreas and was not present in other tissues. Immunofluorescence analysis (FIG. 9) demonstrated that hGH protein was present in the acinar cells, but not in endocrine or connective tissue cells, of the

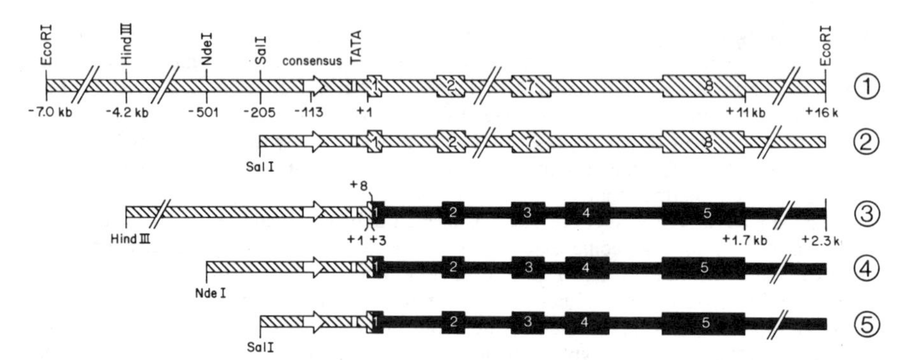

FIGURE 8. A pancreas-specific regulatory element is located near the transcription start site of the elastase I gene. The cloned rat elastase I gene was modified prior to microinjection into mouse egg. Construct 1: Unmodified gene comprising 7 kb of 5′ flanking sequences, 11 kb of exon and intron sequences, and 5 kb of 3′ flanking sequences. Construct 2: Cleavage at a unique *Sal*I site removes all but 205 nucleotides upstream from the start of transcription. Construct 3: Fusion of the 5′ flanking region of the elastase I gene from −4200 to +8 nucleotides to the hGH structural gene at +3. Construct 4: Fusion of −500 to +8 of elastase I to hGH at +3. Construct 5: Fusion of −204 to +8 of elastase I to hGH at +3.

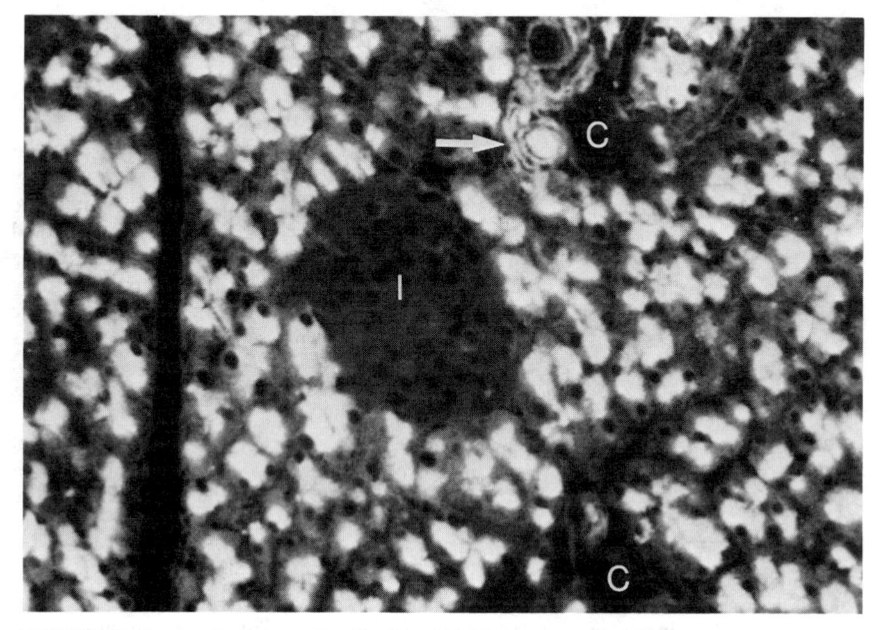

FIGURE 9. Immunofluorescent localization of hGH in the pancreas of a transgenic mouse. Sections were derived from the pancreas of a transgenic mouse bearing rat elastase I-hGH fusion genes containing 0.5 kb of elastase I upstream sequences (FIG. 8, construct 4). Pancreatic sections were treated with rabbit anti-hGH and then with fluorescein isothiocyanate-labeled, goat anti-rabbit immunoglobulin.[24] Acinar cells fluoresce intensely whereas islet cells (I) and cells of connective tissue (C) do not. The presence of immunofluorescence within an intralobular duct (arrow) demonstrated secretion of hGH.

TABLE 3. Rat Elastase I and hGH mRNA Levels in Transgenic Mice

Gene Construct Number	Gene Construct[a]	Number of Transgenic Mice	Number of Transgenic Mice Expressing the Introduced Genes	Transgene mRNA Levels[b] (Molecules per Cell)	
				Pancreas	Other Tissues
1	Intact elastase	7	6	41 000 (2400-120 000)	<40[c]
2	Trimmed elastase	4	4	3400 (1900-4100)	<10
3	4.2-kb Elastase I-hGH fusion	8	6	9900 (1200-28 000)	<10
4	0.5-kb Elastase I-hGH fusion	7	5	15 600 (1500-39 000)	<10
5	0.2-kb Elastase I-hGH fusion	6	4	26 000 (3300-51 000)	<10

[a] See FIGURE 8 for constructs.
[b] Average mRNA levels for transgenic mice that expressed the introduced genes. Ranges are shown in parentheses.
[c] One exception: 300 rat elastase I mRNAs per liver cell were found in one transgenic mouse.

pancreas of transgenic mice bearing a elastase I-hGH fusion gene. Thus, expression was directed to the appropriate cell type within the tissue.

These fusion gene experiments permit a whole-animal biological assay for expression in nonpancreatic tissues. Although synthesis and secretion of hGH by the acinar cells of the pancreas would result in the efficient hydrolysis of the hormone in the intestine and not affect growth, ectopic expression of hGH would dramatically increase growth,[25] even at low levels of expression.[26] None of the transgenic mice with the elastase I-hGH fusion gene, however, grew detectably larger than their normal littermates. Therefore, the absence of enhanced growth in animals bearing the elastase I-hGH fusion genes also indicates that expression does not occur in nonpancreatic tissues. Testing for increased growth effectively assays far more tissues than could be dissected and analyzed by hybridization for mRNA levels. Any cell type in the animal that secretes into a compartment that has access to the circulation can be excluded as having significant expression of the elastase I-hGH transgene.

```
                                   -112                        -31           +1
Elastase I            ......CTTTCATGTCACCTGTGCTTTTCCC.......CGTATAAAGAGGG......CA...
                                   -211                        -31           +1
Chymotrypsin B          .......TCAGGGCACCTGTCCTTTTCCC.......GACATAAAAAGAG......CA...
                                   -95
Elastase II         ......CCCTTT ATTCCAC TGGGCTTT...           -28           +1
                                ...CCACCTTGCGTACTCC........GATATAAACAAAG......CA...
                                   -73
                                   -180                        -31           +1
Trypsin I            ......CCCTTGT   CACCTGTAGGTCTCC........GCTATAAAGGAAG......TA...
                                   -211
Trypsin II              ........GTTTCCAC TGGTTTG...            -31           +1
                     ....CCTTGTCCTTATCAC TG................GGTATAAAAGCAA......TA...
                                   -159
```

FIGURE 10. Regions of homology among the 5′ flanking sequences of pancreatic elastases I and II,[14] chymotrypsin B,[7] and trypsins I and II.[27] The elastase II and trypsin II genes appear to have two conserved regions. Each "+1" indicates the first nucleotide of the mRNAs.

NATURE OF THE PANCREATIC REGULATORY SEQUENCE

The −205 to +8 region of the elastase I gene that is sufficient to direct pancreas-specific expression contains the start site of transcription, the promoter, and a 25-nucleotide sequence recognizably conserved among several pancreas-specific genes (FIG. 10). The conserved sequence is likely an important component of the regulatory region. It is included within a region necessary for selective expression of chymotrypsin gene constructs in pancreatic acinar cells in culture.[15] Octanucleotide sequences within the conserved regions resemble the core sequence of the simian virus 40 enhancer,[14] and preliminary results[15] suggest that the chymotrypsin regulatory sequence has enhancer-like properties and functions selectively in pancreatic acinar cells in culture.

cis-Acting regulatory information sufficient for pancreas-specific expression in animals is contained within the short elastase I gene sequence between −205 and +8. This gene region confers transcriptional specificity in transgenic ice bearing the elastase

I-hGH fusion gene because pancreatic nuclei, but not hepatic nuclei, synthesize hGH RNA.[28] Moreover, the transcription of the elastase I-hGH fusion gene correlates with a pancreas-specific DNase I hypersensitive site within the elastase I regulatory region. If pancreas-specific *trans*-acting factors must bind to activate transcription, then this 213-bp fragment contains all the necessary DNA sequence information for binding. Furthermore, if pancreas-specific alterations in chromatin structure are necessary to activate the gene selectively in the pancreas, that information also must be contained within this narrow upstream gene region. If there are important constraints on nuclear architecture, such as positioning the active gene at or near the nuclear membrane or in association with the nuclear matrix in a pancreas-specific manner, then this fragment must contain that information as well.

ACKNOWLEDGMENTS

We thank Mary Yagle and Kinyua Gikonyo for the technical assistance and Myrna Trumbauer for the DNA microinjection.

REFERENCES

1. NEURATH, H. 1984. Science **224:** 350-357.
2. ASHLEY, P. L. & R. J. MACDONALD. 1985. Biochemistry **24:** 4520-4527.
3. VAN NEST, G. A., R. J. MACDONALD, R. K. RAMAN & W. J. RUTTER. 1980. J. Cell Biol. **86:** 784-794.
4. CHIRGWIN, J. M., A. E. PRZYBYLA, R. J. MACDONALD & W. J. RUTTER. 1979. Biochemistry **24:** 5294-5299.
5. AVIV, H. & P. LEDER. 1972. Proc. Natl. Acad. Sci. USA **69:** 1408-1412.
6. BAILEY, J. M. & N. DAVIDSON. 1976. Anal. Biochem. **70:** 75-85.
7. BELL, G. I., C. QUINTO, M. QUIROGA, P. VALENZUELA, C. S. CRAIK & W. J. RUTTER. 1984. J. Biol. Chem. **259:** 14265-14270.
8. MACDONALD, R. J., G. H. SWIFT, C. QUINTO, W. SWAIN, R. L. PICTET, W. NIKOVITS & W. J. RUTTER. 1982. Biochemistry **21:** 1453-1463.
9. MACDONALD, R. J., S. J. STARY & G. H. SWIFT. 1982. J. Biol. Chem. **257:** 9724-9732.
10. SWIFT, G. H., J.-C. DAGORN, P. L. ASHLEY, S. W. CUMMINGS & R. J. MACDONALD. 1982. Proc. Natl. Acad. Sci. USA **79:** 7263-7267.
11. MACDONALD, R. J., S. J. STARY & G. H. SWIFT. 1982. J. Biol. Chem. **257:** 14582-14585.
12. SWIFT, G. H., R. E. HAMMER, R. J. MACDONALD & R. L. BRINSTER. 1984. Cell **38:** 639-646.
13. MCKNIGHT, G. S. & R. D. PALMITER. 1979. J. Biol. Chem. **254:** 9050-9058.
14. SWIFT, G. H., C. S. CRAIK, S. J. STARY, C. QUINTO, R. G. LAHAIE, W. J. RUTTER & R. J. MACDONALD. 1984. J. Biol. Chem. **259:** 14271-14278.
15. WALKER, M. D., T. EDLUND, A. M. BOULET & W. J. RUTTER. 1983. Nature **306:** 557-561.
16. ROBINS, D. M., I. PAEK, P. H. SEEBURG & R. AXEL. 1982. Cell **29:** 623-631.
17. PALMITER, R. D. & R. L. BRINSTER. 1985. Cell **41:** 343-345.
18. BRINSTER, R. L., H. Y. CHEN, M. E. TRUMBAUER, M. K. YAGLE & R. D. PALMITER. 1985. Proc. Natl. Acad. Sci. USA **82:** 4438-4442.
19. STELLWAG, E. J. & A. E. DAHLBERG. 1980. Nucleic Acids Res. **8:** 299-317.
20. ALWINE, J. C., D. J. KEMP & G. R. STARK. 1977. Proc. Natl. Acad. Sci. USA **74:** 5350-5354.
21. DURNAM, D. M. & R. D. PALMITER. 1983. Anal. Biochem. **131:** 385-393.

22. Scheele, G. A. & G. E. Palade. 1975. J. Biol. Chem. **250:** 5375-5385.
23. O'Farrell, P. Z., H. M. Goodman & P. H. O'Farrell. 1977. Cell **12:** 1133-1141.
24. Ornitz, D. M., R. D. Palmiter, R. E. Hammer, R. L. Brinster, G. H. Swift & R. J. MacDonald. 1985. Nature **313:** 600-603.
25. Palmiter, R. D., R. L. Brinster, R. E. Hammer, M. E. Trumbauer, M. G. Rosenfeld, N. C. Birnberg & R. M. Evans. 1982. Nature **300:** 611-615.
26. Palmiter, R. D., G. Norstedt, R. E. Gelinas, R. E. Hammer & R. L. Brinster. 1983. Science **222:** 809-814.
27. Craik, C. S., Q.-L. Choo, G. H. Swift, C. Quinto, R. J. MacDonald & W. J. Rutter. 1984. J. Biol. Chem. **259:** 14255-14264.
28. Ornitz, D. M., R. D. Palmiter, A. Messing, R. E. Hammer, C. A. Pinkert & R. L. Brinster. 1986. Cold Spring Harbor Symp. Quant. Biol. **50:** 399-409.

Regulation of Lactate Dehydrogenase Gene Expression by cAMP-dependent Protein Kinase Subunits[a]

RICHARD A. JUNGMANN,[b] ANDREAS I. CONSTANTINOU,[c] STEPHEN P. SQUINTO,[c] JOANNA KWAST-WELFELD, AND JOHN S. SCHWEPPE

Department of Molecular Biology
Northwestern University Medical School
Chicago, Illinois 60611

INTRODUCTION

It is now well established that adenosine 3′,5′-monophosphate (cAMP) serves as an intracellular mediator for the action of several glycoprotein, polypeptide, and catecholamine hormones. Among the multitude of physiologic events regulated by cAMP, the cyclic nucleotide mediates the inductive effect of hormones on the biosynthesis of specific proteins in a developmental and differentiation-specific pattern.[1-3] Among several possible induction mechanisms it is thought that cAMP-mediated processes of development and differentiation are controlled, directly or indirectly by changes in the patterns of expression of structural genes. As a result, the spectrum of mRNA molecules synthesized, processed, and translated differs among cell types and in many cases is correlated directly in the form of specific protein products with a particular cellular phenotype.

Rigorous proof that the cAMP-mediated induction process proceeds via a transcriptional mechanism, rather than alteration of posttranscriptional events such as mRNA processing and turnover, altered rates of synthesis/degradation, and/or posttranslational modification of the protein, has been provided in only a few systems. This paper presents a current view of our understanding of the molecular mechanisms responsible for regulation of gene function by cAMP. Specifically, it will summarize recent research carried out in our laboratory dealing with an investigation of the cAMP regulation of the lactate dehydrogenase (LDH) genes. Additionally, we will compare our findings with studies carried out by other investigators dealing with the

[a]The studies described in this paper were supported in part by Grant GM23895 from the National Institutes of Health, by a grant from the American Heart Association, and by the Research and Education Fund, Northwestern University.

[b]Address for correspondence: Department of Molecular Biology, Northwestern University Medical School, 303 East Chicago Avenue, Chicago, Illinois 60611.

[c]Recipient of a National Postdoctoral Research Award from the National Institutes of Health.

inductive effects of cAMP on several other structural genes in the hope of identifying a general and universal mechanism of cAMP-mediated protein induction. We will address ourselves to the following questions: 1) Are posttranscriptional events regulated by cAMP? 2) Which proteins are induced by a transcriptional mechanism? 3) Is cAMP a positive as well as negative modulator of gene regulation? 4) Do cAMP-dependent protein kinase (cADepPK) subunits mediate the effects of cAMP on mRNA synthesis? 4) What are the molecular events that take place during cAMP-mediated transcriptional regulation?

GENE SYSTEMS REGULATED BY cAMP

Posttranscriptional Regulatory Effects

From a historical point of view, regulation of the rate of biosynthesis of several metabolic enzymes by cAMP was the first regulatory mechanism to be unequivocally demonstrated by several investigators.[1-3] These early studies pointed to a posttranscriptional mechanism of enzyme induction, but, due to methodological problems (a lack of methods for determining the levels of low-abundance mRNA species), the role of cAMP in gene regulation remained obscure. It was only through the use of inhibitors of RNA synthesis that a regulatory role of cAMP at the level of transcription was suggested.[1-3] None of these studies, however, provided direct evidence for an effect of cAMP on structural gene transcription. Only after the development of cell-free translation systems and after the development of methodological advances in molecular genetics, which allowed the efficient construction and cloning of DNA probes complementary to relatively rare mRNA species, has it been possible to investigate the induction mechanism at the level of the mRNA.

Assessment of functional mRNA levels through cell-free translation assay and measurement of the actual levels of mRNA molecules through hybridization analysis led to the discovery of several cell systems in which cAMP exerted a marked modulatory effect consisting of increased as well as decreased mRNA levels (TABLE 1). The distinction made here between assessment of the functional level, that is, translational activity, of mRNA and the net level of mRNA molecules determined by hybridization assay with a specific cDNA probe is not a trivial one. Although any alteration of the actual number of mRNA molecules per cell should also be reflected in an alteration of equal magnitude of the *in vitro* translational activity of the cellular mRNA, any discrepancies between these parameters that might arise could result from additional cAMP-modulated controls: the presence of regulatory factors in the poly(A)$^+$ RNA fraction, for instance, which would allow for more effective utilization of the induced mRNA, or the induction of an mRNA species with altered nontranslated sequences, which would result in inherent structural modifications and template utilization of the mRNA.

One of our findings illustrates the importance of the distinction mentioned above. In isoproterenol-stimulated rat C6 glioma cells the rate of accumulation of LDH A subunit mRNA followed a time course identical to the one determined by *in vitro* translation of poly(A)$^+$ RNA (FIG. 1). However, whereas Northern blot and kinetic hybridization analyses determined an increase of about 2.5-fold in the number of LDH mRNA molecules, *in vitro* translation assay of the poly(A)$^+$ RNA from stimulated glioma cells showed an increase of about 8-fold in the functional activity of LDH

TABLE 1. Systems in which cAMP Modulates mRNA of Specific Gene Products

Cell Systems	Gene Product	Effect on mRNA Levels or Translational Activity	Reference
Adipose 3T3 cells	Glycerophosphate dehydrogenase	Decrease[a]	4
	Fatty acid synthetase	Decrease[a]	4
	Malic enzyme	Decrease[a]	4
Dictyostelium discoideum	Discoidin I	Decrease[b]	5
	Developmentally regulated mRNAs	Increase[b]	6
Rat liver	Tyrosine aminotransferase	Increase[a]	7,8
Rat H4IIE hepatoma cells	Tyrosine aminotransferase	Increase[a]	9
Rat liver	Phosphoenolpyruvate carboxykinase	Increase[a]	10-12
Rat liver	Phosphoenolpyruvate carboxykinase	Increase[b]	13,14
Chicken kidney	Phosphoenolpyruvate carboxykinase	Increase[b]	15
Bovine adrenocortical cells	Cytochrome P_{450}	Increase[a]	16
Rat pituitary cells	Prolactin	Increase[b]	17
Rat C6 glioma cells	Lactate dehydrogenase A subunit	Increase[a,b]	18,19
Mouse Hepa-2 cells	Albumin	Increase[b]	20
Mouse L cells	Alkaline phosphatase	Increase[a]	21
Pheochromocytoma cells	Tyrosine hydroxylase	Increase[b]	22
Rat liver	Glucokinase	Decrease[a]	23
Rat liver	L-Type pyruvate kinase	Decrease[a]	24
Rat hepatocytes	Ornithine aminotransferase	Increase[b]	25
Rat liver	Carbamoylphosphate synthetase	Increase[b]	26

[a]Effect determined by assay of translational activity of poly(A)$^{+}$ mRNA.
[b]Analysis of mRNA level by hybridization assay to a specific cDNA.

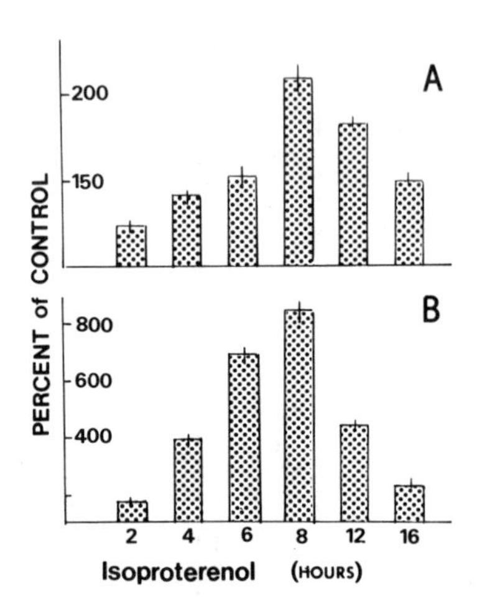

FIGURE 1. Isoproterenol-mediated changes of rat C6 glioma cell LDH A mRNA levels. Confluent rat C6 glioma cells were stimulated with 10^{-5} M isoproterenol. After the indicated time periods, cells were harvested and poly$(A)^+$ RNA was isolated. LDH A mRNA was quantitated (**A**) by Northern blot analysis[19] or (**B**) by translational assay in a reticulocyte lysate system.[18] Each bar represents the mean ± SEM of four determinations.

mRNA. These findings not only indicate an increase of the actual number of induced LDH mRNA molecules but that the newly induced mRNA exhibits a higher translational activity per molecule. Although the molecular basis for this apparent discrepancy has to be elucidated, the induction of a transcriptionally more efficient mRNA species results, in addition to raising mRNA levels, in a more effective amplification of the hormonal inductive signal.

In order to understand the comprehensive mechanism of the cAMP inductive effect, one must also consider a potential regulatory action at the level of the stability of the mRNA. The number of molecules of a given mRNA species in a cell can be assumed to result from the opposing reactions of synthesis and degradation, and either process may be modulated by cAMP. Ideally, turnover studies should be conducted without inhibitors of RNA synthesis because the presence of such inhibitors could complicate results. Methods that allow the half-life of mRNA to be determined involve the quantitation of the rate of change (rate of decay of radioactive mRNA) from an initial induced to a final deinduced steady state level that is equivalent to the turnover rate.[27] This method does not use inhibitors. Stringent experimental conditions, however, such as an effective chase period to avoid reutilization of isotope and the preparation of highly purified mRNA (selected by hybridization to a specific cDNA probe), must be applied in order for the method to be useful.

We have recently reported an example of a cAMP-mediated modulation of the mRNA half-life for the LDH A subunit mRNA in the rat C6 glioma cell.[27] In these cells the decay kinetics of LDH mRNA from isoproterenol-stimulated cells identified two components: a fast-decaying one, with a half-life identical to noninduced LDH mRNA ($t_{1/2}$ = 45 min) (representing about 35% of the total LDH mRNA population), and a slower-decaying one ($t_{1/2}$ = 2.5 hr) (representing the new population of induced LDH mRNA) (TABLE 2). The existence of two LDH mRNA components with different metabolic stabilities in induced C6 glioma cells suggests that the noninduced and induced LDH mRNAs differ structurally, and thus supports the notion mentioned above that such a structural difference may account for the induction of an LDH mRNA of higher translational capacity than would be expected on the basis of the more limited increase of LDH mRNA molecules identified by hybridization analysis. Characterization of several other mRNA species has shown that multiple mRNAs, coding for the same protein sequence, are transcribed from the same structural gene.[28]

These mRNAs differ in the length and nature of their untranslated sequences, and initiation of their transcription is caused by different promoters. Studies conducted along these lines with cAMP-induced mRNAs may be a fruitful area for investigation.

Transcriptional Regulatory Effects

In several different cell and organ systems, cAMP, various cAMP analogues, or cAMP-generating hormones regulate the transcription of specific structural genes. The evidence for this was obtained by quantitation of transcription by a run-off assay: isolated nuclei were derived from unstimulated and stimulated cells; the mRNA chains in these nuclei and the cDNA probes complementary to the mRNA chains were subjected to specific hybridization analysis; and the lengths to which the mRNA chains had become elongated were determined (TABLE 3). In the isoproterenol- or dibutyryl cAMP-stimulated rat C6 glioma cell line, we have quantitated LDH A subunit mRNA transcription by the *in vitro* run-off transcription assay, as well as by assessing LDH mRNA synthesis in intact cells, by applying pulse-labeling with [^{3}H]uridine.[27] Analysis of cAMP-stimulated LDH mRNA synthesis by both methods showed an increase of about 3.6-fold in the basal rate of LDH mRNA transcription, and both methods revealed a close correspondence of the kinetics of LDH mRNA synthesis by RNA polymerase II (FIG. 2). When the degree to which LDH mRNA transcription is increased is considered in context with the increased half-life of LDH mRNA after stimulation, one can calculate that the LDH mRNA concentration should increase some 6- to 10-fold 8 hr after stimulation. Determination of the LDH mRNA levels by hybridization analysis (FIG. 1), however, reveals but a 2.5-fold increase. This apparent discrepancy can best be explained by assuming there are two different species of LDH mRNA, as suggested by the turnover studies shown in TABLE 2. A quantitation of the actual number of LDH mRNA molecules per glioma cell before and after isoproterenol stimulation is illustrated in FIGURE 3. If most (about 90%) of the LDH mRNA in unstimulated glioma cells is of the noninduced type ($t_{1/2}$ = 45 min), and if most of the LDH mRNA synthesized by stimulated glioma cells is of the induced type ($t_{1/2}$ = 2.5 hr), a 10-fold increase of induced LDH mRNA (resulting from the concomitant modulation of the rate of synthesis and degradation of LDH mRNA) would result in only a 2.5-fold overall increase of total LDH mRNA.

The identification of an increasing number of structural genes that are transcriptionally regulated by cAMP indicates that mammalian cells may universally employ a transcriptional mechanism, in addition to posttranscriptional and translational mechanisms, to regulate the hormonally determined expression of specific proteins. In

TABLE 2. Half-life of LDH A Subunit mRNA in Isoproterenol-induced and Noninduced Rat C6 Glioma Cells

Stimulation	$t_{1/2}$ (min)	Percentage of Total LDH mRNA Population
None	45	100
Isoproterenol	45	35
	180	65

TABLE 3. Systems in which cAMP Regulates the Induction of a Specific Gene by a Transcriptional Mechanism

Cell System	Gene Product	Effect on mRNA Transcription	Reference
Dictyostelium discoideum	Discoidin I	Decrease	29
Rat H4IIE hepatoma cells	Phosphoenolpyruvate carboxykinase	Increase	30
Rat kidney	Phosphoenolpyruvate carboxykinase	Increase	31
Rat liver	Phosphoenolpyruvate carboxykinase	Increase	32
Rat C6 glioma cells	Lactate dehydrogenase A subunit	Increase	27
Rat pituitary cells and HTC cells	Prolactin	Increase	33
GH4 cells	Prolactin	Increase	34
Rat liver	Tyrosine aminotransferase	Increase	35
Rat thyroid	Thyroglobulin	Increase	36
Human neuroblastoma cells	Provasoactive intestinal polypeptide (VIP)	Increase	37
Rat liver	Aldolase B	Decrease	38
Rat liver	L-Type pyruvate kinase	Decrease	38

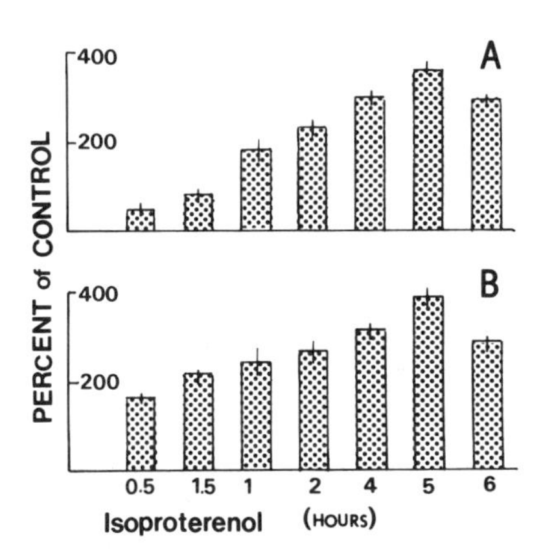

FIGURE 2. Isoproterenol-mediated changes of LDH A mRNA transcription in rat C6 glioma cells. Confluent rat C6 glioma cells were stimulated with 10^{-5} M isoproterenol for the time periods indicated. (**A**) For *in vitro* quantitation of LDH A mRNA synthesis, nuclei were isolated, and a run-off assay was performed.[27] (**B**) *In vivo* LDH A mRNA synthesis was assessed after pulse-labeling of cells with [^{3}H]uridine.[27] Each bar represents the mean ± SEM of four determinations.

addition to the transcriptionally regulated genes shown in TABLE 3, other obvious candidates for transcriptional regulation may include some of the gene products cited in TABLE 1 for which regulatory effects of cAMP on mRNA abundance have been identified.

Interestingly, there are several genes where transcriptional down-regulation occurs as the consequence of cAMP action (TABLE 3). Transcription of discoidin I, aldolase B, and L-type pyruvate kinase is rapidly decreased under conditions of increased cAMP levels. Although these negative regulatory effects are the direct consequence of cAMP action, insulin has been shown to block the inductive effect of cAMP in at least one experimental system. In the rat H4IIE hepatoma cell, insulin reversibly inhibits the cAMP-induced transcription of the phosphoenolpyruvate carboxykinase gene.[30] The molecular mechanisms of these effects are not known, but it is obvious that any general model of transcriptional control by cAMP will have to account for both positive as well as negative induction effects of cAMP and for the role of effector agents that modulate cAMP action.

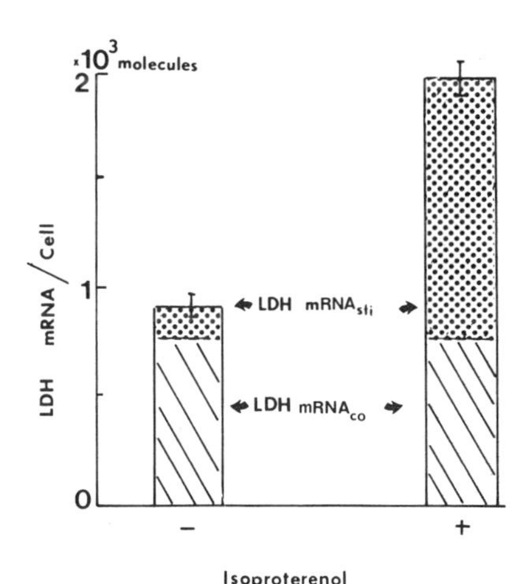

FIGURE 3. Effect of isoproterenol on the number of LDH A mRNA molecules per rat C6 glioma cell. Confluent rat C6 glioma cells were stimulated for 8 hr with 10^{-5} M isoproterenol. The total number of LDH A mRNA molecules are represented by each complete bar. The diagonally striped bars indicate the assumed concentration of the noninduced LDH A mRNA$_{co}$ population; the dotted bars, the assumed concentration of the induced LDH A mRNA$_{sti}$ population.

MECHANISM OF TRANSCRIPTION CONTROL BY cAMP

When evaluating the molecular mechanism of transcription control by cAMP, it is helpful to recall that there are several features that appear to be characteristic for the cAMP-mediated mechanism of gene control in eukaryotic cells. In contrast to steroid hormone- or thyroid hormone-mediated protein synthesis, the induction process initiated by cAMP is extremely rapid. Several cell systems present ideal experimental situations in which a comparison of the induction kinetics of a single gene product, inducible by both cAMP as well as a steroid hormone, can be carried out. For instance, in rat liver glucagon as well as glucocorticoids are capable of inducing tyrosine aminotransferase. However, tyrosine aminotransferase mRNA synthesis increases rapidly and transiently in response to glucagon (or cAMP), whereas there is a delay in the transcriptional response to dexamethasone, followed by a more persistent but slower rate of increase of enzyme synthesis.[35] Another example of differential induction kinetics consists of the glucocorticoid and cAMP regulation of the enzyme phosphoenolpyruvate carboxykinase.[30,31] This evidence suggests there are different induction mechanisms for both types of effector agents.

Although intense efforts have been made by many investigators, efforts which were aided in part by the identification of steroid hormone-inducible and relatively abundant gene products, and which have resulted in an elucidation of some of the initial molecular steps of steroid hormone action, there has been little progress in elucidating the cAMP-mediated inductive mechanisms. During the process of gene regulation by cAMP, one gene (or a few genes) must be selected from many genomic structural gene sequences. The selection and activation process may involve a direct interaction with cAMP, or may involve interactions with cAMP-regulated factors at various pretranscriptional and transcriptional steps, that is, at the DNA level, directly, through specific structural and functional modifications of chromatin and/or the action of cAMP upon DNA-binding proteins that regulate the binding to, and transcription of, the gene by RNA polymerase II. The relative rapidity of the cAMP-mediated induction process, together with findings that *de novo* protein synthesis (of mediatory regulatory factors) is not mandatory, suggests that a transient allosteric and/or covalent structural and functional modification of regulatory factor(s) may be involved. Although the cAMP regulatory action in prokaryotic enzyme induction processes has given rise to much speculation that a similar mechanism might be operable in eukaryotic systems, no direct evidence is available to support such a notion. For these reasons, and others discussed below, a pivotal role for cADepPK subunits in cAMP-initiated enzyme induction has to be seriously considered.

Subunits of cADepPK and Gene Regulation

Based on the evidence available at present, it is generally accepted that the physiologic effects of cAMP in eukaryotic cells are mediated primarily, if not exclusively, by activation and dissociation of cADepPK isozymes (types I and II) into their regulatory (RI and RII) and catalytic (C) subunits, processes that are followed by the phosphorylative and functional modification of regulatory proteins. Although there is no direct evidence that cADepPK mediates the effects of cAMP on gene expression, several findings are compatible with such a mechanism. For instance, genetic evidence of the involvement of cADepPK in enzyme induction has been obtained by Insel *et al.*[39] and Steinberg and Coffino,[40] investigators who have studied mutants of the mouse lymphoma S49 cell line. But not only has there been a successful demonstration of the involvement of cADepPK in enzyme induction by genetic means, there is now evidence utilizing biochemical approaches of a role of cADepPK in enzyme induction.[41–44]

On the premise that cADepPK mediates the inductive effect of cAMP, a gene-regulatory role of cADepPK, then, would require either 1) the presence of a resident nuclear holoenzyme that may be activated by an influx of cAMP into the nucleus, or 2) the translocation of gene-regulatory phosphoproteins into the nucleus, or 3) the translocation of cADepPK subunits from the extranuclear space to the nucleus. A more orthodox explanation would attribute the inductive effect to the activation of a resident nuclear cADepPK holoenzyme, but most of the experimental data that have accumulated are inconsistent with such a mechanism.[45] We favor a more dynamic model of nuclear cADepPK subunit modulation, one that has received considerable experimental support. Several years ago, we suggested that a modulation of nuclear cADepPK subunit levels could be achieved through a cAMP-mediated temporary (reversible) translocation of cADepPK subunits from the extranuclear space to the nuclear compartment.[45] There are several experimental observations that correlate cADepPK subunit translocation with specific processes such as cAMP-mediated regulation of cell proliferation,[46,47] LDH mRNA induction in rat C6 glioma cells,[27] and phosphoenolpyruvate carboxykinase mRNA induction in rat H4IIE hepatoma cells. In the rat H4IIE hepatoma cell, we recently studied nuclear cADepPK subunit modulation after dibutyryl cAMP stimulation under conditions that lead to the induction of phosphoenolpyruvate carboxykinase mRNA transcription. In addition to the demonstration that dibutyryl cAMP rapidly increased nuclear catalytic subunit activity, precise quantitation of cADepPK subunits by enzyme-linked immunoabsorbent assay showed a marked increase of the number of nuclear RII and C subunits but not of the RI subunit (TABLE 4).

TABLE 4. Quantitation of Nuclear cADepPK Subunits in Rat H4IIE Hepatoma Cells by Enzyme-linked Immunosorbant Assay[a]

	Subunit (μg of Protein per 10^4 Nuclei)		
Stimulation	Catalytic Subunit C	Regulatory Subunit RI	Regulatory Subunit RII
None	4.1 ± 0.7 (3)	ND	0.72 ± 0.3 (4)
Dibutyryl cAMP (10^{-5} M)	8.5 ± 0.9 (3)	ND	2.0 ± 0.5 (3)

[a]Hepatoma cells were treated for 10 min with dibutyryl cAMP. Each result is the mean ± SEM for the number of experiments indicated in parentheses. ND: Not detected.

These data do not only demonstrate unequivocally an actual increase of the number of nuclear RII and C subunits, strongly supporting a translocation mechanism, but show the presence of both subunits in a nonstoichiometric relationship (about five molecules of C for each molecule of RII). The markedly higher concentration of C, in relation to RII, in both unstimulated and stimulated cells is incompatible with the transfer or presence of holoenzyme in the nucleus.

Function of Subunits at the Nuclear Level

Phosphorylation of nonhistone chromosomal proteins has been implicated as playing a potentially important role in determining the availability of genetic sequences for transcription.[45] If phosphorylative modification of chromosomal proteins is related to the regulation of gene expression, one would expect to observe a correlation between the modulation of nuclear cADepPK levels and changes in specific gene expression. Indeed, correlations between chromosomal protein phosphorylation and alterations of gene activity have been reported in many different experimental systems.[45] Unfortunately, most of the reports in which such correlations were observed noted that the methodological shortcomings of the time prevented the precise characterization of induced low-abundance gene products. The lack of such a characterization did not enable investigators to make a distinction between the stimulation of general RNA synthesis and specific mRNA induction.

More recently, several studies have unequivocally identified a phosphorylative modification of nuclear proteins during specific gene-regulatory events. Nuclear proteins whose degree of phosphorylation is increased in a cAMP-mediated mechanism include histone H1 in the glucagon-stimulated rat liver;[42] histones H1-1, H1-2, H1-3, and H3 and RNA polymerase II subunits in rat C6 glioma cells;[48,49] histone H1 during induction of thyroglobulin;[50] a 19 000-dalton peptide in forskolin-treated pituitary cells;[51,52] and a 76 000-dalton protein during dibutryryl cAMP-induced regression of mammary tumor growth.[46] In each case there has been a good correlation between the phosphorylative modifications and specific regulatory events such as the inductions

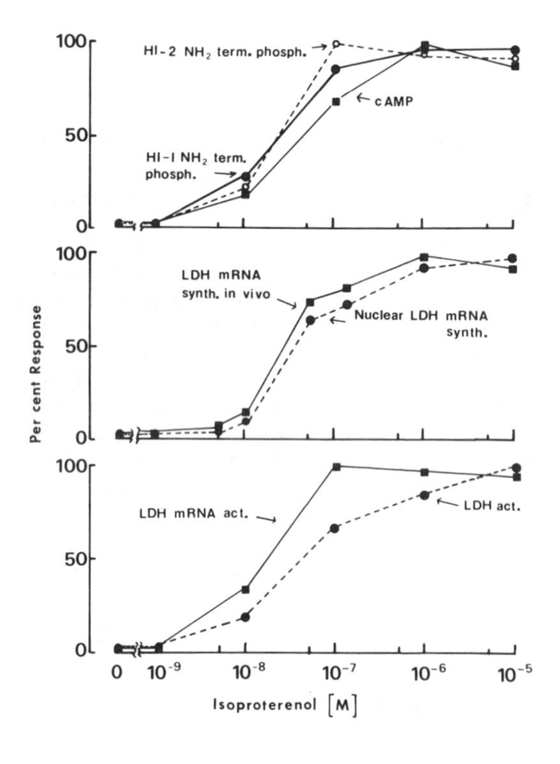

FIGURE 4. Dose relationship of isoproterenol-mediated molecular events during the LDH A mRNA induction phase (cAMP formation,[48] histone H1-1 and histone H1-2 aminoterminal phosphorylations,[48] LDH A mRNA synthesis *in vitro*[27] and *in vivo*,[27] LDH mRNA translational activity,[18] and LDH-5 enzymatic activity[18]).

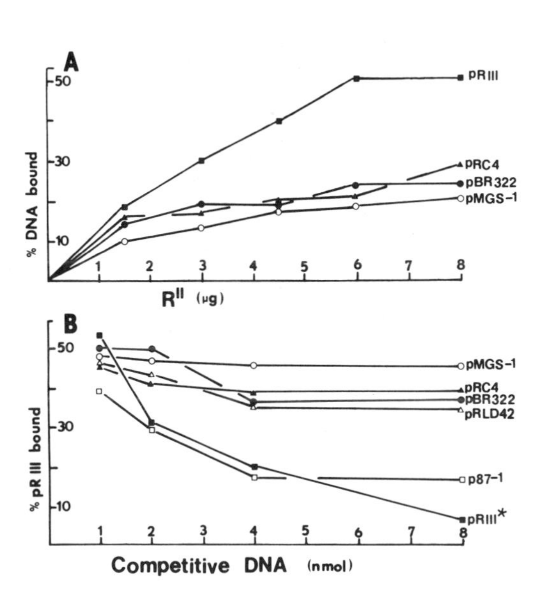

FIGURE 5. Assessment of the relative binding affinity of the regulatory subunit RII for various DNA templates. (**A**) Binding of increasing amounts of rat liver RII to ^{32}P-labeled DNA templates (1.4 nmoles each) was assessed by the nitrocellulose filter binding assay as described by Nagata *et al.*[55] (**B**) Binding of 8 μg of rat liver RII to ^{32}P-labeled pRIII was performed in the presence of increasing amounts of competitor DNAs. The following competitor DNAs were used: pRIII*: *Eco*RI/*Hind* III restriction fragment of pλpC112.R3 containing about 300 bp of the 5′ flanking region including the cAMP-responsive promoter sequence[56] (pλpC112.R3 was kindly provided by Dr. Daryl Granner, Vanderbilt University); p87-1: 3-kbp circular plasmid containing about 600 bp of the 5′ flanking region, the transcription initiation site, and additional structural sequence of the rat LDH A subunit gene (Jungmann *et al.*, unpublished results); pRLD42: LDH A cDNAl;[19] pRC4: genomic cytochrome *c* DNA cloned into pBR322;[57] pMGS-1: genomic mouse β-globin DNA cloned into pBR322.[58]

of tyrosine aminotransferase,[42] LDH,[48,49] thyroglobulin,[50] and prolactin[51,52] and the regulation of cellular growth.[46,47] FIGURE 4 illustrates the functional correlations between the various biochemical events that occur during LDH induction; that is, it illustrates the dose relationships of isoproterenol-mediated cAMP formation, histone H1 aminoterminal phosphorylation, and LDH mRNA transcription, and the relationship between LDH mRNA levels and LDH activity in rat C6 glioma cells. The similarity of the maximal and half-maximal stimulation values indicates a close dose-response relationship for the various processes listed and suggests a functional correlation between them.

Although the nuclear association and cAMP-mediated modulation of catalytic subunit levels is important for the structural covalent modification of nuclear proteins, no function of the regulatory subunits at the nuclear level has yet been identified. Based on the partial sequence homology between the RII subunit and the prokaryotic cAMP catabolite gene-activator protein,[53] it has been speculated that a functional similarity exists between these two proteins in regulating structural gene sequences. The first concrete evidence of an important role for the regulatory subunits, other than one of inhibiting catalytic subunit activity, was provided by our recent findings[54] that the rat liver RII subunit possesses topoisomerase activity. Furthermore, we have identified that the toposiomerase activity of RII is expressed only in the presence of cAMP and is altered by changes in the degree of phosphorylation of RII. This strongly

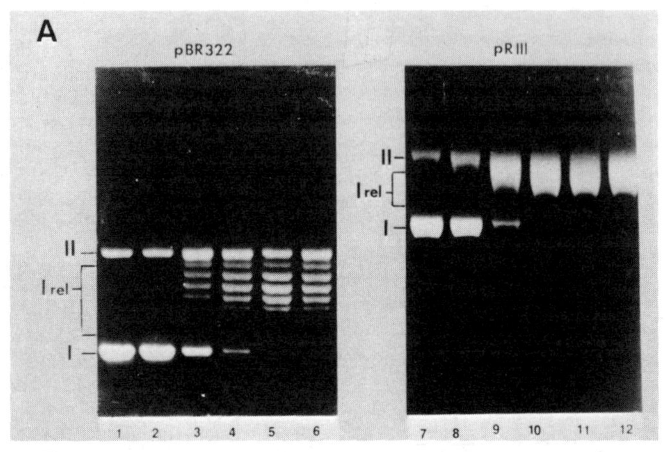

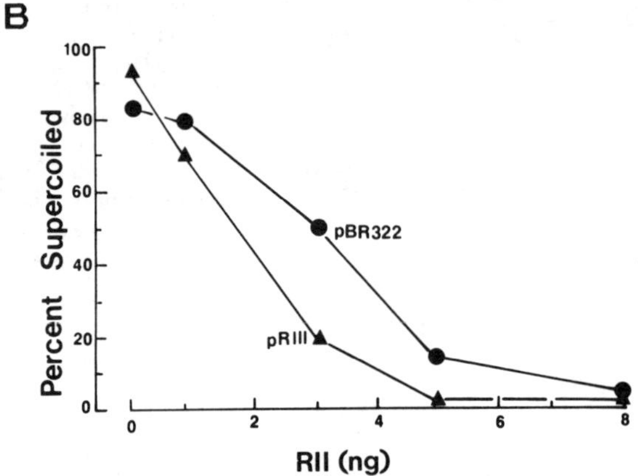

FIGURE 6. Effect of RII concentration on the relaxation of pBR322 and pRIII. (**A**) RII was serially diluted and assayed for DNA relaxing activity with 35 nmoles of negatively supercoiled pBR322 or pRIII as described by us.[54] The reactions contain the following amounts of RII: 0 ng (lanes 1 and 7); 1 ng (lanes 2 and 8); 3 ng (lanes 3 and 9); 5 ng (lanes 4 and 10); 8 ng (lanes 5 and 11); 10 ng (lanes 6 and 12). Reaction products were separated by agarose gel electrophoresis. (**B**) Quantitative assessment of the decreased amounts of supercoiled pBR322 and pRIII as a function of increasing RII concentrations. Lanes 1-6 and 7-12 in **A** were scanned with a Zeineh soft-laser-beam densitometer integrated with a Hewlett Packard computer model 3390A to calculate the percentage of supercoiled DNA remaining in the reaction mixture. I: supercoiled DNA; I_{rel}: relaxed topological isomers (closed circular DNA); II: nicked circular DNA.

suggests a cAMP-mediated functional interaction of RII with DNA and the potential ability of nuclear RII to affect the topology and transcription of cAMP-regulated structural gene sequences.

Because a selective interaction of RII with cAMP-regulated (in contrast to cAMP-unresponsive) promoter sequences was clearly suggested, we have studied the possibility of a selective binding of RII to cAMP-regulated promoter sequences in addition to an evaluation of the rate of relaxation of these DNA templates. We have quantitated the relative binding affinity of the regulatory subunit RII for various recombinant plasmids, some of which carry promoter sequences that are not expected to be regulated by cAMP (pMGS-1, pRC4), and two plasmids (p87-1, pRIII) that carry inserts containing the cAMP-regulated promoter sequences of the LDH and the phosphoenolpyruvate carboxykinase genes, respectively. FIGURE 5A shows that when equimolar concentrations of plasmid are used, RII has a markedly higher binding affinity for pRIII than it has for pBR322, pRC4, or pMGS-1. This selectivity of binding is confirmed by the competitive binding assay shown in FIGURE 5B. We studied the binding of RII to pRIII, which contains the cAMP-responsive phosphoenolpyruvate carboxykinase promoter sequence, and which was labeled with ^{32}P, in the absence and then in the presence of increasing concentrations of competitor DNAs. Only plasmids containing cAMP-regulated promoter sequences (p87-1, pRIII) effectively reduced the binding of RII to ^{32}P-labeled pRIII. Plasmids without cAMP-regulated promoters (pMGS-1, pRC4, pRLD42, pBR322) had little or no competitive activity.

FIGURE 6 graphically illustrates the relaxing effect of RII on negatively supercoiled DNA templates. At equimolar concentrations of DNA, less RII was required to achieve the same percentage of pRIII relaxation as compared to pBR322. The time courses of relaxation of various DNA templates by RII are shown in FIGURE 7. Under identical experimental conditions pRIII was a more efficiently relaxed template than either pBR322, pMGS-1, or pRC4. Although much work remains to be carried out on this aspect, these data suggest a selective binding affinity of RII to plasmids with cAMP-regulated promoters and the selective utilization of these DNA templates as substrates for the RII topoisomerase.

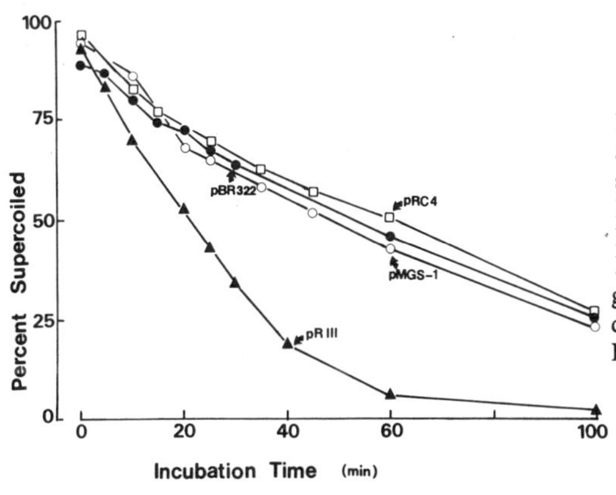

FIGURE 7. Time course of relaxation of negatively supercoiled DNAs by the regulatory subunit RII. Negatively supercoiled plasmid DNA (35 nmoles) was incubated with rat liver RII (7 fmoles) for the time periods indicated. Reaction products were separated by agarose gel electrophoresis and quantitated by densitometry as described in the legend of FIGURE 6.

SUMMARY

The studies described in this report suggest a rather complex, albeit incomplete, sequence of molecular events that we believe form part of the cascade of reactions through which a series of hormones, via cAMP, regulates the expression of specific gene products. The majority of our own studies relate to cAMP-mediated induction of LDH. Some, if not all, of the molecular steps discussed in this paper may ultimately be recognized as part of a universal mechanism by which cAMP controls gene expression in higher eukaryotes.

The idea of a functional role for cAMP-dependent protein kinase subunits in cAMP-mediated gene control has already had experimental support,[45] but our identification of the regulatory subunit RII as a topoisomerase now more firmly points to a complex function for the kinase in regulating gene function at the DNA level. We look forward to the elucidation of the function of those nuclear proteins that serve as substrate for the catalytic subunit of cAMP-dependent protein kinase. Further studies related to the molecular interaction of RII with chromosomal DNA should be a fruitful area for future research.

ACKNOWLEDGMENTS

We are grateful to Drs. D. Granner, R. Roeder, and R. Scarpulla for generous gifts of genomic clones.

REFERENCES

1. Wicks, W. D. 1974. Adv. Cyclic Nucleotide Res. **4:** 335-438.
2. Nimmo, H. G. & P. Cohen. 1977. Adv. Cyclic Nucleotide Res. **8:** 145-266.
3. Rosenfeld, M. G. & A. Barrieux. 1979. Adv. Cyclic Nucleotide Res. **11:** 205-264.
4. Spiegelman, B. M. & H. Green. 1981. Cell **24:** 503-510.
5. Berger, E. A., D. M. Bozzone, M. B. Berman, J. A. Morgenthaler & J. M. Clark. 1985. J. Cell. Biochem. **27:** 391-400.
6. Chung, S., S. M. Landfear, D. D. Blumberg, N. S. Cohen & H. F. Lodish. 1981. Cell **24:** 785-797.
7. Ernest, M. J. & P. Feigelson. 1978. J. Biol. Chem. **253:** 319-322.
8. Noguchi, T., M. Diesterhaft & D. Granner. 1978. J. Biol. Chem. **253:** 1332-1335.
9. Culpepper, J. A. & A. Y.-C. Liu. 1983. J. Biol. Chem. **258:** 13812-13819.
10. Iynedjian, P. B. & R. W. Hanson. 1977. J. Biol. Chem. **252:** 655-662.
11. Kioussis, D., L. Reshef, H. Cohen, S. M. Tilghman, P. B. Iynedjian, F. J. Ballard & R. W. Hanson. 1978. J. Biol. Chem. **253:** 4327-4332.
12. Beale, E. G., C. S. Katzen & D. K. Granner. 1981. Biochemistry **20:** 4878-4883.
13. Cimbala, M. A., W. H. Lamers, K. Nelson, J. E. Monahan, H. Yoo-Warren & R. W. Hanson. 1982. J. Biol. Chem. **257:** 7629-7636.
14. Beale, E. G., J. L. Hartley & D. K. Granner. 1982. J. Biol. Chem. **257:** 2022-2028.
15. Hod, Y., S. M. Morris & R. W. Hanson. 1984. J. Biol. Chem. **259:** 15603-15608.
16. Kramer, R. E., W. E. Rainey, B. Funkenstein, A. Dee, E. R. Simpson & M. R. Waterman. 1984. J. Biol. Chem. **259:** 707-713.
17. Maurer, R. A. 1982. Endocrinology **110:** 1957-1963.
18. Derda, D. F., M. F. Miles, J. S. Schweppe & R. A. Jungmann. 1980. J. Biol. Chem. **255:** 11112-11121.
19. Miles, M. F., P. Hung & R. A. Jungmann. 1981. J. Biol. Chem. **256:** 12545-12552.
20. Brown, P. C. & J. Papaconstantinou. 1979. J. Biol. Chem. **254:** 9379-9384.
21. Firestone, G. L. & E. C. Heath. 1981. J. Biol. Chem. **256:** 1396-1403.
22. Lewis, E. J., A. W. Tank, N. Weiner & D. M. Chikaraishi. 1983. J. Biol. Chem. **258:** 14632-14637.

23. SIBROWSKI, W. & H. J. SEITZ. 1984. J. Biol. Chem. **259:** 343-346.
24. MUNNICH, A., J. MARIE, G. REACH, S. VAULONT, M.-P. SIMON & A. KAHN. 1984. J. Biol. Chem. **259:** 10228-10231.
25. MERRILL, M. J., M. M. MUECKLER & H. C. PITOT. 1985. J. Biol. Chem. **260:** 11248-11251.
26. DE GROTT, C. J., A. J. VAN ZONNEVELD, P. G. MOOREN, D. ZONNEVELD, A. VAN DEN DOOL, A. J. W. VAN DEN BOGAERT, W. H. LAMERS, A. F. M. MOORMAN & R. CHARLES. 1984. Biochem. Biophys. Res. Commun. **124:** 882-888.
27. JUNGMANN, R. A., D. C. KELLEY, M. F. MILES & D. M. MILKOWSKI. 1983. J. Biol. Chem. **258:** 5312-5318.
28. TOSI, M., R. A. YOUNG, O. HAGENBUCHLE & U. SCHIBLER. 1981. Nucleic Acids. Res. **9:** 2313-2323.
29. WILLIAMS, J. G., A. S. TSANG & H. MAHBUBAMI. 1980. Proc. Natl. Acad. Sci. USA **77:** 7171-7175.
30. SASAKI, K., T. P. CRIPE, S. R. KOCH, T. L. ANDREONE, D. P. PETERSEN, E. G. BEALE & D. K. GRANNER. 1984. J. Biol. Chem. **259:** 15242-15251.
31. MEISNER, H., D. S. LOOSE & R. W. HANSON. 1985. Biochemistry **24:** 421-425.
32. LAMERS, W. H., R. W. HANSON & H. M. MEISNER. 1982. Proc. Natl. Acad. Sci. USA **79:** 5137-5141.
33. MAURER, R. 1981. Nature **294:** 94-97.
34. MURDOCH, G. M., R. FRANCO, R. E. EVANS & M. G. ROSENFELD. 1983. J. Biol. Chem. **258:** 15329-15335.
35. HASHIMOTO, S., W. SCHMID & G. SCHULTZ. 1984. Proc. Natl. Acad. Sci. USA **81:** 6637-6641.
36. VAN HEUVERSWYN, B., C. STREYDIO, H. BROCAS, S. REFETOFF, J. DUMONT & G. VASSART. 1984. Proc. Natl. Acad. Sci. USA **81:** 5941-5945.
37. HAYAKAWA, Y., K.-I. OBATA, N. ITOH, N. YANAIHARA & H. OKAMOTO. 1984. J. Biol. Chem. **259:** 9207-9211.
38. VAULONT, S., A. MUNNICH, J. MARIE, G. REACH, A.-L. PICHARD, M.-P. SIMON, C. BESMOND, P. BARBRY & A. KAHN. 1984. Biochem. Biophys. Res. Commun. **125:** 135-141.
39. INSEL, P. A., H. R. BOURNE, P. COFFINO & G. M. TOMKINS. 1975. Science **190:** 896-899.
40. STEINBERG, R. A. & P. COFFINO. 1979. Cell **18:** 719-726.
41. BONEY, C., D. FINK, D. SCHLICHTER, K. CARR & W. D. WICKS. 1983. J. Biol. Chem. **258:** 4911-4918.
42. LANGAN, T. A. 1978. Methods Cell Biol. **19:** 127-142.
43. WICKS, W. D., J. KOONTZ & K. WAGNER. 1975. J. Cyclic Nucleotide Res. **1:** 49-58.
44. WAGNER, K., M. D. ROPER, B. H. LEICHTLING, J. WIMALASENA & W. D. WICKS. 1975. J. Biol. Chem. **250:** 231-239.
45. JUNGMANN, R. A. & E. G. KRANIAS. 1977. Int. J. Biochem. **8:** 819-830.
46. LAKS, M. S., J. J. HARRISON, G. SCHWOCH & R. A. JUNGMANN. 1981. J. Biol. Chem. **256:** 8775-8785.
47. CHO-CHUNG, Y. S. 1980. J. Cyclic Nucleotide Res. **6:** 163-177.
48. HARRISON, J. J., G. SCHWOCH, J. S. SCHWEPPE & R. A. JUNGMANN. 1982. J. Biol. Chem. **257:** 13602-13609.
49. LEE, S. K., J. S. SCHWEPPE & R. A. JUNGMANN. 1984. J. Biol. Chem. **259:** 14695-14701.
50. LAMY, F., R. LECOCQ & J. DUMONT. 1977. Eur. J. Biochem. **73:** 529-535.
51. MURDOCH, G. H., M. G. ROSENFELD & R. M. EVANS. 1982. Science **218:** 1315-1317.
52. WATERMAN, M., G. H. MURDOCH, R. M. EVANS & M. G. ROSENFELD. 1985. Science **229:** 267-269.
53. WEBER, I. T., K. TAKIO & T. A. STEITZ. 1982. Proc. Natl. Acad. Sci. USA **79:** 7679-7684.
54. CONSTANTINOU, A. I., S. P. SQUINTO & R. A. JUNGMANN. 1985. Cell **42:** 429-437.
55. NAGATA, K., R. A. GUGGENHEIMER & J. HURWITZ. 1984. Proc. Natl. Acad. Sci. USA **80:** 6177-6181.
56. WYNSHAW-BORIS, A., T. G. LUGO, J. M. SHORT, R. E. K. FOURNIER & R. W. HANSON. 1984. J. Biol. Chem. **259:** 12161-12169.
57. SCARPULLA, R. C., K. M. AGNE & R. WU. 1981. J. Biol. Chem. **256:** 6480-6486.
58. LUSE, D. S. & R. G. ROEDER. 1980. Cell **20:** 691-699.

Molecular Genetic Analysis of cAMP-dependent Protein Kinase

MICHAEL M. GOTTESMAN[a] AND
ROBERT FLEISCHMANN

Laboratory of Molecular Biology
National Cancer Institute
National Institutes of Health
Bethesda, Maryland 20892

IRENE ABRAHAM

Cell Biology Department
The Upjohn Company
Kalamazoo, Michigan 29001

The mechanism by which cAMP and hormones that alter cAMP levels exert many of their pleiotropic effects on mammalian cells is still not fully understood at the molecular level. Two alternative hypotheses have been proposed. The first suggests that all of the effects of cAMP can be attributed to activation of the enzyme cAMP-dependent protein kinase (cADepPK) (the biochemical evidence for this point of view is reviewed in reference 1); the second is that some of these effects can be attributed to a direct interaction of cAMP-binding proteins (the regulatory subunits of the protein kinase, termed RI and RII) with DNA. This latter mechanism is known to be the way in which cAMP regulates metabolism in prokaryotes by interaction with the bacterial protein CRP (for a review, see reference 2). Recent published evidence has suggested that the 5′ flanking regions of eukaryotic genes whose levels of expression are regulated by cAMP contain sequence homology with CRP DNA binding sites found in *Escherichia coli*,[3,4] raising the possibility of a direct interaction of a mammalian cAMP-binding protein with DNA. Such an interaction has recently been suggested *in vitro* for RII, the phosphorylated form of which appears to have a cAMP-dependent DNA topoisomerase activity.[5] Although the relevance of these observations for the mechanism of cAMP effects in intact cells has not been proved, these observations suggest that a reexamination of the mechanism by which cAMP affects cells is in order.

We have taken a genetic approach to the analysis of cAMP effects in cultured cells. For these studies, we have used cultured Chinese hamster ovary (CHO) cells, a fibroblast-like cell line that responds to cAMP by changing from an oval to an elongated morphology,[6,7] by a reduction in growth rate,[7] by a decrease in transport of glucose and amino acids,[8] and by increases in activity of a variety of enzymes,

[a] Address for correspondence: National Cancer Institute, National Institutes of Health, Building 37, Room 2E18, Bethesda, Maryland 20892.

including transglutaminase, cyclic nucleotide phosphodiesterase, and ornithine decarboxylase.[8-11] We have isolated mutants resistant to the growth-inhibitory effects of cAMP analogues and/or cholera toxin,[7] and have found that essentially all of these mutants have specific deficiencies in the levels or activities of the enzyme cADepPK.[11]

STRATEGIES FOR USING SOMATIC CELL GENETICS TO STUDY cAMP-DEPENDENT PROTEIN KINASE

Our long-range strategy for the use of these CHO cAMP-resistant mutants has been as follows:

1) We would like to define the nature of those mutations in CHO cells that make cells resistant to the growth-inhibitory effects of cAMP. In essentially all cases, we have found that alterations in either the catalytic (C) subunit or the RI subunit of cADepPK render cells resistant to the growth-inhibitory effects of cAMP. All other known effects of cAMP on CHO cells are also blocked in these mutants (for reviews, see references 11 and 12). This result demonstrates that a fully intact cADepPK system is not necessary for normal growth of CHO cells, and that mutations in this system can block all known responses to cAMP in CHO cells. Similar conclusions have been reached by several other laboratories working with cAMP-resistant mouse S49 lymphoma cells,[13] mouse Y-1 adrenal cells,[14] mouse J774.2 macrophages,[15] mouse neuroblastoma cells,[16] and mouse Cloudman melanoma cells.[17,18]

2) We have demonstrated the dominant nature of many of our cADepPK CHO mutants by analysis of the phenotype of somatic cell hybrids.[7] This result has encouraged us to attempt DNA-mediated gene transfer of DNA encoding the defective catalytic or regulatory subunits of two of our mutants. We have found that both of these genes can be transferred into CHO cells by DNA-mediated gene transfer.[19,20] The transfer of these genes results in a cAMP-resistant phenotype indistinguishable from that of the cAMP-resistant DNA donor. These results indicate that it should be possible to clone a mutant gene and use this cloned gene to confer cAMP resistance on any cell line that is a recipient for DNA in the transfer procedure.

3) Having cloned DNAs encoding dominant mutant regulatory and catalytic subunits of cADepPK from CHO cells, it should be possible to transfer these mutant subunits into differentiated cell lines in which cAMP exerts transcriptional control over expression of specific genes. Examples of these genes include phosphoenolpyruvate carboxykinase (PEPCK),[21-23] tyrosine aminotransferase,[24] alkaline phosphatase,[25] and lactic dehydrogenase-5.[26] Using this approach, it should be feasible to determine whether cADepPK is an obligatory intermediate in the induction of each of these enzymes by cAMP. An alternate approach, once these cAMP-responsive genes have been fully cloned, or their promoters have been linked to a reporter gene such as the chloramphenicol acetyltransferase (CAT) gene,[27] would be to introduce them into the cAMP-resistant mutant CHO cells or other cell types carrying mutant cADepPK genes. The disadvantage of this alternative approach is that cAMP-dependent expression of these genes may depend on other factors present only in differentiated cells.

4) The final goal of these studies is to introduce genes encoding dominant mutant cADepPK subunits into intact animals, such as transgenic mice. As the technology develops, it should be possible to direct expression of these genes to specific tissues. Such studies would allow the independent determination of the role of cADepPK in the development and function of a variety of endocrine tissues. The development of

genetic models in intact organisms is the ultimate goal of the molecular endocrinologist who works in somatic cell genetics, and such models may soon be within our grasp.

This paper constitutes a progress report concerning our long-term efforts to isolate and clone mutant cADepPK subunits and to use them to generate animal models of cAMP-resistance. We shall summarize what we know about the mutants we have isolated, describe our successful efforts to transfer mutant cADepPK genes by DNA-mediated gene transfer, and describe the isolation of a cosmid clone carrying most of the gene for the RI subunit of cADepPK.

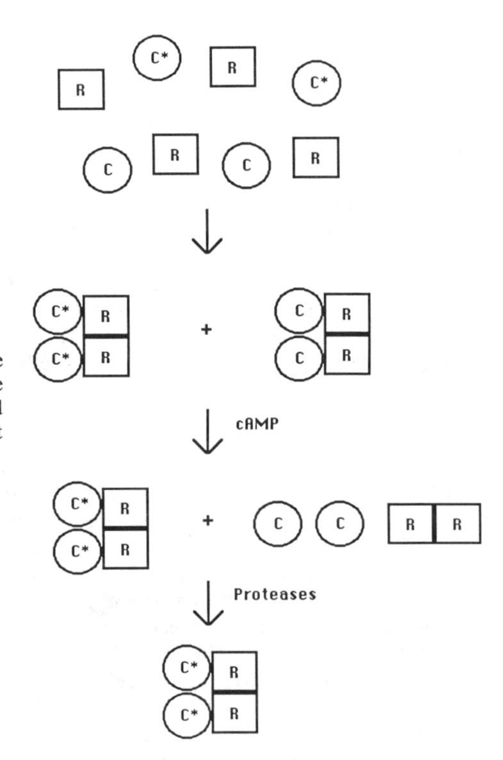

FIGURE 1. Model to explain dominance of the C subunit mutant 10215. See the text for details. A similar model would explain the dominance of the RI mutant 10248.

CHARACTERIZATION OF MUTANTS APPROPRIATE AS DNA DONORS FOR DNA-MEDIATED GENE TRANSFER

We have isolated over 30 independent CHO mutants resistant to the growth-inhibitory effects of cAMP. The vast majority of these mutants carry dominant mutations as indicated by the cAMP-resistance of somatic cell hybrids between cAMP-resistant cells and wild-type CHO cells.[7] Extracts of seven of these mutants have been analyzed by DEAE chromatography to determine whether they contain the two cADepPK activities found in extracts from wild-type CHO cells.[12] Four of the mutants we have analyzed are dominant and lack type II cADepPK activity. Three recessive mutants have also been analyzed; all of these lack type I cADepPK activity, and one of these also has very little type II kinase activity. These recessive mutants appear to fall into two distinct complementation groups,[28] and a mutant from one of these complementation groups has no detectable levels of the C subunit.[29] Whether this mutant carries a recessive regulatory mutation (defective promoter or lack of an activator for expression of the C subunit gene) or has a structural mutation in the C subunit gene has not yet been determined.

The dominant mutants are of more interest for the studies we are reporting here. Two of the four dominant mutants have been analyzed in some detail. One, mutant 10215, has been shown to have a defective C subunit with an altered K_m for ATP, and an apparent increased affinity of the mutant C subunit for RI, so that cAMP binding to RI does not dissociate the RI-C complex.[30] Another mutant, 10248, appears to have a defective RI subunit, as indicated by altered binding of cAMP and its photoaffinity labeling analogue, 8-N_3[^{32}P]cAMP. This mutant appears to contain the wild-type RI subunit as well, but the mutant RI subunit is preferentially associated with C.[31] For both mutants 10215 and 10248, the mutant RI-C complex is not dissociated by levels of cAMP normally achieved within cells.

One interesting characteristic of extracts from both of these mutants is the reduced level of type II cADepPK when compared to wild-type cells. A detailed analysis[28,31] has indicated that both mutants contain normal levels of RII (the regulatory subunit normally found in type II kinase), yet this RII is present as a dimer and is not complexed to C. One hypothesis to explain this phenomenon is that all of the free C in the cell is associated with RI in both of these mutants and that there is no C available to bind RII. Thus, a defect in either RI or C that results in increased stability of the RI-C complex would tend, over time, to eliminate RII-C (type II protein kinase) from the cell. This conclusion is based on the assumption that type II cADepPK would be dissociated normally by cAMP and that free C would be rapidly degraded by proteases in the cell, or become associated with RI in a nondissociable complex. This model is presented schematically in FIGURE 1, in which only the case where C is mutated is shown. The same model would hold for mutants such as 10248 in which RI is affected.

DNA-MEDIATED GENE TRANSFER OF MUTANT REGULATORY AND CATALYTIC SUBUNIT GENES

We have used a two-step protocol to transfer the genes for cAMP resistance from the dominant cAMP-resistant mutants 10248 and 10215 to CHO cells, and to thus make these CHO cells cAMP resistant.[32] Initial experiments in CHO cells indicated that the overall ratio of the number of cells that acquired the genomic thymidine kinase gene to the total number of cells with which transfer was attempted was low (approximately 10^{-6} to 10^{-7}).[33] Our first efforts to transfer genes for cAMP resistance into CHO cells were unsuccessful because the background frequency of phenotypically cAMP-resistant clones is high. The two-step protocol we finally used eliminated this problem of a high background frequency by selecting, in the first step, cells that had taken up DNA and then, in the second step, those DNA recipients that were cAMP resistant.[32]

Details of the procedure we used are summarized in FIGURES 2 & 3. In the first step, a total of 5×10^6 cells (2.5×10^5 cells/100-mm tissue culture dish) were treated

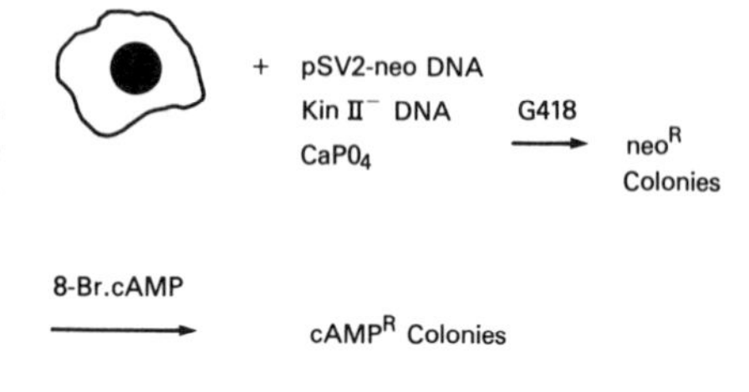

FIGURE 2. Two-step protocol for DNA-mediated transfer of dominant cADepPK mutations. Kin II-DNA is DNA from mutant 10215 or mutant 10248. Details are given in the text.

with 1 μg of pSV$_2$*neo* plasmid and 20 μg of high-molecular-weight genomic DNA from either wild-type cells or one of the mutants. Cells which had taken up pSV$_2$*neo* and genomic DNA were selected in the neomycin analogue G418. The ratio of the number of G418-resistant clones to the total number of cells was approximately 7×10^{-4}. These clones, of which there were about 3500 for each of the three transformations, were grown for 10 days and were then trypsinized and pooled. These clones, once their cells had been pooled, constituted a library of CHO cells carrying pSV$_2$*neo* sequences and genomic DNA fragments from the donor cells. The cAMP-resistant colonies among these clones were selected by plating the clones in agar containing 1 mM 8-Br-cAMP as previously described for the isolation of cAMP-resistant mutants.[12] When DNA from wild-type CHO cells was used as the donor DNA, only a few colonies, which were very small, appeared in agar. These were subsequently shown to be cAMP sensitive when growth was measured in monolayer culture in the presence and in the absence of 8-Br-cAMP. In contrast, when DNA from either of the mutant cells was used as the donor DNA, true cAMP-resistant clones of normal size were found: the ratio of cAMP-resistant cells to the total number of cells was 8×10^{-6} for the RI mutant (10248) and 3.3×10^{-5} for the C mutant (10215). Presumably, these numbers represent daughter clones derived from at least one cloned, cAMP-resistant cell out of the 3500 G418-resistant cells. Overall, the ratio of cells that acquired the phenotype for cAMP-resistance to the total number of cells, or, the frequency of transfer, was $>2 \times 10^{-7}$ for both RI and C mutant genes.

What is the biochemical evidence that we have actually transferred genes encoding mutant RI and C subunits? We have examined the phenotype of cADepPK activity in both the cAMP-resistant transformants by separating type I and type II protein kinases on DEAE columns. In each case, the DEAE profile of a given transformant was identical to that of the parent cell line (FIGURE 4). For either transformant, the type I enzyme that is present shows a reduced response to stimulation by cAMP in *in vitro* tests. In the case of the transformant from the RI mutant (11564), we have

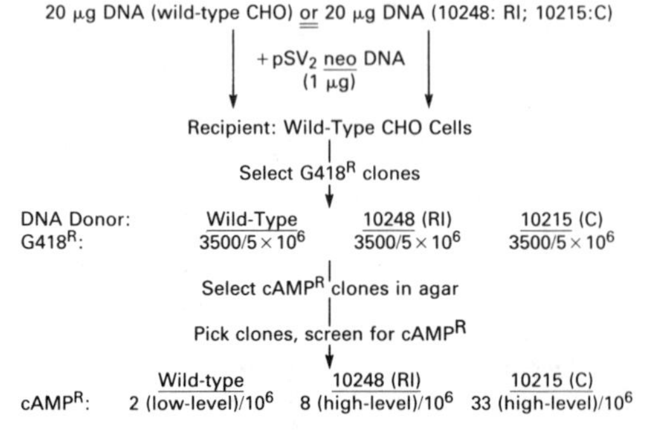

FIGURE 3. Summary of frequencies of DNA-mediated transfer of cAMP resistance from mutants 10215 and 10248.

also demonstrated that the transferred mutant subunit binds 8-N_3-cAMP with reduced affinity.[32] DNA from cells of this clone contains at least one additional RI gene as indicated by Southern blots and slot blots.[19] Both mutants fail to phosphorylate a specific 52000-dalton cellular protein in response to cAMP treatment,[34] and their corresponding transformants show the same defect. The important phenotypes of the mutants 10215 (C subunit) and 10248 (RI subunit) and their corresponding transformants (11586 and 11564, respectively) are shown in TABLE 1.

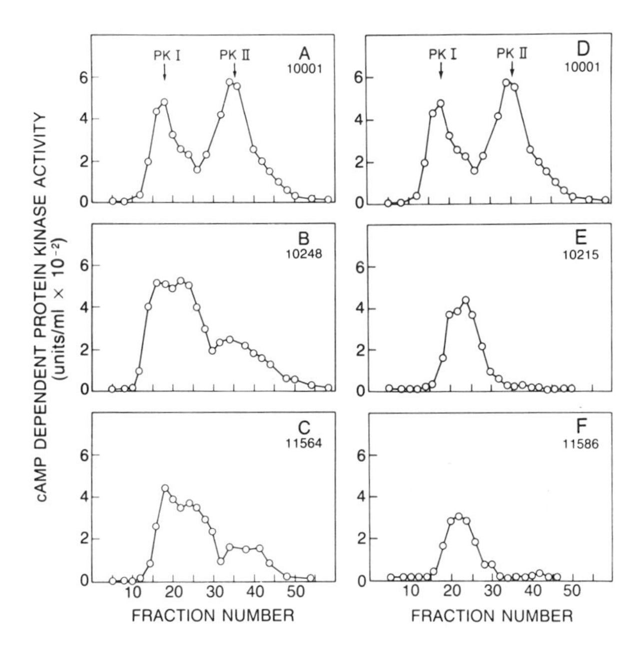

FIGURE 4. DEAE chromatography of cADepPK activities from wild-type, mutant, and cAMP-resistant clone extracts. The cADepPK activity was measured as transfer of ^{32}P from [γ-^{32}P]ATP to calf thymus histone in the presence of 10^{-5} M cAMP. Activity in the absence of cAMP was subtracted as background. **A** & **D**: wild-type extracts; **B**: mutant 10248; **C**: transformant 11564 from mutant 10248; **E**: mutant 10215; **F**: transformant 11586 from mutant 10215.

PREPARATION OF A COSMID LIBRARY CONTAINING THE GENE FOR THE MUTANT RI SUBUNIT

Having demonstrated the dominant nature of two of our cADepPK mutants by gene transfer experiments, we were interested in isolating the genes encoding the mutant subunits in these cell lines. Because of the availability of a cDNA probe for the RI gene,[35] we began working with the gene for the RI subunit of cADepPK. Our goal was to determine whether gene regulation at the transcriptional level acts either directly or indirectly through the interaction of cAMP with the RI subunit of cADepPK by introducing into a variety of cell lines a cloned gene encoding the dominant-acting mutant RI subunit with altered cAMP binding.

TABLE 1. Comparison of the Phenotypes of Mutants 10215 and 10248 and Their cAMP-resistant Transformants

	Wild-type Parent	Mutant 10215	Transformant 11586	Mutant 10248	Transformant 11564
Resistance to growth inhibition	No	Yes	Yes	Yes	Yes
Resistance to morphological change	No	Yes	Yes	Yes	Yes
Presence of abnormal type I kinase activity	No	Yes	Yes	Yes	Yes
Reduced type II kinase activity	No	Yes	Yes	Yes	Yes
Phosphorylation of 52 000-dalton protein	Yes	No	No	No	No
Affected subunit	None	C	C	RI	RI

In order to obtain a movable genetic element containing the mutant type I regulatory subunit, we chose to clone the RI CHO mutant gene from a CHO genomic cosmid library. Cosmid libraries allow the cloning of large segments of genomic DNA (up to 40 kb) in a single recombinant molecule.[36–40] This method increases the likelihood of obtaining all the coding information as well as the promoter region in a single recombinant molecule. Cloning was performed using the cosmid vector pSV_{13}*cos* developed by B. Howard and M. McCormick of the National Institutes of Health (FIG. 5). The vector pSV_{13}*cos* contains the genes for resistance to the drugs chloramphenicol and ampicillin, as well as the gene for the bacterial enzyme Eco *gpt*. Expression of this enzyme in stably transfected mammalian cell lines allows the cells to grow in MXHAT medium.[41] As outlined in FIGURE 6, high-molecular-weight genomic DNA was isolated from the cAMP-resistant CHO cell line 10248. Partial digestion of the DNA with the restriction enzyme *Pst*I was done in order to obtain *Pst*I-cut genomic DNA approximately 40 kb long. The cosmid cloning vector pSV_{13}*cos* was linearized with *Pst*I and treated with calf intestinal phosphatase in order to minimize self-ligation. The recombinant molecules were produced by ligation of partial

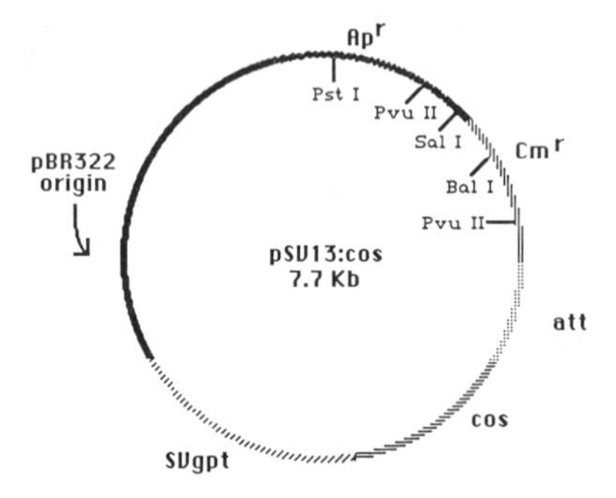

FIGURE 5. Important features of the cosmid shuttle vector pSV_{13}*cos*. This cosmid cloning vector was developed by B. Howard and M. McCormick of the National Institutes of Health. It contains a unique *Pst*I site for insertion of genomic DNA, the pBR322 origin of replication for growth in *E. coli,* the genes for chloramphenicol and ampicillin resistance (Cm^r and Ap^r, respectively) for selection in *E. coli,* the Eco *gpt* gene under control of the SV_{40} promoter (SV *gpt*) for selection in mammalian cells, and the λ cos site for *in vitro* packaging.

digests of *Pst*I-cut 40-kb-long genomic DNA with *Pst*I-linearized pSV_{13}*cos* vector DNA in the presence of T4 ligase. The recombinant DNA molecules were then packaged into λ phage heads using the Amersham *in vitro* packaging kit. *E. coli* HB101 was made competent for λ phage infection by growth overnight in LB broth supplemented with 0.2% maltose. The bacteria were infected with the λ phage containing the

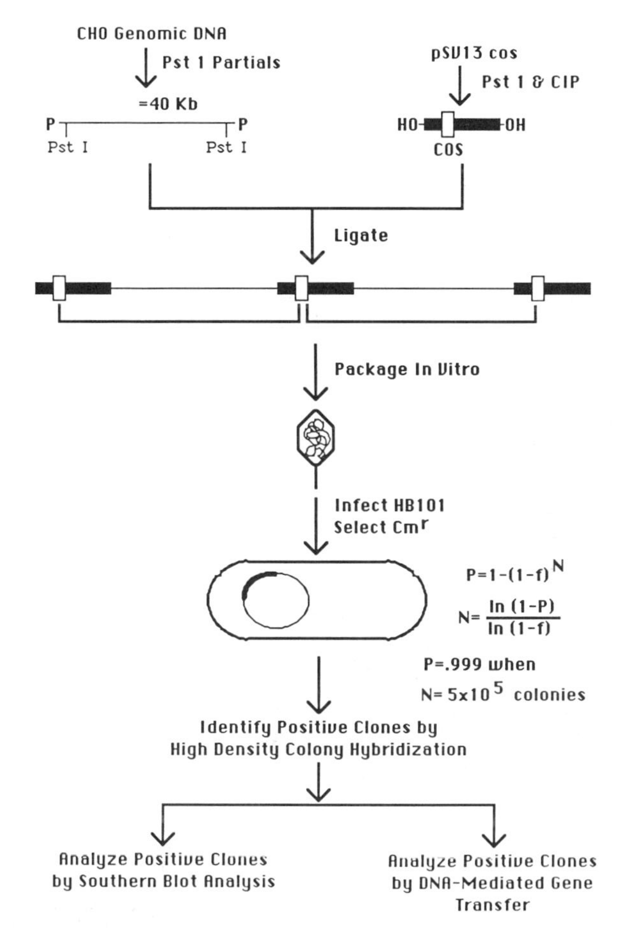

FIGURE 6. Outline of steps involved in preparation of a cosmid library from mutant CHO cells. CIP: calf intestinal phosphatase; P: probability of having a single copy gene in the library; f: fractional proportion of genome in a single recombinant.

recombinant DNA molecules and grown overnight on LB plates containing chloramphenicol. Approximately 5.0×10^5 chloramphenicol-resistant bacterial colonies are necessary in order to have a cosmid library in which a single copy gene has at least a 99.9% chance of being represented. The chloramphenicol-resistant bacterial colonies were collected to constitute our CHO cosmid library. Aliquots of the library were frozen for later screening with the appropriate cDNA probe.

IDENTIFICATION OF COSMID CLONES ENCODING THE RI GENE

The CHO cosmid library was screened for the mutant RI gene using the technique of high-density colony hybridization.[42] The RI cDNA (pRI) isolated from a bovine testes cDNA library was used as a probe (the pRI was generously provided by G. S. McKnight of the University of Washington).[35] Approximately 2×10^5 bacterial colonies were screened from our CHO cosmid library using a ^{32}P-labeled, nick-trans-

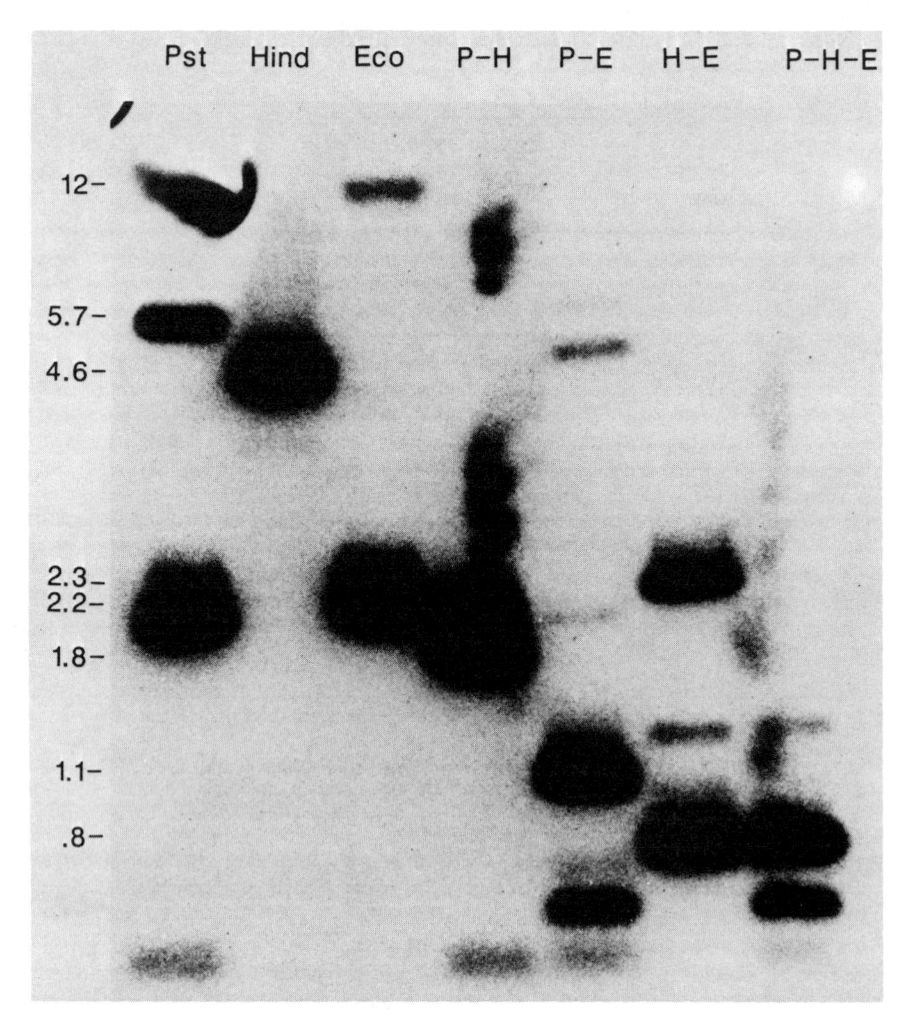

FIGURE 7. Southern blot of pcosRI-1 DNA digested with *Pst*I, *Hind*III, *Eco*RI, or combinations of these enzymes. The probe was a nick-translated 770-bp *Pst*I insert from the pR1 cDNA clone (gift of G. S. McKnight).[35] Data such as these were used to generate the map shown in FIGURE 8. P: *Pst*I; H: *Hind*III; E: *Eco*RI.

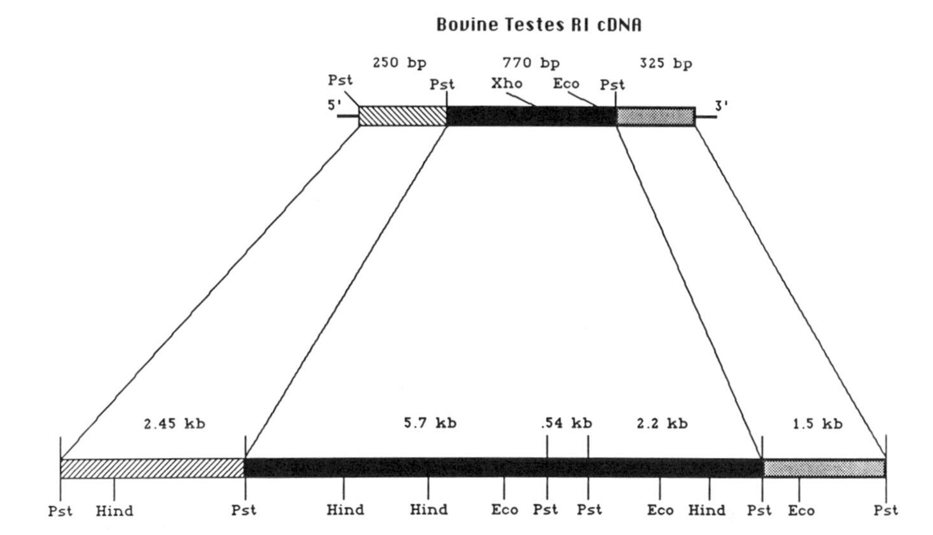

FIGURE 8. Restriction maps of RI cDNA and CHO genomic pcosRI DNA.

lated, 770-bp insert from the bovine testes RI cDNA. Three positive clones (pcosRI-1, pcosRI-2, and pcosRI-3) have been identified. These three clones appear to be identical as shown by partial restriction mapping and Southern analysis using both the internal 770-bp and 250-bp 5′ inserts from the pRI clone as a probe. FIGURE 7 is a Southern blot in which pcosRI-1 DNA was digested with three different restriction enzymes alone and in combination. The digested cosmid clone was probed with a 770-bp insert from the pRI cDNA plasmid. Restriction fragments of the same size were detected by the probe in both genomic and cosmid clone DNA (data not shown). The size of the hybridized restriction fragments indicates there is a gene of at least 12.4 kb encoding the structural part of the CHO RI protein. FIGURE 8 is a partial restriction map of the three pcosRI clones based on the restriction fragments shown in FIGURE 7, as well as Southern blots with the 250-bp 5′ *Pst*I fragment of the pRI cDNA and the entire pRI plasmid. A 2.45-kb *Pst*I fragment hybridizes with the 250-bp 5′ *Pst*I fragment of the pRI cDNA. This 2.45-kb *Pst*I fragment is further upstream than any other restriction fragment containing coding sequences for the RI subunit. The internal 770-bp insert of the pRI cDNA hybridizes to 5.7-, 2.2-, and 0.54-kb *Pst*I fragments, a 4.7-kb *Hind*III fragment, and 2.3- and 2.2-kb *Eco*RI fragments. The remaining 325-bp 3′ insert of the cDNA hybridizes to a 1.5-kb *Pst*I fragment.

A stretch of 57 nucleotides beginning at the *Xho* site in the middle of the 770-bp pRI cDNA insert has been sequenced in the 5′-to-3′ direction.[35] This nucleotide sequence corresponds to a stretch of 19 amino acids beginning at amino acid position 213 of the bovine skeletal muscle RI protein.[43] This sequence of amino acids is part of the cAMP-binding site I of the RI protein. This region is of special interest because our RI cosmid clones may encode an RI protein subunit with altered cAMP-binding activity. The intense hybridization of the 770-bp pRI insert to several restriction fragments from our cosmid clone should allow us to use these fragments to determine the nucleotide sequence encoding the cAMP-binding regions of the mutant RI protein in the cAMP-resistant CHO cell line 10248. These restriction fragments include the 2.2-kb *Pst*I fragment and an overlapping 1.85-kb *Pst*I-*Hind*III fragment (FIG. 7).

ASSESSING pcosRI FUNCTION

Presumably both the wild-type and mutant alleles are present in our CHO cosmid library, so one way to define which allele we have is to assess the functional properties of the three cosmid clones isolated thus far. In order to test for the mutant allele the cosmid clones were transfected into wild-type cAMP-sensitive CHO cells by DNA-mediated gene transfer using a $CaPO_4$ precipitation technique.[32] Stable cAMP-resistant transformants were selected by growth in MXHAT medium. Cells that grow in MXHAT medium are gpt^+ as a result of the stable integration of the cosmid DNA. The gpt^+ clones were then tested for the phenotype of cAMP resistance by assessing colony morphology and growth in soft agar in the absence and in the presence of 1 mM of 8-Br-cAMP. As yet no stable clones expressing the phenotype for cAMP resistance have been identified. This indicates that we have isolated the wild-type gene or that the isolated gene is not in a sufficiently complete form to allow expression.

In order to test (in collaboration with B. Schimmer of the University of Toronto) whether our cosmid clones represent the wild-type allele, cosmid DNA was cotransfected with pSV_2*neo* DNA into cADepPK mutant mouse adrenal Y-1 cells. In this system the presence of the wild-type allele of the RI subunit should be dominant and is expected to result in a rounded morphology of the Y-1 cells in the presence of 8-Br-cAMP.[44] Stable G418 drug-resistant clones were selected and tested for the presence of the wild-type allele of the RI subunit. No morphologic change was observed in the presence of 8-Br-cAMP, indicating that our cosmid clone did not encode a complete wild-type allele. These results suggest that although we have isolated several cosmid clones from CHO cells containing extensive homology with the bovine testes cDNA of the RI subunit, we have not yet isolated a cosmid clone containing either the complete wild-type or mutant alleles. Our clones may be missing either a segment of the gene encoding a structural aspect of the RI subunit or a segment of the regulatory promoter region. Chromosome-walking to find these missing segments is underway. An alternate explanation is that the intact gene and regulatory regions have been cloned, but that these regions are not expressed after transfer.

SUMMARY AND CONCLUSIONS

Genetic evidence has been obtained indicating that the responsiveness of CHO cells to cAMP requires an intact cADepPK system. DNA carrying mutant RI and C subunit genes has been transferred to wild-type cells. Recipient cells carrying the DNA have been detected by selecting for expression of the phenotype for cAMP resistance. We are in the process of cloning a functional mutant RI subunit to use as a moveable genetic element that can confer cAMP resistance on recipient cells. This cloned gene should allow us to test hypotheses concerning the mechanism of cAMP regulation of transcription in mammalian cells.

REFERENCES

1. BEAVO, J. A., P. J. BECHTEL & E. G. KREBS. 1975. Adv. Cyclic Nucleotide Res. **5:** 241-251.
2. PASTAN, I. & S. ADHYA. 1976. Bacteriol. Rev. **40:** 527-551.
3. WYNSHAW-BORIS, A., T. G. LUGO, J. M. SHORT, R. E. K. FOURNIER & R. W. HANSON. 1984. J. Biol. Chem. **259:** 12161-12169.
4. NAGAMINE, Y. & E. REICH. 1985. Proc. Natl. Acad. Sci. USA **82:** 4606-4610.
5. CONSTANTINOU, A. I., S. P. SQUINTO & R. A. JUNGMANN. 1985. Cell **42:** 429-437.
6. HSIE, A. W. & T. T. PUCK. 1971. Proc. Natl. Acad. Sci. USA **68:** 358-361.
7. GOTTESMAN, M. M., A. LECAM, M. BUKOWSKI & I. PASTAN. 1980. Somat. Cell Genet. **6:** 45-61.
8. LECAM, A., M. M. GOTTESMAN & I. PASTAN. 1980. J. Biol. Chem. **255:** 8103-8108.
9. LICHTI, U. & M. M. GOTTESMAN. 1982. J. Cell. Physiol. **113:** 433-439.
10. MILHAUD, P. G., P. J. A. DAVIES, I. PASTAN & M. M. GOTTESMAN. 1980. Biochim. Biophys. Acta **630:** 476-484.
11. GOTTESMAN, M. M. 1985. *In* Molecular Cell Genetics. M. M. Gottesman, Ed.: 711-743. John Wiley & Sons. New York, NY.
12. GOTTESMAN, M. M. 1983. *In* Methods in Enzymology. J. D. Corbin & J. G. Hardman, Eds. Vol. **99:** 197-206. Academic Press. New York, NY.
13. STEINBERG, R. A., P. H. O'FARRELL, U. FRIEDRICH & P. COFFINO. 1977. Cell **10:** 381-391.
14. DOHERTY, P. J., J. TASO, B. P. SCHIMMER, M. C. MUMBY & J. A. BEAVO. 1982. J. Biol. Chem. **257:** 5877-5883.
15. ROSEN, N., J. PISCITELLO, J. SCHNECK, R. J. MUSCHEL, B. R. BLOOM & O. M. ROSEN. 1979. J. Cell. Physiol. **98:** 125-136.
16. SIMANTOV, R. & L. SACHS. 1975. J. Biol. Chem. **250:** 3236-3242.
17. PAWELEK, J. M. 1979. J. Cell. Physiol. **98:** 619-626.
18. GOTTESMAN, M. M. 1980. Cell **22:** 329-330.
19. ABRAHAM, I., S. BRILL, J. HYDE, R. FLEISCHMANN, M. CHAPMAN & M. M. GOTTESMAN. 1985. J. Biol. Chem. **260:** 13934-13940.
20. ABRAHAM, I., S. BRILL, M. CHAPMAN, J. HYDE & M. M. GOTTESMAN. 1986. J. Cell. Physiol. **127:** 89-94.
21. WICKS, W. 1969. J. Biol. Chem. **244:** 3941-3950.
22. BEALE, E. G., J. L. HARTLEY & D. K. GRANNER. 1982. J. Biol. Chem. **257:** 2022-2028.
23. CHRAPKIEWICZ, N. B., E. G. BEALE & D. K. GRANNER. 1982. J. Biol. Chem. **257:** 14428-14432.
24. ERNEST, M. J. & P. FEIGELSON. 1978. J. Biol. Chem. **253:** 319-322.
25. FIRESTONE, G. & E. HEATH. 1981. J. Biol. Chem. **256:** 1396-1403.
26. DERDA, D., M. MILES, J. SCHWEPPE & R. JUNGMANN. 1980. J. Biol. Chem. **255:** 11112-11121.
27. GORMAN, C. M., L. F. MOFFAT & B. H. HOWARD. 1982. Mol. Cell. Biol. **2:** 1044-1051.
28. SINGH, T. J., C. ROTH, M. M. GOTTESMAN & I. PASTAN. 1981. J. Biol. Chem. **256:** 926-932.
29. MURTAUGH, M. P., A. L. STEINER & P. J. A. DAVIES. 1982. J. Cell Biol. **95:** 64-72.
30. EVAIN, D., M. M. GOTTESMAN, I. PASTAN & W. B. ANDERSON. 1979. J. Biol. Chem. **254:** 6931-6937.
31. SINGH, T. J., J. HOCHMAN, R. VERNA, M. CHAPMAN, I. ABRAHAM, I. PASTAN & M. M. GOTTESMAN. 1985. J. Biol. Chem. **260:** 13927-13933.
32. ABRAHAM, I. 1985. *In* Molecular Cell Genetics. M. M. Gottesman, Ed.: 181-210. John Wiley & Sons. New York, NY.
33. ABRAHAM, I., J. S. TYAGI & M. M. GOTTESMAN. 1982. Somat. Cell Genet. **8:** 23-39.
34. LECAM, A., J.-C. NICOLAS, T. SINGH, F. CABRAL, I. PASTAN & M. M. GOTTESMAN. 1981. J. Biol. Chem. **256:** 933-941.
35. LEE, D. C., D. F. CARMICHAEL, E. G. KREBS & G. S. MCKNIGHT. 1983. Proc. Natl. Acad. Sci. USA **80:** 3608-3612.
36. HOHN, B. 1979. *In* Methods in Enzymology. R. Wu, Ed. Vol. **68:** 299-309. Academic Press. New York, NY.

37. HOHN, B. & J. COLLINS. 1980. Gene **11:** 291-298.
38. ISH-HOROWICZ, D. & J. F. BURKE. 1981. Nucleic Acids Res. **9:** 2989-2998.
39. MANIATIS, T., E. F. FRITSCH & J. SAMBROOK, Eds. 1982. *In* Molecular Cloning: A Laboratory Manual: 295-308. Cold Spring Harbor Press. Cold Spring Harbor, NY.
40. LAU, Y. F. & Y. W. KAN. 1983. Proc. Natl. Acad. Sci. USA **80:** 5225-5229.
41. HOWARD, B. H. & M. MCCORMICK. 1985. *In* Molecular Cell Genetics. M. M. Gottesman, Ed.: 211-233. John Wiley & Sons. New York, NY.
42. HANAHAN, D. & M. MESELSON. 1983. *In* Methods in Enzymology. R. Wu, L. Grossman & K. Moldave, Eds. Vol. **100:** 333-342. Academic Press. New York, NY.
43. TITANI, K., T. SASAGAWA, L. H. ERICSSON, S. KUMAR, S. B. SMITH, E. G. KREBS & K. A. WALSH. 1984. Biochemistry **23:** 4193-4199.
44. SCHIMMER, B. P., J. TSAO & M. KNAPP. 1977. Mol. Cell. Endocrinol. **8:** 135-145.

Multihormonal Regulation of Phosphoenolpyruvate Carboxykinase Gene Transcription

The Dominant Role of Insulin[a]

DARYL K. GRANNER, KAZUYUKI SASAKI, AND DAVID CHU

Department of Molecular Physiology and Biophysics
Vanderbilt University Medical Center
Nashville, Tennessee 37232

INTRODUCTION

The most important goal of our laboratory is to understand how several hormones, each of which presumably has a unique mechanism of action, provide an integrated biological response. We have concentrated our efforts on the regulation of the enzyme phosphoenolpyruvate carboxykinase (PEPCK) (EC 4.1.1.32) in cultured H4IIE cells, a clone of Reuber H35 rat hepatoma cells. PEPCK catalyzes the conversion of oxalacetate to phosphoenolpyruvate and is a rate-limiting gluconeogenic enzyme. Cyclic AMP, which mediates the action of glucagon, and glucocorticoids increase the rate of synthesis of this protein and enhance gluconeogenesis.[1,2] Insulin decreases the rate of synthesis of PEPCK and correspondingly decreases gluconeogenesis.[1,3] By coupling measurements of PEPCK synthesis with assays capable of quantitating the activity and the amount of mRNA coding for PEPCK ($mRNA^{PEPCK}$), we were able to exclude the following possibilities as means of hormonal control: alterations of $mRNA^{PEPCK}$ translational efficiency, alterations in the rate of $mRNA^{PEPCK}$ egress from the nucleus, and changes in the rate of degradation of $mRNA^{PEPCK}$ in the nucleus or cytoplasm.[2,4–6] Our interpretation of these studies was that cAMP, glucocorticoids, and insulin affect $mRNA^{PEPCK}$ production. The purpose of the studies described in this paper was to determine whether each of these molecules regulated PEPCK synthesis at the level of PEPCK gene transcription and then to explore how they acted in concert.

Steroid hormones are thought to regulate metabolic processes by affecting the rate of transcription of specific genes. The first step in this action involves the binding of the ligand to intracellular receptors. The hormone-receptor complex binds in a site-specific manner to DNA, and thereby regulates the transcription of specific genes.

[a] This research was supported by Grant AM 35107 and Grant AM 07061 from the Vanderbilt Diabetes Research Training Center.

Until very recently hormones that bind to plasma membrane receptors were not thought to regulate gene expression, but evidence that such is the case is rapidly accumulating. In most instances the internalization of the plasma membrane hormone-receptor complex is not a prerequisite for action. This implies that another molecule or process mediates the intracellular action of this large class of hormones. In many cases the interaction of peptide hormones with their specific receptors results in the activation of adenylate cyclase, which, in turn, increases cAMP within the cell. Cyclic AMP regulates the mRNAs that code for several proteins (including tyrosine aminotransferase, PEPCK, lactate dehydrogenase, prolactin, albumin, alkaline phosphatase, haptoglobin, and discoidin) in eukaryotic cells.[7] In several of these instances, cAMP, or one of its various analogues, regulates gene transcription. Such regulations may be quantitated by measuring RNA elongation in isolated nuclei.

TABLE 1. Messenger RNAs Regulated by Insulin

Intracellular Enzymes
PEPCK[a]
Fatty acid synthase
Pyruvate kinase[a]
Glucokinase
Glycerol phosphate dehydrogenase[a]
Glyceraldehyde phosphate dehydrogenase[a]
Secreted Proteins/Enzymes
Albumin[a]
Amylase
α_{2U}-Globulin
Growth hormone
Proteins Involved in Reproduction
Ovalbumin[a]
Casein[a]
Structural Proteins
δ-Crystallin

[a] Insulin regulates the rate of specific mRNA transcription.

Although insulin influences the rate of synthesis of many proteins, the molecular details of this action and the intracellular mediator of this effect have not been elucidated. An effect of insulin on ribosomes or translation factors, or both, could explain how insulin enhances total protein synthesis in a tissue,[8,9] but it is difficult to explain selective effects using this mechanism, especially since stimulatory and inhibitory responses have been noted. Insulin stimulates RNA synthesis and RNA polymerase activity,[10,11] an indication that the hormone may regulate protein synthesis at a pretranslational site. Observations that the messenger RNAs that code for several proteins (TABLE 1) change in response to insulin treatment support this hypothesis. These selective mRNA changes could result from effects of insulin on several processes including an effect on the rate of cytoplasmic mRNA degradation, on mRNA transport from the nucleus to cytoplasm, or on the synthesis, modification, processing, or degradation of the specific mRNA in the nucleus.

We have studied the regulation of PEPCK in H4IIE cells in an effort to learn how insulin regulates the synthesis of a specific protein. Concentrations of insulin in the pM to nM range decrease H4IIE cell cytoplasmic $mRNA^{PEPCK}$ and PEPCK synthesis rapidly: this effect is mediated through the insulin receptor, and it appears to be due to decreased synthesis of $mRNA^{PEPCK}$ from the PEPCK gene.[6,12]

In this report we present an analysis of the individual and combined effects cAMP analogues, dexamethasone, and insulin have on transcription of the PEPCK gene in H4IIE hepatoma cells. In addition we present data that indicate phorbol esters mimic the action of insulin on PEPCK gene transcription. This may provide a clue about insulin action, and about the multihormonal regulation of PEPCK.

EXPERIMENTAL PROCEDURES

Materials

Swim's S-77 culture medium, hormones, chemicals, enzymes, and radionuclides were obtained from standard sources and were used as described previously.[7]

Cell Culture

H4IIE cells were obtained from Dr. Van Potter of the University of Wisconsin, and monolayer cultures were grown to confluency in Corning 150-cm^2 flasks, using conditions described previously.[7]

Quantitation of $mRNA^{PEPCK}$ and PEPCK Gene Transcription

Total cell poly$(A)^+$ RNA was isolated and quantitated using an adaptation of the dot blot hybridization technique described previously.[7] The cDNA probe was a 1300-bp *Sma*I-*Sph*I fragment isolated from pPC116, a cloned recombinant DNA plasmid[13] and labeled by nick-translation. The results were analyzed by a densitometer scan of an autoradiogram. To quantitate PEPCK gene transcription cells were treated with hormones or effectors, nuclei were isolated, and the elongation of nascent RNA transcripts on the PEPCK gene was measured by a modification[7] of a procedure described previously.[14] Isolated nuclei were incubated in a mixture containing 0.4 mM each of ATP, CTP, GTP, and [α-^{32}P]UTP. The reaction was terminated by digestion with DNase I and proteinase K; the samples were treated twice with an equal volume of phenol-chloroform (1:1; v/v) and then twice with ether; and the RNA was precipitated by adding ethanol. Hybridization of radiolabeled of PEPCK mRNA was to nitrocellulose-filter-bound pBR322 or pλPC112.R3 as control and PEPCK-specific probes, respectively. The pλPC112.R3 probe is a recombinant plasmid with a 5.8-kb *Eco*RI fragment of genomic DNA inserted into the *Eco*RI site of pBR322. All of the insert represents PEPCK DNA that is transcribed into precursor $mRNA^{PEPCK}$. The isolation and characterization of this plasmid is described elsewhere.[13] Hybridization was carried out for 2-3 days at 45° C in the presence of 6000-7000 cpm [^{3}H]cRNA, prepared as previously described,[7] to monitor hybridization efficiency. The filters were washed, incubated with RNases A and T1, and washed again. The RNA was then eluted, and radioactivity was measured in a scintillation spectrometer. The results express the amount of RNA^{PEPCK} transcribed relative to the total amount of RNA transcribed in ppm. The general features of this assay, including the calculations for the amounts of $mRNA^{PEPCK}$ transcribed, are described in detail in a previous paper.[7]

RESULTS

Effect of Cyclic Nucleotide Analogues on PEPCK Gene Transcription

We first assessed the effect dibutyryl cAMP (Bt_2cAMP) has on PEPCK gene transcription (TABLE 2). The basal rate of incorporation of isotope into elongating RNA^{PEPCK} transcripts was 93 ppm, but within 20 min after adding Bt_2cAMP there was an eightfold increase of PEPCK gene transcription (the maximal rate was reached within 40 min after the addition of the cyclic nucleotide). It is important to note that there is no effect of Bt_2cAMP on total RNA synthesis in these cells, hence the stimulatory effect of the nucleotide on PEPCK gene transcription is quite specific.

Although Bt_2cAMP is an effective inducer of PEPCK gene transcription, the results varied considerably between experiments. Thus a nonmetabolizable analogue, 8-(4-chlorophenylthio)-cAMP (8-CPT-cAMP), was used in most experiments. In the experiment illustrated in FIGURE 1A, the maximal rate of transcription was obtained 30 min after the addition of 8-CPT-cAMP. This was maintained for about 3-4 hr. Then there was a gradual decline to a new steady state, which was maintained at approximately 2-3 times the basal rate for at least 72 hr. The amount of total cellular $mRNA^{PEPCK}$ changed commensurate with, but slightly slower than, the rate of transcription, as would be expected if transcription were considered the principal site of $mRNA^{PEPCK}$ regulation. FIGURE 1B represents a composite of the results of several different experiments designed to examine the early time course of induction. Transcription of the PEPCK gene doubled within 5 min after the addition of 8-CPT-cAMP, and the peak rate of transcription occurred after 30 min.

TABLE 2. Effect of Bt_2cAMP on PEPCK Gene Transcription

Bt_2cAMP Treatment (min)	Total RNA Synthesis ($cpm \times 10^{-6}$)	$mRNA^{PEPCK}$ Synthesis (ppm)
0	18.1 ± 0.4	93 ± 8
20	18.0 ± 1.0	765 ± 51
40	18.4 ± 1.1	1004 ± 175

NOTE: The effect of Bt_2cAMP on PEPCK gene transcription was quantitated as described in the text. The data reflect total RNA synthesis and mRNA PEPCK synthesis in H4IIE cells incubated for 0, 20, or 40 min in the presence of a 0.5 mM concentration of Bt_2cAMP. Adapted from Granner *et al.*[7]

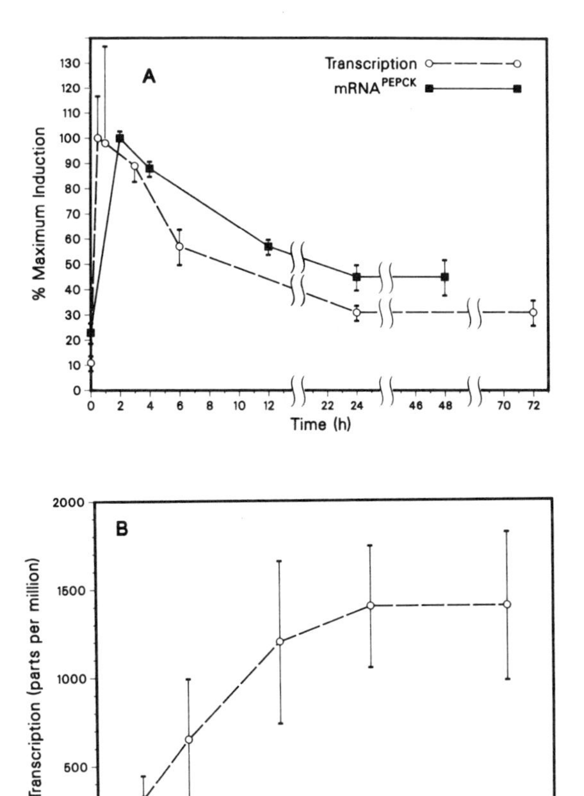

FIGURE 1. Time course of 8-CPT-cAMP induction. Cells were treated with 0.1 mM 8-CPT-cAMP for the times indicated. PEPCK gene transcription was measured using the run-off assay, and $mRNA^{PEPCK}$ was measured using dot blot hybridization. Each result in **A** is shown as the mean ± SD of the percent of maximal induction for duplicate samples, and each curve represents a separate experiment. The maximal mean value for PEPCK gene transcription was 1030 ppm, and for the amount of $mRNA^{PEPCK}$ it was 2.72 integrator units (see Beale *et al.*[4] for an explanation of this measurement). Each result in **B** is a mean ± SD from six assays and shows the rate of gene transcription at various times after the addition of 8-CPT-cAMP to H4IIE cells. Details of the assay and calculations are described in Sasaki *et al.*[7]

Effect of a Glucocorticoid Hormone on PEPCK Gene Transcription

TABLE 3 depicts the effect of the synthetic glucocorticoid dexamethasone on PEPCK gene transcription. In experiment 1 we quantitated both the amount of PEPCK protein and PEPCK gene transcription. Dexamethasone at 5×10^{-8} M caused a fourfold increase of PEPCK protein in H4IIE cells, and there was a corresponding increase in PEPCK gene transcription (from 108 to 605 ppm). This increase

TABLE 3. Regulation of PEPCK Gene Transcription by Dexamethasone

Experiment Number	Treatment	PEPCK (nmols/mg protein)	$mRNA^{PEPCK}$ Synthesis (ppm)
1	None	0.71 ± 0.03	108 ± 5
	Dexamethasone (50 nM)	262 ± 0.11	605 ± 10
2	None	—	123
	Dexamethasone (50 nM)	—	305
	Dexamethasone (500 nM)	—	640

NOTE: The effect of dexamethasone on PEPCK gene transcription in H4IIE cells was quantitated in two separate experiments. In experiment 1, 50 nM dexamethasone was added to H4IIE cells for 48 hr, and PEPCK protein, expressed in nmols/mg total protein, was quantitated using a specific radioimmunoassay. Transcription elongation was quantitated as described in the text, using parallel cultures of H4IIE cells incubated in 500 nM dexamethasone for 30 min. In experiment 2, the effect of 50 or 500 nM dexamethasone on transcription was quantitated as described in the text. Adapted from Granner *et al.*[7]

in transcription is sufficient to account for the increased accumulation of PEPCK protein. The second experiment shown in TABLE 3 reveals that this is a concentration-dependent phenomenon. The half-maximal effect of dexamethasone on PEPCK gene transcription was achieved at 50 nM, a concentration similar to that required to half-maximal occupation of the glucocorticoid receptor in these cells. Not shown is the fact that this maximal rate is achieved within 30 min, and that the hormone caused no change in total RNA transcription. Therefore dexamethasone enhances PEPCK protein production by selectively and rapidly increasing transcription of this gene.

Effect of Insulin on PEPCK Gene Transcription

Having shown that both of the positive regulators of gluconeogenesis stimulate PEPCK gene transcription, we next turned our attention to an analysis of the effect of insulin on this process. In the experiment illustrated in TABLE 4, H4IIE cells kept in serum-free medium had a basal rate of PEPCK gene transcription of 177 ppm. The addition of insulin at 10 pM caused a substantial inhibition of this process, and at 1 nM insulin, it was reduced to 25 ppm. This is the level of incorporation noted

TABLE 4. Effect of Insulin on PEPCK Gene Transcription

Insulin Concentration	$mRNA^{PEPCK}$ Synthesis (ppm)
0	177 ± 12
10 pM	71 ± 3
1 nM	25 ± 14

NOTE: H4IIE cells were placed in serum-free medium for 24 hr, and then insulin at either 10 pM or 1 nM was added. Sixty minutes later PEPCK gene transcription was quantitated according to the procedure described in the text and in Sasaki *et al.*[7]

when H4IIE cells are incubated in the presence of α-amanitin at 1 μg/ml, a concentration that completely inhibits class II gene transcription. Hence in this experiment insulin at 1 nM completely abolished transcription of the PEPCK gene.

This experiment shows that insulin can inhibit PEPCK gene transcription in the absence of other hormones, but the relatively low basal rate of transcription in such cells precludes an accurate analysis of the effect of various concentrations of insulin on the process, or of an analysis of the rapidity of the response. To study these parameters of insulin action we first maximized the rate of transcription of the PEPCK gene by treating the H4IIE cells with 8-CPT-cAMP.

Effect of Insulin on PEPCK Gene Transcription: Time Course

H4IIE cells were first exposed to 8-CPT-cAMP, which caused a rapid, ninefold increase in transcription within 30 min after 8-CPT-cAMP was added (FIG. 2A). Transcription continued at the maximal rate of about 1500 ppm for an additional 60 min, then decreased to 800 ppm 150 min after the addition of 8-CPT-cAMP. Addition of insulin 30 min after the cyclic nucleotide was added resulted in a rapid decrease in PEPCK gene transcription. Detectable inhibition was noted at 5 min, and within 60 min after adding insulin the rate of PEPCK gene transcription was reduced to the basal level. The $t_{1/2}$ of this inhibition is thus approximately 12-15 min, notably faster than the decrease in cytoplasmic $mRNA^{PEPCK}$.[2,4,6,15] Two hours after insulin was added transcription had decreased to 100 ppm, a value less than the uninduced rate of 150 ppm. Transcription was also reduced to the basal rate when 5 nM insulin was added to H4IIE cells previously treated with 50 μM forskolin. Insulin thus inhibits PEPCK gene transcription when cAMP is added to the cells exogenously in the form of analogues, or when endogenous cAMP is increased.

The repression of $mRNA^{PEPCK}$ by insulin is exerted quickly[6] and, given the rapid response of $mRNA^{PEPCK}$ during feeding and fasting cycles,[16] should be readily reversible. Other studies show that this is the case. Removal of insulin by a mild acid wash results in an almost immediate restoration of basal PEPCK gene transcription and unresponsiveness to 8-CPT-cAMP.[7]

Effect of Physiologic Concentrations of Insulin on PEPCK Gene Transcription

H4IIE cells are remarkably sensitive to insulin.[12] Thus it is not surprising that 1 pM insulin caused a significant decrease in transcription (FIG. 2B). The half-maximal decrease occurred with 2-5 pM insulin, and 10 nM insulin reduced transcription to the level measured in untreated cells. Proinsulin also decreased PEPCK gene transcription, but approximately 30-50-fold higher concentrations were needed to achieve an inhibition equivalent to that obtained with insulin. These insulin concentration effects are virtually identical to those obtained when either $mRNA^{PEPCK}$ or the rate of synthesis of PEPCK are measured in H4IIE cells.[6,12] This provides additional evidence in support of the hypothesis that transcription is the primary site at which insulin regulates the amount of $mRNA^{PEPCK}$ and subsequent enzyme synthesis. The

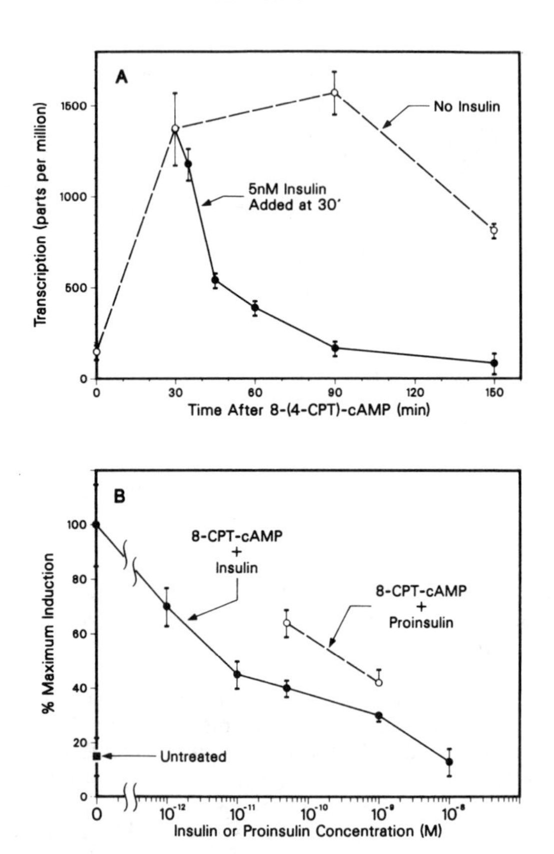

FIGURE 2. Effects of 8-CPT-cAMP, insulin, and proinsulin on $mRNA^{PEPCK}$ transcription as a function of concentration and time. In the experiment illustrated in **A**, nuclei were isolated from H4IIE cells that had been exposed to 0.1 mM 8-CPT-cAMP for the time indicated, or from cells to which 5 nM insulin was added 30 min after the addition of cyclic nucleotide. The transcriptional rate of the PEPCK gene was assayed as described in the text; each result is a mean; and the results represent the ± SD from triplicate assays. In the experiment illustrated in **B**, H4IIE cells were exposed to 0.1 mM 8-CPT-cAMP for the first 90 min and to various concentrations of insulin or proinsulin for the final 60 min. Each point represents the mean of three transcription assays. The rate of transcription is expressed as the percentage of maximum induction. Details are described in Sasaki *et al.*[7]

30-50-fold difference in sensitivity between insulin and proinsulin supports our previous contention that this effect is mediated by the insulin receptor.[12]

Effects of 8-CPT-cAMP and Insulin: Relationship to Protein Synthesis

The hormonal induction of several different mRNAs is prevented when protein synthesis is inhibited.[17,18] This is generally interpreted to mean that the hormone acts by inducing a regulatory protein that has a rapid turnover time, rather than by acting directly on the gene in question.[19] The 10-fold decrease in mRNA^{PEPCK} caused by insulin in H4IIE cells (from 0.32 ± 0.08 integrator units to 0.03 ± 0.01) is not affected by cycloheximide, emetine, or puromycin at concentrations (10 μM, 10 μM, and 1 mM, respectively) that inhibit protein synthesis by greater than 95%, as assessed by incorporation of [^{3}H]leucine into trichloroacetic acid-precipitable material. Under these conditions the inhibitors have no effect on total RNA concentration.

We then tested for similar effects in the RNA transcript elongation assay. At early times (up to 30 min) transcription of the PEPCK gene is unaffected by cycloheximide. From the data presented above, one might also expect that cycloheximide would not block the effect that insulin exerts on transcription on the PEPCK gene. That this is so is illustrated in TABLE 5. Cycloheximide had no effect on transcription in either untreated or insulin-plus-8-CPT-cAMP-treated cells, nor did it affect total RNA transcription under these conditions. These observations indicate that the regulation of PEPCK gene transcription by cAMP and insulin does not require the synthesis of a protein with a rapid turnover time.

Effect of the Phorbol Ester Phorbol Myristate Acetate on mRNA^{PEPCK}

In a study of possible effects of calcium and phospholipid metabolites in PEPCK gene transcription we discovered that phorbol myristate acetate is a potent inhibitor of this process. H4IIE cells were treated with or without 8-CPT-cAMP for 3 hr with different concentrations of phorbol myristate acetate (FIG. 3). 8-CPT-cAMP increased the amount of mRNA^{PEPCK} by about threefold as compared with the basal value. Increasing concentrations of phorbol myristate acetate decreased the amount of

TABLE 5. The Inability of Cycloheximide to Influence the Effects 8-CPT-cAMP and Insulin Have on PEPCK Gene Transcription

Treatment	Total RNA Synthesis (cmp × 10^{-6})	mRNA^{PEPCK} Synthesis (ppm)
None		
Without cycloheximide	1.44 ± 0.16	186 ± 30
With cycloheximide	1.55 ± 0.14	195 ± 22
8-CPT-cAMP		
Without cycloheximide	1.14 ± 0.15	1597 ± 88
With cycloheximide	1.63 ± 0.18	1196 ± 112
8-CPT-cAMP and Insulin		
Without cycloheximide	1.39 ± 0.26	447 ± 28
With cycloheximide	1.34 ± 0.17	383 ± 36

NOTE: H4IIE cells were prestimulated with 0.1 mM 8-CPT-cAMP for 30 min and then were incubated for an additional 60 min in the presence or absence of 10 μM cycloheximide with or without 8-CPT-cAMP. Insulin, where indicated, was added to a final concentration of 5 nM 30 min before the cells were harvested. Nuclei were isolated and transcription was assayed as described above. Each result is expressed as the mean ± SD of triplicate assays. Adapted from Sasaki *et al.*[7]

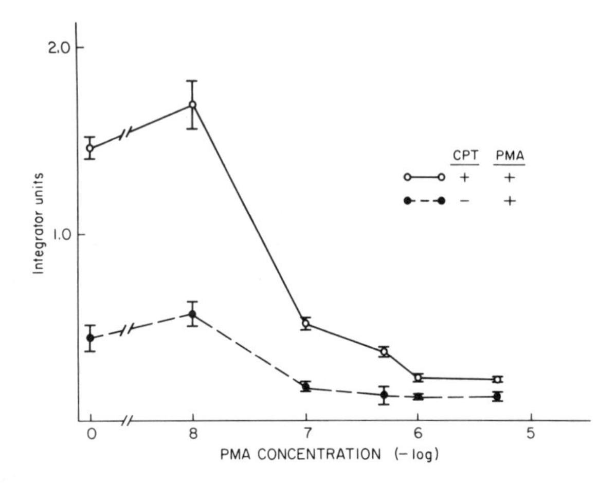

FIGURE 3. Phorbol myristate acetate represses mRNA^{PEPCK} in H4IIE cells. H4IIE cells were treated for 3 hr with or without 0.1 mM 8-CPT-cAMP and various concentrations of phorbol myristate acetate. The mRNA^{PEPCK} levels were determined as described above. Each result is expressed in relative integrator units and represents the mean ± SEM of four samples.

mRNA^{PEPCK} in basal and induced cells. The half-maximal effect occurred at about 50 nM, and phorbol myristate acetate exerted its maximal effect at a 1 μM concentration.

The inhibition of mRNA^{PEPCK} by phorbol myristate acetate also occurs rapidly, as illustrated in FIGURE 4. Levels of mRNA^{PEPCK} decrease within 30 min after treatment with phorbal myristate acetate, and a 50% reduction occurs within 1 hr.

Effect of Hormone Combinations on PEPCK Gene Transcription: Dominant Role of Insulin and Phorbol Esters

It has been suggested that cAMP is the primary regulator of PEPCK synthesis in rat liver and that insulin decreases PEPCK synthesis by reducing the intracellular concentration of cAMP,[20] an effect presumably due to enhanced hydrolysis of cAMP through increased phosphodiesterase activity.[21] The induction of PEPCK by a cyclic nucleotide that cannot be metabolized, or by an agent other than cAMP, should be unaffected by insulin if insulin acts in this manner. We therefore investigated insulin action in H4IIE cells in which mRNA^{PEPCK} was induced by 8-CPT-cAMP, a fully active but nonmetabolizable derivative, or by dexamethasone, a glucocorticoid that induces enzymes in a cAMP-independent manner.[22] The relative rate of transcription of mRNA^{PEPCK}, which was 108 ppm in untreated H4IIE cells, increased to 606 ppm in cells treated with dexamethasone, to 1355 ppm in cells treated with 8-CPT-cAMP, and to 1619 ppm in cells treated with the combination of 8-CPT-cAMP plus dexamethasone (TABLE 6). Insulin at 5 nM decreased mRNA^{PEPCK} transcription in otherwise untreated H4IIE cells to 70 ppm, and in cells treated with dexamethasone, 8-CPT-cAMP, and 8-CPT-cAMP plus dexamethasone to 58 ppm, 365 ppm, and 520 ppm, respectively. The dominant role of insulin is apparent in all combinations, yet there is a fundamental difference in the responses. In this and other similar experiments in which a 60-min exposure to hormones was employed, insulin did not always completely reduce transcription to the basal level when a cAMP analogue was also present. None of the individual hormone, or the various combinations, had any demonstrable effect on the synthesis of total RNA, so again these effects were very specific.

We also used the nuclear run-off transcription assay to study the effect of phorbol myristate acetate on PEPCK gene expression (TABLE 7). Dexamethasone increased the relative rate of transcription of PEPCK 6-fold, 0.1 mM 8-CPT-cAMP increased it more than 10-fold, and the combination of both increased it even further. However, 1 μM PMA totally inhibited the transcription of the PEPCK gene in the presence of 8-CPT-cAMP, dexamethasone, or the combination. Again, there was no detectable change in the amount of total RNA transcribed in the presence of the different effectors.

DISCUSSION

Few examples of regulation of the transcription of mammalian genes by peptide hormones or their intracellular mediators have been reported, and examples of multihormonal regulation of transcription of a single gene involving peptides are even rarer.[7] This paper summarizes evidence that shows that cAMP and glucocorticoids exert a rapid stimulatory effect on transcription of the PEPCK gene in H4IIE cells whereas insulin inhibits transcription of this gene and overrides the stimulatory actions of cAMP and glucocorticoids.

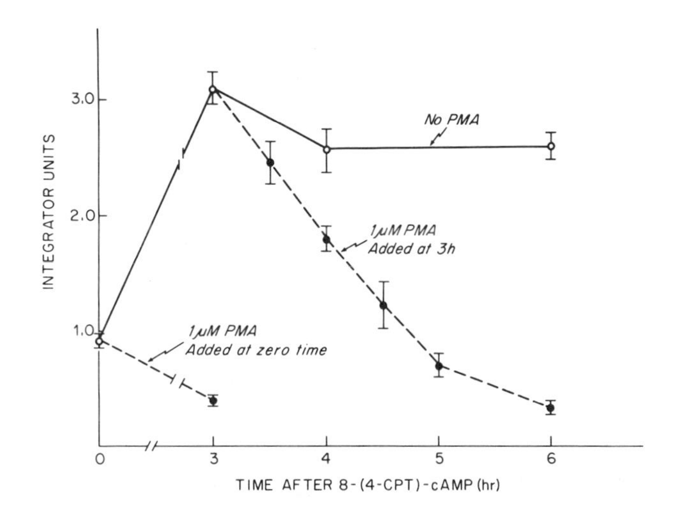

FIGURE 4. Rapidity of the effect of phorbol myristate acetate on $mRNA^{PEPCK}$. Culture conditions, isolation of extracts, and cytodot hybridization were as described above. Cytoplasmic extracts were isolated from H4IIE cells that had been exposed to 0.1 mM 8-CPT-cAMP for the times indicated, or from cells to which 1 μM phorbol myristate acetate was added 30 min after the cyclic nucleotide treatment. Each result is expressed as the mean ± SEM of four samples.

TABLE 6. Effect of Hormone Combinations on PEPCK Gene Transcription: Dominant Role of Insulin.

Primary Treatment	mRNA^{PEPCK} Synthesis (ppm) No Insulin	Insulin
Control	108 ± 3	70 ± 6
Dexamethasone	606 ± 3	58 ± 23
8-CPT-cAMP	1355 ± 43	365 ± 6
Dexamethasone and 8-CPT-cAMP	1619 ± 131	520 ± 10

NOTE: H4IIE cells were treated for 60 min with 0.1 mM 8-CPT-cAMP, 0.5 μM dexamethasone, or a combination of both, in the absence or presence of 5 nM insulin. Nuclei were then isolated and transcription was quantitated as described in the text. Each result represents the mean ± SD of triplicate assays. Adapted from Sasaki *et al.*[7]

These experiments provide clues as to the mechanism of action of these hormones. The fact that dexamethasone and 8-CPT-cAMP give an additive response (TABLE 6) suggests, not surprisingly, that they act by different mechanisms. Additional evidence for this is provided by the companion experiment in which insulin abolished transcription in the steroid-treated cells within 60 min whereas transcription in H4IIE cells treated with 8-CPT-cAMP or 8-CPT-cAMP plus dexamethasone, although significantly inhibited, was still readily detectable.

The major focus of our recent research has been the mechanism of action of insulin on gene transcription. The effect of insulin on PEPCK gene transcription is 1) achieved at physiologic concentrations of the hormone; 2) mediated through the insulin receptor; 3) specific; 4) seen in the absence of continuous protein synthesis; and 5) readily reversible.

Studies of the regulation of mRNA^{PEPCK} synthesis provided the first evidence of an effect of insulin on gene transcription, and would be of interest even if this were the only example of this action of the hormone. In recent years several additional examples have been reported, and it appears that the regulation of PEPCK synthesis represents a prototype of a general regulatory role of insulin. The mRNAs known to be regulated by insulin (many at the level of transcriptional control) are listed in TABLE 1, and this list is expanding at a rapid rate (see Granner and Andreone[23] for references). Indeed it has more entries than one for proteins whose activities are modulated by insulin-mediated changes in phosphorylation.

Several of these mRNAs direct the synthesis of enzymes that have a well-established metabolic connection to insulin (that is, PEPCK, glucokinase, fatty acid synthetase, pyruvate kinase, and tyrosine aminotransferase) whereas others represent major secretory proteins (albumin, amylase, growth hormone, and α_{2U}-globulin) in which the metabolic role of insulin is obscure. The production of ovalbumin and casein is involved in reproductive function in birds and mammals, and δ-crystallin is a structural protein. Insulin therefore regulates mRNAs that represent a spectrum of different actions in several tissues including liver, pancreas, oviduct, mammary gland, and the lens.

At least six of these gene products, the mRNAs that direct the synthesis of glycerol phosphate dehydrogenase, glyceraldehyde phosphate dehydrogenase, pyruvate kinase, albumin, ovalbumin, and casein, are regulated at the level of transcription. Other examples of effects of insulin on gene transcription should be forthcoming. It will be interesting to see whether transcriptional control is involved in the effect insulin has on "total" protein synthesis in liver, muscle, and adipose tissue.

The question of how a hormone regulates the transcription of a specific gene perplexes all investigators involved in this aspect of hormone action. Unfortunately, the general mechanism of class II gene (mRNA) transcription in mammalian cells is poorly understood, but this is an extremely active area of investigation and basic principles should be elucidated in the near future. There are probably many factors that control the rate and fidelity of transcript initiation, just as there are probably several factors involved in elongation, termination, and release of the mRNA molecules. The fact that insulin inhibits the basal rate of transcription of the PEPCK gene, as well as that stimulated by two different effectors, suggests an action at or near RNA polymerase II initiation; some evidence for this exists (Sasaki and Granner, unpublished observations). The action of insulin must not be exerted on RNA polymerase II itself, nor on any other general transcription factor, since the effect is very specific and total gene transcription is not affected.

The rapidity of the effect (FIG. 2A) and the fact that protein synthesis is not required (TABLE 5) suggests that this action of insulin, like so many others, may involve the covalent modification of a transcription regulatory factor(s) that somehow selectively inhibits the transcription of the PEPCK gene. Furthermore, it is possible that insulin, cAMP, and glucocorticoids act through a common mechanism, perhaps a protein phosphorylation-dephosphorylation system, perhaps one that requires specific protein kinases and phosphoprotein phosphatases. Such a mechanism is well established for cAMP. Insulin, whether it works via activation of the tyrosine kinase on its receptor and a subsequent phosphorylation cascade, or via a unique intracellular mediator, may act in an analogous manner. The observations regarding the action of the phorbol ester phorbol myristate acetate on PEPCK gene transcription are interesting because these actions resemble those of insulin in every respect (compare TABLES 6 & 7). Phorbol esters presumably activate protein kinase C, so this could result in activation (or inactivation) of the substrate affected by insulin and, in an opposite manner, by cAMP. Finally, one must consider the action of glucocorticoid hormones in this context. Although the actions of this hormone do not directly involve cAMP,[22] recent evidence suggests that the glucocorticoid receptor can be phosphorylated,[24] perhaps this modification alters the activity of the receptor. The concerted action of

TABLE 7. Effect of Hormone Combinations on PEPCK Gene Transcription: Dominant Role of Phorbol Myristate Acetate

Primary Treatment	mRNA^{PEPCK} Synthesis (ppm) Without PMA	With PMA
Control	113 ± 23	NT
Dexamethasone	686 ± 51	20 ± 7
8-CPT-cAMP	1112 ± 30	23 ± 1
Dexamethasone and 8-CPT-cAMP	1328 ± 91	17 ± 6

NOTE: Nuclei were isolated from H4IIE cells that had been treated for 60 min with 0.1 mM 8-CPT-cAMP, 0.5 μM dexamethasone, or a combination of both, in the presence or absence of 1 μM phorbol myristate acetate (PMA). Each result represents the mean ± SEM of triplicate assays.

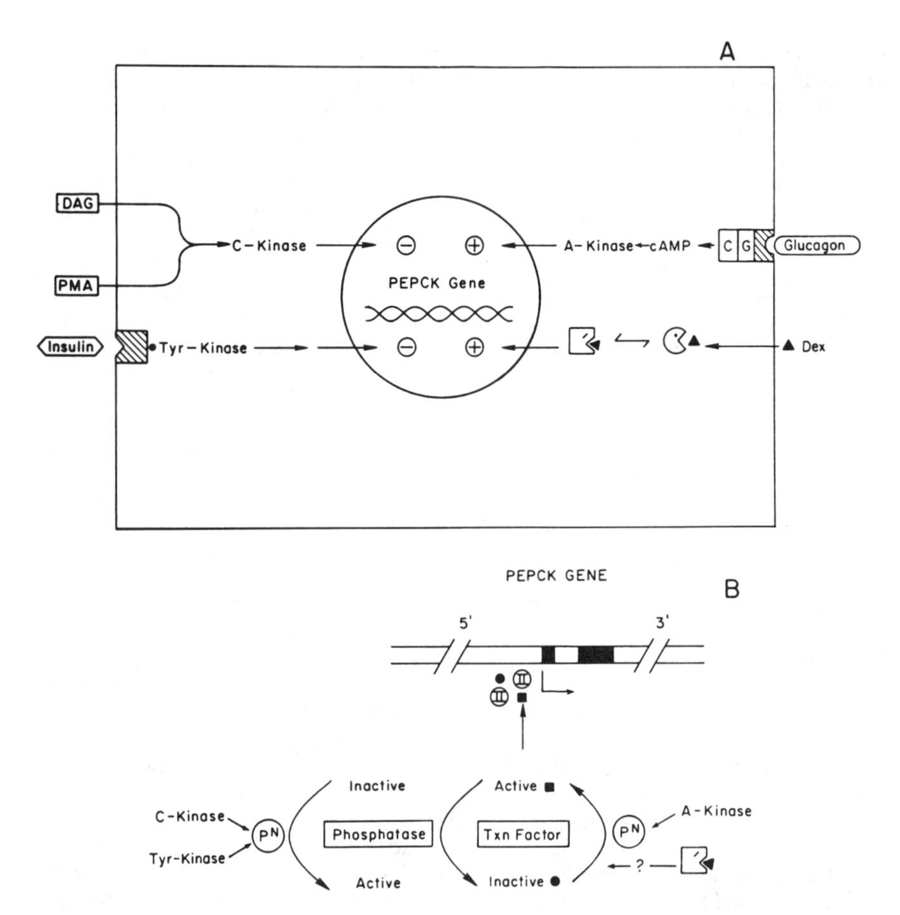

FIGURE 5. A hypothetical model for multihormonal regulation of PEPCK gene transcription. **A** illustrates how four different hormones, each presumably acting by a different mechanism, might regulate transcription from the PEPCK gene. **B** illustrates a highly speculative possibility, namely that all of these hormones might converge on a single protein involved in PEPCK gene transcription and alter its activity state.

these hormones could thus be explained if there exists a protein that regulates transcription of the PEPCK gene in proportion to its extent of phosphorylation (or dephosphorylation). A model illustrating these features is shown in FIGURE 5.

CONCLUSION

Considerable progress has been made toward understanding how insulin regulates the rate at which a specific protein is synthesized within a target tissue. There are now several examples of effects of insulin on mRNA metabolism, and in some of these instances this hormone influences the rate of transcription of specific genes. Based on the rate at which additional examples are appearing, we suggest that this regulation of gene transcription and the modulation of enzyme activity through covalent modification are equally important manifestations of insulin action.

ACKNOWLEDGMENTS

We would like to recognize our colleagues who played an integral role in the research described in this article: Terry Andreone, Elmus Beale, Cathy Caldwell, Nancy Chrapkiewicz, Tim Cripe, James Hartley, Stephen Koch, and Dan Petersen.

REFERENCES

1. GUNN, J., S. TILGHMAN, R. HANSON & F. BALLARD. 1975. Effects of cyclic adenosine monophosphate, dexamethasone and insulin on phosphoenolpyruvate carboxykinase synthesis in Reuber H-35 hepatoma cells. Biochemistry **14:** 2350-2357.
2. BEALE, E., C. KATZEN & D. K. GRANNER. 1981. Regulation of rat liver phosphoenolpyruvate carboxykinase (GTP) messenger ribonucleic acid activity by N^6, $O^{2'}$-dibutyryladenosine 3′,5′-monophosphate. Biochemistry **20:** 4878-4883.
3. BEALE, E., T. ANDREONE, S. KOCH, M. GRANNER & D. GRANNER. 1984. Insulin and glucagon regulate cytosolic phosphoenolpyruvate carboxykinase (GTP) mRNA in rat liver. Diabetes **33:** 328-332.
4. BEALE, E. G., J. L. HARTLEY & D. K. GRANNER. 1982. N^6, $O^{2'}$-dibutyryl cyclic AMP and glucose regulate the amount of messenger RNA coding for hepatic phosphoenolpyruvate carboxykinase (GTP). J. Biol. Chem. **257:** 2022-2028.
5. CHRAPKIEWICZ, N. B., E. B. BEALE & D. K. GRANNER. 1982. Induction of the messenger ribonucleic acid coding for phosphoenolpyruvate carboxykinase in H4IIE cells. J. Biol. Chem. **257:** 14428-14432.
6. GRANNER, D., T. ANDREONE, K. SASAKI & E. BEALE. 1983. Inhibition of transcription of the phosphoenolpyruvate carboxykinase gene by insulin. Nature **305:** 545-549.
7. SASAKI, K., T. R. CRIPE, S. R. KOCH, T. L. ANDREONE, D. D. PETERSEN, E. G. BEALE & D. K. GRANNER. 1984. Multihormonal regulation of phosphoenolpyruvate carboxykinase gene transcription: The dominant role of insulin. J. Biol. Chem. **259:** 15242-15251.
8. JEFFERSON, L. S. 1980. Role of insulin in the regulation of protein synthesis. Diabetes **29:** 487-496.
9. KORC, M., Y. IWAMOTO, H. SANDARAN, J. A. WILLIAMS & I. D. GOLDFINE. 1981. Insulin action in pancreatic acini from streptozotocin-treated rats. I. Stimulation of protein synthesis. Am. J. Physiol. **240:** G56-G62.
10. HORVAT, A. 1980. Stimulation of RNA synthesis in isolated nuclei by an insulin-induced factor in liver. Nature **286:** 906-908.
11. GRISWOLD, M. D. & J. MERRYWEATHER. 1982. Insulin stimulates the incorporation of ^{32}P into ribonucleic acid in cultured Sertoli cells. Endocrinology **111:** 661-667.
12. ANDREONE, T. L., E. BEALE, R. BAR & D. K. GRANNER. 1982. Insulin decreases phosphoenolpyruvate carboxykinase (GTP) mRNA activity by a receptor-mediated process. J. Biol. Chem. **257:** 35-38.
13. BEALE, E. G., N. B. CHRAPKIEWICZ, H. SCOBLE, R. J. METZ, R. L. NOBLE, D. P. QUICK, J. E. DONELSON, K. BIEMANN & D. K. GRANNER. 1985. Structure of the rat cytosolic phosphoenolpyruvate carboxykinase protein, messenger RNA, and gene. J. Biol. Chem. **260:** 10748-10760.

14. McKnight, G. S. & R. D. Palmiter. 1979. Transcription regulation of the ovalbumin and conalbumin genes by steroid hormones in chick oviduct. J. Biol. Chem. **254:** 9050-9058.
15. Nelson, K., M. Cimbala & R. W. Hanson. 1980. Regulation of phosphoenolpyruvate carboxykinase (GTP) mRNA turnover in rat liver. J. Biol. Chem. **255:** 8509-8514.
16. Tilghman, S. M., F. J. Ballard & R. W. Hanson. 1976. Hormonal regulation of phosphoenolpyruvate carboxykinase (GTP) in mammalian tissues. *In* Gluconeogenesis: Its Regulation in Mammalian Species. R. W. Hanson & M. A. Mehlman, Eds.: 47-87. John Wiley & Sons. New York, NY.
17. McKnight, G. S. 1978. The induction of ovalbumin and conalbumin mRNA by estrogen and progesterone in chick oviduct explant cultures. Cell **14:** 403-413.
18. Chen, C.-L. C. & P. Feigelson. 1979. Cycloheximide inhibition of hormonal induction of α_{2U}-globulin mRNA. Proc. Natl. Acad. Sci. USA **76:** 2669-2673.
19. Yamamoto, K. R. & B. M. Alberts. 1976. Steroid receptors: Elements for modulation of eukaryotic transcription. Annu. Rev. Biochem. **45:** 721-746.
20. Cimbala, M. A., W. H. Lamers, K. Nelson, J. E. Monahan, H. Yoo-Warren & R. W. Hanson. 1982. Rapid changes in the concentration of phosphoenolpyruvate carboxykinase mRNA in rat liver and kidney. J. Biol. Chem. **257:** 7629-7636.
21. Loten, E. G. & J. G. T. Sneyd. 1970. An effect of insulin on adipose tissue adenosine 3′,5′-cyclic monophosphate phosphodiesterase. Biochem. J. **120:** 187-193.
22. Granner, D. K., L. Chase, G. D. Aurbach & G. M. Tomkins. 1968. Tyrosine aminotransferase: Enzyme induction independent of adenosine 3′,5′-monophosphate. Science **162:** 1018-1020.
23. Granner, D. K. & T. L. Andreone. 1985. Insulin modulation of gene expression. *In* Diabetes Metabolism Reviews. R. DeFronzo, Ed. Vol. **1**(Nos. 1 & 2): 139-170. John Wiley & Sons. New York, NY.
24. Housley, P. R. & W. B. Pratt. 1983. Direct demonstration of glucocorticoid receptor phosphorylation by intact L cells. J. Biol. Chem. **258:** 4630-4635.

The Relationship between Structure and Function in cAMP-dependent Protein Kinases[a]

SUSAN S. TAYLOR, LAKSHMI D. SARASWAT, JEAN A. TONER, AND JOSÉ BUBIS[b]

Department of Chemistry
University of California, San Diego
La Jolla, California 92093

Since the discovery of cAMP and its principal mode of action in eukaryotic cells, it has been clear that protein phosphorylation plays an important role in the regulation of cellular processes.[1,2] The recent discovery that several oncogenes code for protein kinases that specifically phosphorylate tyrosine residues has expanded the role attributed to protein phosphorylation considerably.[3,4] Because these oncogenes have cellular homologues as well,[5,6] it is likely that protein phosphorylation plays a role not only in some forms of oncogenesis, but also in normal growth regulation. The finding that several growth factor receptors, such as the epidermal growth factor receptor and the insulin receptor, also have intrinsic protein kinase activity that is specific for tyrosine[7,8] has expanded even further the scope of the processes that are regulated by protein phosphorylation.

Of the many protein kinases that have been discovered recently, the catalytic (C) subunit of cAMP-dependent protein kinase remains the best characterized at the biochemical, molecular, and structural levels. The functional sites for the C subunit include an ATP-binding site, two Mg^{2+}-binding sites, a peptide-binding site, and a recognition site for binding the regulatory subunit. Several approaches have been used to characterize these functional sites. Kinetic mapping with ATP analogues has defined general features of the ATP-binding site[9] whereas affinity labeling has provided a mechanism for identifying specific amino acid residues that contribute to ATP binding.[10] Synthetic peptides have been used to characterize the peptide-recognition site,[11] and once again affinity labeling has been used to identify specific residues that are associated with peptide binding.[12] NMR has provided further information about both the Mg-ATP- and the peptide-recognition sites.[13] In addition, the overall sequence of the protein has been elucidated.[14] Sequence comparisons of the C subunit with other protein kinases such as the transforming protein from Rous Sarcoma Virus, $pp60^{src}$,[15] clearly identify a common homologous core that can be recognized as a catalytic "domain" in all protein kinases.

[a] This work was supported in part by Grant GM-19301 from the United States Public Health Service and Grant NP-419 from the American Cancer Society.

[b] Supported by the Consejo Nacional de Investigaciones Científicas y Tecnológicas, Caracas, Venezuela.

The regulatory (R) subunit of cAMP-dependent protein kinase also has been characterized extensively. It is a protein with a well-defined domain structure that includes multiple functional sites. In eukaryotic cells, the R subunit appears to represent a family of proteins that can be antigenically distinguished[16,17] and presumably play different roles in terms of compartmentalization, differentiation of cell types, regulation of gene expression, and protein-protein interactions. These roles and the reasons for this diversity are still not fully understood, but it is becoming apparent that the R subunit may have additional functions beyond its role as an inhibitor of the C subunit.[18,19] Although the R subunit does not belong to such a broad family of related proteins as the C subunit, it is clearly homologous to the catabolite gene-activator protein (CAP) in *Escherichia coli,* a protein that binds cAMP and plays a direct role in the regulation of gene expression.[20] The homologies with CAP, for which a crystal structure is available,[21] have provided a good model for the cAMP-binding domains of the R subunit that is consistent with affinity labeling results.[22]

Our laboratory has used a variety of approaches to probe the structure of both the R and C subunits of cAMP-dependent protein kinase. Approaches such as affinity labeling and monoclonal antibody specificity have been expanded to include preliminary X-ray crystallographic analysis of the C subunit, gene cloning of the R and C subunits, expression of these cloned genes, and directed mutagenesis of the R subunit.

CATALYTIC SUBUNIT

The various members of the protein kinase family differ substantially in size, in the ligand responsible for activation, and in subcellular location and membrane association. They also have clear differences in both the target amino acid that is recognized as a phosphate acceptor and in the amino acid-recognition site that surrounds the phosphorylated residue.[11,23] Nevertheless, each has conserved regions that include, in particular, those sites that are associated with the binding of Mg-ATP. The only amino acid that has been shown to be directly associated with ATP binding in the C subunit of cAMP-dependent protein kinase is Lys 72 that was covalently modified by fluorosulfonylbenzoyl 5′-adenosine (FSBA).[10] Other analogues have subsequently been used to map this ATP-binding site. For example, 8-N_3-ATP has been used as an affinity analogue of ATP that presumably targets a region of the protein that is in close proximity to the C-8 position of the adenine ring. Although a specific modified amino acid has not been identified yet, one of the major tryptic peptides that is covalently labeled following photoactivation of the C subunit and 8-N_3-ATP has been isolated and corresponds to a peptide in the NH_2-terminal region of the protein.

Another approach for identifying residues that are essential for catalysis has been to treat the C subunit with dicyclohexyl carbodiimide (DCCD), which selectively targets carboxylic acid residues. The C subunit is inhibited by DCCD, and both Mg-ATP and the R subunit are capable of protecting against this inhibition (FIG. 1). When peptide maps of the free C subunit or of the C subunit treated with DCCD in the presence of Mg-ATP are compared with maps of DCCD-inhibited protein, one major peptide is found to be consistently modified in the DCCD-treated enzyme. Preliminary results indicate that this peptide corresponds to residues 166-189 in the amino acid sequence. This portion of the sequence is as follows: Asp-Leu-Lys-Pro-Glu-Asn-Leu-Leu-Ile-Asp-Gln-Gln-Gly-Tyr-Ile-Gln-Val-Thr-Asp-Phe-Gly-Phe-Ala-

Lys. It contains four potential carboxylic acid residues, and efforts are now underway to determine if one of the four is responsible for the inhibition of catalytic activity.

If the overall homologies in sequence are compared for a variety of kinases, a common core can be readily identified (FIG. 2). Closer examination of these homologies reveals that there are specific regions within this core that are more highly conserved, as well as several residues that are invariant. Lys 72, the residue modified by FSBA, is one of these invariant residues. Although most of the other kinases, such as the

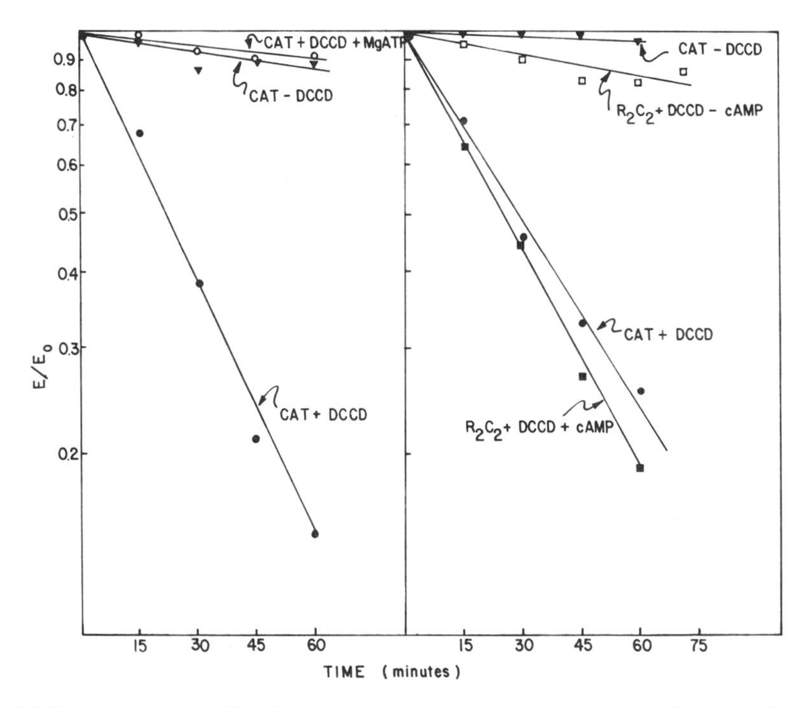

FIGURE 1. Inactivation of the C subunit by DCCD. Reaction mixtures contained the following in 0.15 ml of 50 mM 2-(N-morpholino)ethane sulfonic acid/NaOH, pH 6.0: 0.8 μM C subunit, $\pm$1.2 mM Mg-ATP, and $\pm$0.86 mM DCCD (left panel); 0.8 μM C subunit, $\pm$33 μM cAMP, and $\pm$1.0 mM DCCD (right panel). CAT: the free C subunit; R_2C_2: the holoenzyme. Reactions were initiated by the addition of DCCD after preincubation for 5 min at 37° C. Samples were withdrawn at 15-min intervals and assayed for kinase activity as described previously.[10] The holoenzyme was prepared using RII purified from porcine heart.

oncogenes, have not been purified in amounts that permit careful enzymological or structural analysis, Kamps *et al.* were able to demonstrate that the homologous lysine residue in $pp60^{src}$, Lys 295, was specifically modified by FSBA and that this modification led to inhibition of enzymatic activity.[31] The essential role of this residue has been confirmed further by directed mutagenesis of $pp60^{src}$ where conversion of Lys 295 to a methionine abolished kinase activity, transformation, and tumorigenicity.[32] A structural comparison of FSBA and ATP indicates that FSBA probably modifies a residue that is close to the γ-phosphate of ATP. Even though FSBA lacks the negative charges of ATP, the reactive fluorosulfonyl group is approximately the same

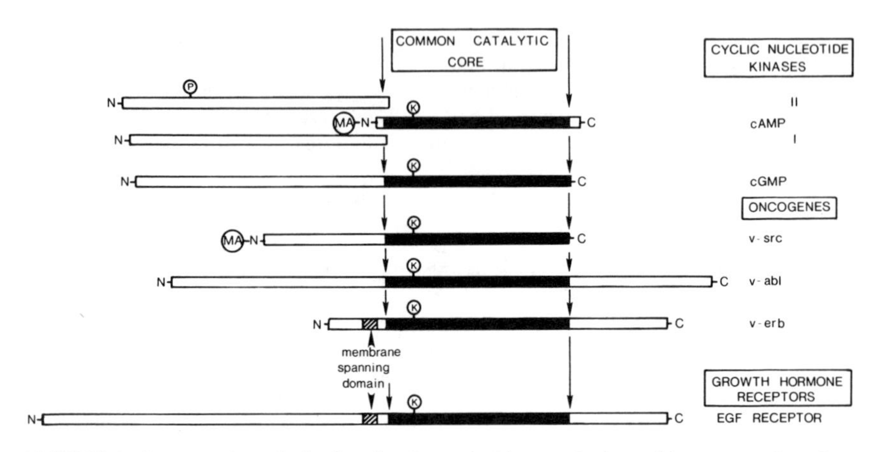

FIGURE 2. Sequence homologies in related protein kinases. Amino acid sequences have been elucidated for the RI, RII, and C subunits of cAMP-dependent protein kinases[24,25] and for cGMP-dependent protein kinase.[26] The sequences for v-src, v-abl, v-erb, and the epidermal growth factor (EGF) receptor have been deduced from DNA sequences.[27-30] The regions in black represent the homologous segment that is related to the C subunit of cAMP-dependent protein kinase. Within this segment the invariant lysine residue, Lys 72 in the C subunit, is indicated (K). N: the NH_2-terminal; C: the COOH-terminal; P: the autophosphorylation site; and MA: myristic acid.

distance from the ribose ring as the γ-phosphate of ATP. This proximity may allow this lysine to play a direct role in catalysis as indicated in FIGURE 3.

The peptide modified by DCCD also includes a highly conserved region. One segment of the peptide's sequence in particular, Asp-Phe-Gly, is another invariant region that is seen in every protein that has been demonstrated to have protein kinase activity. This sequence is near the end of the modified peptide. The Asp in this segment is one of the four carboxylic acid residues in this peptide, and one of two that are conserved in all protein kinases.

An earlier segment of the sequence, Gly-Thr-Gly-Ser-Phe-Gly, which includes residues 50-55, is also conserved in all protein kinases. This glycine-rich sequence, Gly-X-Gly-X-X-Gly, is found in many proteins that bind adenine nucleotides. For those proteins where a crystal structure is known, this region is associated with nucleotide binding.[33,34] By analogy, it is reasonable to think that the NH_2-terminal portion of the C subunit is associated with ATP binding—an assumption that is confirmed by affinity labeling. This region may represent another example of the nucleotide folding pattern that has now been found in many nucleotide-binding proteins.[35,36] Those regions that are highly conserved and that are most likely associated with ATP binding are summarized in FIGURE 4.

Much less is known about the peptide-binding site, although, based on clear differences in specificity, this region is likely to be more variable within the larger family of protein kinases. A synthetic peptide analogue has been used to target the peptide recognition site in the C subunit.[12] This led to stoichiometric modification of Cys 199 that is consistent with the peptide-binding site being more closely associated with the COOH-terminal region of the protein. The fact that this cysteine is not conserved again reflects the likelihood that the peptide-recognition site will differ in the serine- and tyrosine-specific kinases. The functional sites of the C subunit have been summarized in FIGURE 3. Those regions that are associated with Mg-ATP binding will most likely be conserved in all protein kinases, including the oncogenes and growth factor receptors, whereas the regions involved in peptide recognition will clearly differ.

In order to more definitively describe the structure of the C subunit, X-ray crystallographic characterization of the protein was initiated. Crystals that diffract to 3.5 Å were grown,[38] and two additional and unique crystal forms were obtained subsequently. One form was grown in the presence of 5′-adenylyl(β,γ-methylene)diphosphate (AMP-PCP), a nonhydrolyzable analogue of ATP, and the other form represents a ternary complex of Mg-ATP, the C subunit, and a 20-residue inhibitor peptide that represents the active segment of the heat-stable protein kinase inhibitor.[39]

REGULATORY SUBUNIT

Although there appears to be one form of the catalytic subunit, there are several forms of cAMP-dependent protein kinase. In general, these are classified as types I and II, based on elution from DEAE-cellulose,[40] and differences in the holoenzymes can be attributed primarily to differences in the R subunits.[41] Type I and II R subunits can be readily distinguished on the basis of molecular weight and autophosphorylation[41,43] and are immulogically distinct.[16,17] Antibodies directed toward the holoenzymes are directed almost exclusively against the R subunits, and in general no cross-reactivity is observed between serum antibodies directed against RI or holoenzyme I and RII or holoenzyme II.

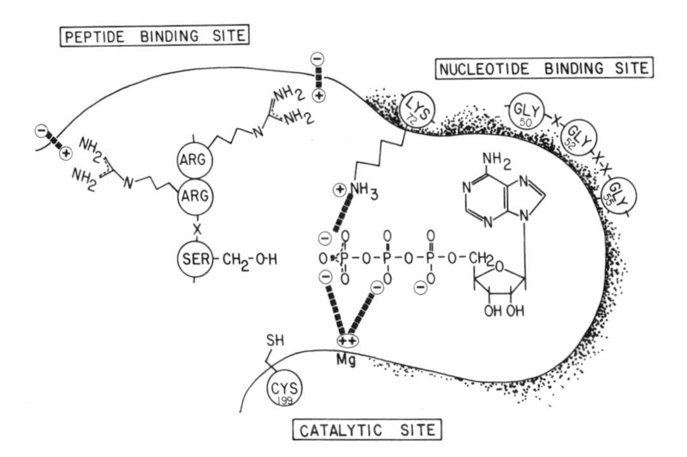

FIGURE 3. A model of the active site of the C subunit of cAMP-dependent protein kinase. The regions that are associated with ATP binding and that are likely to be conserved in all protein kinases are indicated by the shaded region on the right. This region includes Lys 72, the segment from Gly 50 to Gly 55, and the Mg^{2+}-binding site. It is likely that the Asp-Phe-Gly sequence (184-186) will also be included in this region. The peptide-binding site indicated on the left differs in the various kinases. Thus it is not anticipated that essential residues here will be conserved throughout the larger family of protein kinases.

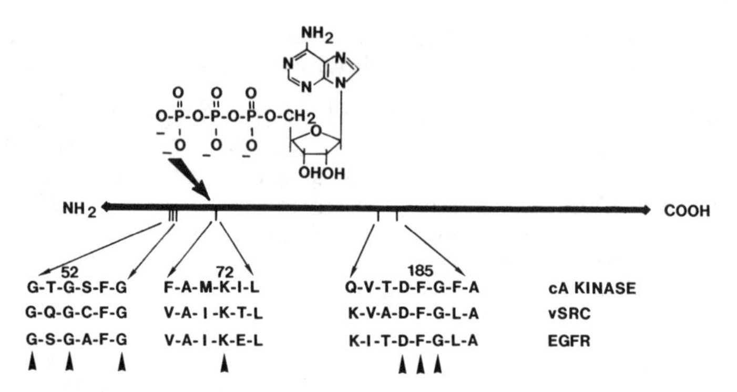

FIGURE 4. Some of the highly conserved regions in protein kinases. Three regions are indicated. Each contains amino acid residues that are invariant in all protein kinases. The invariant residues are indicated with arrows. The glycine-rich sequence seen on the left is common to many nucleotide-binding sites.[33] Lys 72 has been shown to be involved in ATP binding and most likely is in close proximity to the γ-phosphate of ATP.[10,31] The third region containing the D-F-G sequence is contained within a peptide that has been modified by DCCD. These preliminary data suggest that this third region may also be associated with nucleotide binding.

Several lines of evidence suggest that there is a single gene product for RI,[44] whereas the RII classification appears to represent greater diversity. The major form of RII that has been characterized is from heart; however, the major forms of the R subunit in brain can also be classified as being RII subunits based on several criteria such as elution from DEAE-cellulose and autophosphorylation.[45–47] Nevertheless, several features distinguish neural RII from the RII that predominates in heart. It can be readily distinguished qualitatively because the electrophoretic mobility of the protein on SDS-polyacrylamide does not shift in response to autophosphorylation. These differences were further confirmed once it was established that monoclonal antibodies generated to heart RII did not cross-react with the major form of RII in brain.[47] After several antigenic sites had been localized in heart RII,[48,49] it was possible to target specific regions of neural RII where antigenic differences were anticipated. One such region of RII was sequenced, and the results, summarized in FIGURE 5, definitively established that brain RII and heart RII represent unique gene products. Whether the diversity seen in other tissues[50] represents additional gene products remains to be established. The most definitive mechanism for unambiguously distinguishing these multiple forms of RII is with monoclonal antibodies, which appear to be extremely sensitive to sequence differences in the various classes of RII.[51] At this point, it is not clear how many unique forms of RII exist, nor is it apparent why there is such diversity.

Limited proteolysis was used to establish the general domain structure of the R subunits.[52,53] Each R subunit consists of three segments. The COOH-terminal for two-thirds of the protein consists of two homologous, tandem, cAMP-binding domains. These domains, designated A and B, will be described more fully later. The NH_2-terminal segment accounts for approximately one-third of the molecule and contains the major C subunit recognition site as well as the major sites for interaction between the two protomers in the R subunit dimer. A hinge region that is susceptible to limited proteolysis is also contained within this segment.[52–54] Localization of antigenic sites recognized by both monoclonal and serum antibodies has established that the antigenic sites in heart RII are confined to this NH_2-terminal third of the molecule and do not appear to extend into either of the cAMP-binding domains.[47] It is the region that also differs most when RI and RII are compared. The homology between the two tandem, cAMP-binding domains within each subunit is in the range of 40%; the homologies between the A domains and the B domains in RI and RII are closer to 50%.[24-26] On the other hand, the homology in the NH_2-terminal segment is less than 20% and is confined primarily to the C subunit interaction site. The NH_2-terminal regions of RI and RII are both highly charged—roughly one-third of the first 140 residues are charged. However, the net charge in the RI segment is highly positive whereas the corresponding net charge in heart RII is highly negative. Thus, an unusual class of isozymes emerges: each isozyme of this class contains a highly conserved region that is associated primarily with cAMP binding and a much more variable region. Since this variable region is masked in part by the C subunit in the holoenzyme, it follows that this region is exposed when the C subunit is dissociated in the presence of cAMP.

The cAMP-binding domains of the R subunit that account for the remaining two-thirds of the protein are strikingly nonantigenic. This region has been probed most directly by photoaffinity labeling. Photolabeling of RII with 8-N_3-cAMP leads to nearly stoichiometric modification of only one residue[55] whereas two residues are covalently modified in RI.[22] In both proteins a homologous tyrosine residue is labeled. This residue, Tyr 371 in RI and Tyr 381 in RII, is localized near the end of the

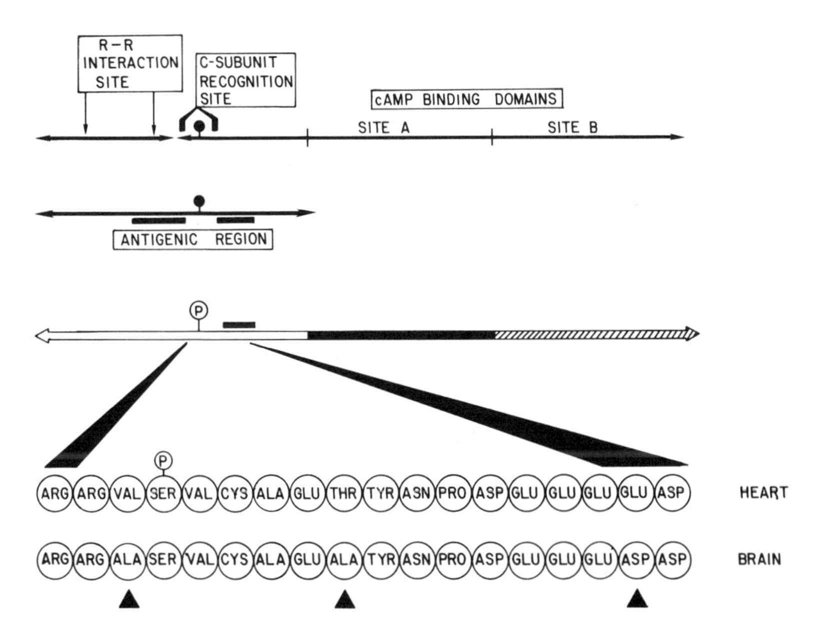

FIGURE 5. Summary of the antigenic regions of RII and the comparative sequences of heart and neural RII flanking the autophosphorylation site. The general domain structure of RII is summarized at the top. The antigenic sites are confined to the NH_2-terminal segment and do not extend into the cAMP-binding domains.[49] Specific sites recognized by two monoclonal antibodies are indicated by bars.[48,49] The amino acid sequence indicated represents a portion of the antigenic site that is recognized by a monoclonal antibody that cross-reacts with bovine heart RII but not with neural RII. Isolation of the peptide and its sequencing were described by Weldon *et al.*[47] The autophosphorylation site is indicated by a darkened circle whereas sequence differences between heart and neural RII are indicated by darkened triangles.

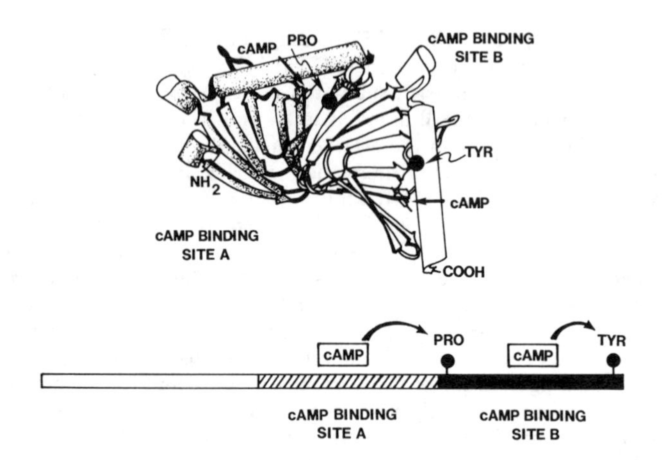

FIGURE 6. Model of the two tandem cAMP-binding sites in the COOH-terminal region of RI. Two cAMP-binding domains, each homologous to the cAMP-binding domain of CAP,[20,21] are positioned in tandem to one another. The domain corresponding to site A is indicated as a shaded ribbon; the domain corresponding to site B is indicated as an open ribbon. The model correlates known sites of covalent modification with 8-N_3-cAMP with cAMP binding in each domain. The bound cAMP is indicated by arrows in each domain as are Pro 271 and Tyr 371, the two covalently modified residues.

second cAMP-binding domain. Interpretation of this photolabeling has been greatly facilitated by the homology of the R subunit with CAP, the major cAMP-binding protein in *E. coli.* Comparison of amino acid sequences has clearly established that the cAMP-binding domains of CAP and RI and RII are homologous.[20] The crystal structure of CAP is known,[21] so the amino acid sequences of the R subunits can be superimposed on top of the amino acid sequence of the CAP structure to determine whether the observed covalent modifications are consistent with cAMP binding in a similar manner in the R subunits and CAP. When the sequence of the cAMP-binding domain B of the R subunit is superimposed on the sequence of the crystal structure of the cAMP-binding domain of CAP, Tyr 381 in RII and Tyr 371 in RI both lie in the C helix of CAP (FIG. 6). It is the long C helix that provides the major region of interaction between the protein and the adenine ring. The orientation of the tyrosine ring is also directed toward the adenine ring. Thus, it seems reasonable that the tyrosine residue is modified by 8-N_3-cAMP bound to the second cAMP-binding domain, and that cAMP is binding in a generally analogous manner in both proteins.

The second residue that is modified in RI is Pro 271, which is also located in the second cAMP-binding domain. If one considers the presumed general location of this proline based on the homology with CAP, however, it would be located in the first β-strand of domain B, which would lie more than 20 Å away from Tyr 371. Thus, it seems unlikely that the same 8-N_3-cAMP molecule could modify both residues. If two tandem cAMP-binding domains are juxtaposed as seen in FIGURE 6, however, it can be hypothesized that Pro 271 is covalently modified by 8-N_3-cAMP bound to domain A and that it represents a contact point between the two domains.

In order to test this hypothesis, we took advantage of the analogue specificity of the two cAMP-binding sites. N^6-substituted analogues of cAMP are known to preferentially displace cAMP from one site whereas C-8-substituted analogues can displace cAMP preferentially from the other site.[56] Analogues that have been shown to have a pronounced site preference were used to selectively alter photolabeling of RI with 8-N_3-cAMP. For example, RI was initially saturated at both sites with 8-N_3-[^{3}H]-cAMP. A nonradioactive N^6-substituted analogue of cAMP, N^6-monobutyryl cAMP, was then added, and dissociation of the radioactively labeled photoaffinity label was allowed to procede until more than 50% of the radioactivity was dissociated. The protein at this point was photolabeled, and the incorporation of radioactivity into Pro 271 and Tyr 371 was determined by HPLC separation of tryptic peptides. As summarized in FIGURE 7, N^6-monobutyryl cAMP selectively abolished labeling of Pro 271 whereas the labeling of Tyr 371 was much less reduced. N^6-substituted analogues, therefore, selectively target site A whereas C-8-substituted analogues target site B (data not shown). These results also established that labeling of the tyrosine residue can be attributed to 8-N_3-cAMP bound to domain B and that labeling of the proline residue clearly results from 8-N_3-cAMP bound to domain A, thus confirming the model proposed in FIGURE 6.

CLONING AND EXPRESSION OF CATALYTIC AND REGULATORY SUBUNITS

Another approach to further our understanding of the structure of cAMP-dependent protein kinases is to clone the genes for both the C and the R subunits. Partial cDNA clones for both the C and RI subunits have been isolated from a bovine

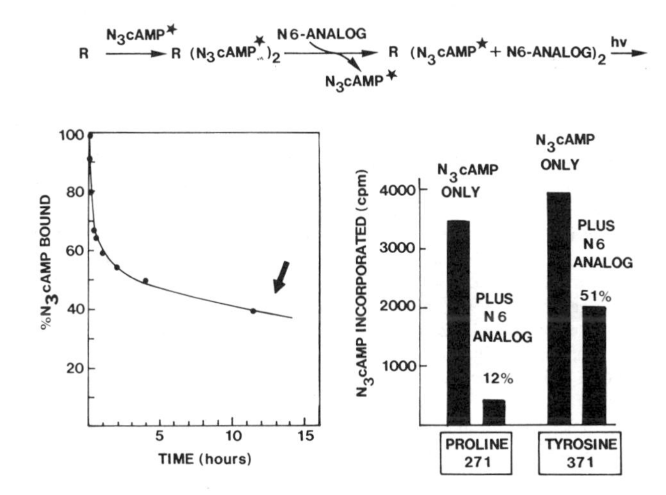

FIGURE 7. Covalent modification of RI with 8-N_3-cAMP following partial dissociation with an N^6-substituted analogue of cAMP. Holoenzyme I (3.7 nmoles) was incubated with 8-N_3-[^{3}H]-cAMP (37 nmoles) (designated with a star) for 16 hr at 4° C to saturate both cAMP-binding sites of RI. The radioactive photoaffinity probe was then dissociated with a 15-fold excess of N^6-monobutyryl cAMP (580 nmoles) for 13 hr at 4° C. When 60% of the 8-N_3-cAMP had dissociated (see the time dissociation curve on the left), the sample was irradiated on ice with ultraviolet light (254 nm) for 10 min. Free C subunit was removed using CM Sepharose, and RI peptides were separated using HPLC after treatment with trypsin as described previously.[22] The radioactivity incorporated into each peptide is indicated in the bar graph on the right.

pituitary cDNA library using oligonucleotide probes. These cDNA clones each represent approximately 75% of the coding region. In order to proceed with the expression of RI, a cDNA clone containing the entire coding region of bovine RI, which was generously provided by G. S. McKnight of the University of Washington, was used. This clone, 62C12, was isolated from a pBR322 library from bovine testes.[44] The eventual objective is to utilize directed mutagenesis to probe the multiple functional sites of RI. The first step was to construct an expression vector for RI. To achieve

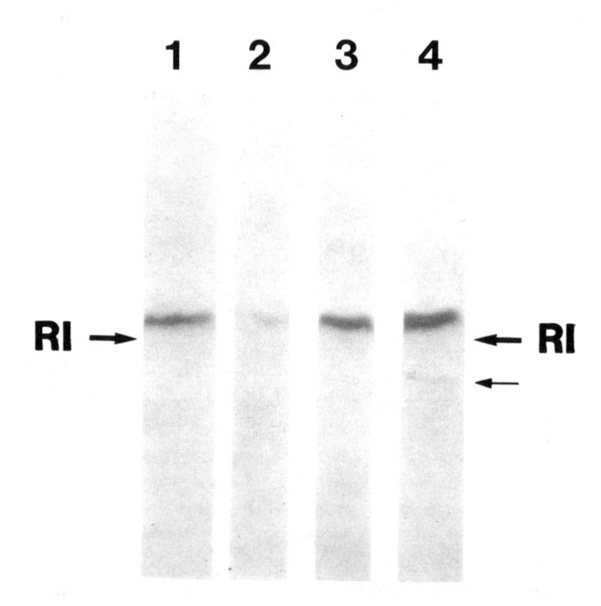

FIGURE 8. Partial purification of RI from *E. coli.* One liter of culture was centrifuged after 24 hr, and the cells were resuspended in 20 mM potassium phosphate, 5 mM EDTA, 5 mM 2-mercaptoethanol, 1 mg/l N-α_p-tosyl-L-lysylchloromethyl ketone, and 15 mg/l phenylmethylsulfonylfluoride (pH 6.2). The suspension was passed through a French pressure cell and centrifuged at 3600 × g for 20 min at 4° C. The pellet that formed was then reextracted and recentrifuged. Proteins from the combined extracts were precipitated with 70% ammonium sulfate and centrifuged at 12 000 × g for 30 min at 4° C. Aliquots were taken out at various steps and subjected to electrophoresis on a 12.5% SDS-polyacrylamide gel.[59] Proteins were electrotransferred onto nitrocellulose, labeled with 8-N_3-[^{32}P]-cAMP in 10 mM Tris-150 mM sodium chloride (pH 7.5), and subjected to autoradiography. Lane 1: the total extract; lane 2: the residual pellet after two extractions; lane 3: the pooled supernatant fractions; lane 4: the redissolved pellet after ammonium sulfate precipitation. The mobility of bovine RI is shown by the large arrows, and the mobility of the proteolytic fragment of RI is shown by the small arrow.

this, the entire coding segment of the RI-cDNA clone was excised according to the procedure described by Saraswat *et al.*[57] After digestion with *Nar*I and partial digestion with *Nco*I, the RI-cDNA clone was isolated and ligated into the pUC7 plasmid.[58] *E. coli* (JM105) was transformed with this ligation mix, and a clone was selected that contained the entire RI insert in the same orientation as the *lac Z* gene.[57]

Because *E. coli* normally contain a relatively high level of endogenous cAMP, a sensitive screening procedure was devised that circumvented the problems associated with these high levels of cAMP. After electrophoresis of the total solubilized cell extract on polyacrylamide gels containing sodium dodecyl sulfate, the proteins were electrotransferred to nitrocellulose. The nitrocellulose was then blocked, incubated with 8-N_3-[^{32}P]-cAMP, washed, and photolyzed. The RI subunit was then visualized by autoradiography. Although this technique may not be quantitative, it does provide a very rapid and sensitive means of assessing relative kinase levels of the R subunit.[57]

As indicated in FIGURE 8, not only was RI expressed, but all of the RI that was synthesized was soluble and could be readily extracted. Negligible photolabeled material remained in the particulate fraction. The amount of R subunit expressed in *E. coli* (JM105) based on comparative photolabeling of RI standards and on direct cAMP binding following ammonium sulfate precipitation was 2-4 mg/liter of culture.

Because the expressed protein uses the start codon provided by the vector, which was confirmed by amino acid sequencing (data not shown), the protein has an additional 10 residues at its NH_2-terminal. Nevertheless, the behavior of the expressed protein was similar to that of RI purified from bovine tissues using several criteria. The expressed R subunit was a dimeric protein that showed interchain disulfide bonding analogous to bovine RI and formed a holoenzyme that was dependent on cAMP for activation. It also bound cAMP with a high affinity to both cAMP-binding sites.[57] This protein has now been purified to homogeneity, and crystallization studies have been initiated. Moreover, directed mutations are being introduced into the protein as a mechanism for further probing the cAMP-binding sites as well as the R-C-interaction sites.

REFERENCES

1. SUTHERLAND, E. W. & T. W. RALL. 1958. J. Biol. Chem. **232:** 1077-1091.
2. KREBS, E. G. & J. A. BEAVO. 1979. Annu. Rev. Biochem. **48:** 923-960.
3. COLLETT, M. S. & R. L. ERIKSON. 1978. Proc. Natl. Acad. Sci. USA **75:** 2021-2024.
4. HUNTER, T. & B. M. SEFTON. 1980. Proc. Natl. Acad. Sci. USA **77:** 1311-1315.
5. HUNTER, T. 1984. Sci. Am. **251:** 70-79.
6. BISHOP, J. M. 1983. Annu. Rev. Biochem. **52:** 301-354.
7. USHIRA, H. & S. COHEN. 1980. J. Biol. Chem. **255:** 8363-8365.
8. KASUGA, M., Y. ZICK, D. BLITHE, M. CRETTAZ & C. KAHN. 1982. Nature **298:** 667-669.
9. HOPPE, J., W. FREIST, R. MARUTSKY & K. G. WAGNER. 1977. Eur. J. Biochem. **80:** 369-373.
10. ZOLLER, M. J., N. C. NELSON & S. S. TAYLOR. 1981. J. Biol. Chem. **256:** 10837-10842.
11. KEMP, B. E., E. BENJAMINI & E. G. KREBS. 1976. Proc. Natl. Acad. Sci. USA **73:** 1038-1042.
12. BRAMSON, H. N., N. THOMAS, R. MATSUIDA, N. C. NELSON, S. S. TAYLOR & E. T. KAISER. 1982. J. Biol. Chem. **257:** 10575-10581.
13. GRANOT, J., A. S. MILDVAN & E. T. KAISER. 1980. Arch. Biochem. Biophys. **205:** 1-17.
14. SHOJI, S., L. H. ERICSSON, K. A. WALSH, E. H. FISCHER & K. TITANI. 1983. Biochemistry **22:** 3702-3709.
15. BARKER, W. C. & M. O. DAYHOFF. 1982. Proc. Natl. Acad. Sci. USA **79:** 2836-2839.
16. FLEISCHER, N., O. M. ROSEN & M. REICHLIN. 1976. Proc. Natl. Acad. Sci. USA **73:** 54-58.
17. KAPOOR, C. L., J. A. BEAVO & A. L. STEINER. 1979. J. Biol. Chem. **254:** 12427-12432.
18. CONSTANTINOU, A. I., S. P. SQUINTO & R. A. JUNGMANN. 1985. Cell **42:** 429-437.
19. JURGENSEN, S. R., P. B. CHOCK, S. S. TAYLOR, J. R. VANDENHEEDE & W. MERLEVEDE. 1985. Proc. Natl. Acad Sci USA **82:** 7565-7569.
20. WEBER, I. T., K. TAKIO, K. TITANI & T. A. STEITZ. 1982. Proc. Natl. Acad. Sci. USA **79:** 7679-7693.
21. MCKAY, D. B., I. T. WEBER & T. A. STEITZ. 1982. J. Biol. Chem. **257:** 9518-9524.
22. BUBIS, J. & S. S. TAYLOR. 1985. Biochemistry **24:** 2163-2170.

23. Hunter, T. 1983. J. Biol. Chem. **257:** 4843-4848.
24. Titani, K., T. Sasagowa, L. H. Ericsson, S. Jumar, S. B. Smith, E. G. Krebs & K. A. Walsh. 1984. Biochemistry **23:** 4193-4199.
25. Takio, K., S. B. Smith, E. G. Krebs, K. A. Walsh & K. Titani. 1984. Biochemistry **23:** 4200-4206.
26. Takio, K., R. D. Wade, S. B. Smith, E. G. Krebs, K. A. Walsh & K. Titani. 1984. Biochemistry **23:** 4207-4213.
27. Schwartz, D., R. Tizzard & W. Gilbert. 1983. Cell **32:** 853-869.
28. Reddy, G. P., M. J. Smith & A. Srinivasan. 1983. Proc. Natl. Acad. Sci. USA **80:** 3623-3626.
29. Yamamoto, T., T. Nishida, N. Miyajima, S. Kawai, T. Ooi & K. Toyoshima. 1983. Cell **35:** 71-78.
30. Urich, A., L. Coussens, J. S. Hayflich, T. J. Dull, A. Gray, A. W. Tam, J. Lee, Y. Yarden, T. A. Libermann, J. Schlessinger, J. Downware, E. L. V. Mayes, N. Whittle, J. D. Waterfield & P. H. Seeburg. 1984. Nature **309:** 418-425.
31. Kamps, M. P., S. S. Taylor & B. M. Sefton. 1984. Nature **310:** 589-592.
32. Snyder, M. & M. J. Bishop. 1985. Mol. Cell. Biol. **5:** 1772-1779.
33. Buehner, M., G. C. Ford, D. Moras, K. W. Olsen & M. G. Rossmann. 1974. J. Mol. Biol. **90:** 25-49.
34. Hol, W. G. J., P. T. Van Duijhne & H. J. C. Berendsen. 1978. Nature **273:** 443-446.
35. Rossmann, M. G., D. Moras & K. W. Olsen. 1974. Nature **250:** 194-199.
36. Holbrooke, J. J., A. Liljas, S. J. Steindel & M. G. Rossmann. 1975. *In* The Enzymes. Vol. **11A:** 191-192.
37. Wierenga, R. K., P. Terpstra & W. G. J. Hol. 1986. J. Mol. Biol. **187:** 101-107.
38. Sowadski, J. M., N. G. Xuong, D. Anderson & S. S. Taylor. 1985. J. Mol. Biol. **182:** 617-620.
39. Cheng, H. C., S. M. Van Patten, A. J. Smith & D. A. Walsh. 1985. Biochem. J. **231:** 655-661.
40. Corbin, J. D. & S. L. Keely. 1977. J. Biol. Chem. **252:** 910-918.
41. Zoller, M. J., A. R. Kerlavage & S. S. Taylor. 1979. J. Biol. Chem. **254:** 2408-2412.
42. Hoffman, F., J. A. Beavo, P. T. Bechtel & E. G. Krebs. 1975. J. Biol. Chem. **250:** 7795-7801.
43. Rosen, O. M. & J. Ehrlichman. 1975. J. Biol. Chem. **250:** 7788-7794.
44. Lee, D. C., D. F. Carmichael, E. G. Krebs & G. S. McKnight. 1983. Proc. Natl. Acad. Sci. USA **80:** 3608-3612.
45. Hart, F. T. & R. Roskoski. 1982. Biochemistry **21:** 5175-5183.
46. Ehrlichman, J., D. Sarkov, N. Fleisher & C. S. Rubin. 1980. J. Biol. Chem. **225:** 8179-7184.
47. Weldon, S. L., M. C. Mumby & S. S. Taylor. 1985. J. Biol. Chem. **260:** 6440-6448.
48. Weldon, S. L., M. C. Mumby, J. A. Beavo & S. S. Taylor. 1983. J. Biol. Chem. **258:** 1129-1135.
49. Weldon, S. L. & S. S. Taylor. 1985. J. Biol. Chem. **260:** 4203-4209.
50. Robinson-Steiner, A. M., S. J. Beebe, S. R. Rannels & J. D. Corbin. 1984. J. Biol. Chem. **259:** 10596-10605.
51. Mumby, M. C., S. L. Weldon, C. W. Scott & S. S. Taylor. 1985. Pharm. Therap. **28:** 367-387.
52. Corbin, J. D., P. H. Sugden, L. West, D. A. Flockhart, T. M. Lincoln & D. McCarthy. 1978. J. Biol. Chem. **253:** 3997-4003.
53. Potter, R. L. & S. S. Taylor. 1979. J. Biol. Chem. **254:** 2413-2418.
54. Takio, K., K. A. Walsh, H. Neurath, S. B. Smith, E. G. Krebs & K. Titani. 1980. FEBS Lett. **114:** 83-91.
55. Kerlavage, A. R. & S. S. Taylor. 1980. J. Biol. Chem. **255:** 8483-8488.
56. Rannels, S. R. & J. D. Corbin. 1981. J. Biol. Chem. **256:** 7871-7877.
57. Saraswat, L. D., M. Filutowicz & S. S. Taylor. 1985. J. Biol. Chem. **261:** in press.
58. Vieira, J. & J. Mession. 1982. Gene **19:** 259-268.
59. Laemmli, U. K. 1970. Nature **227:** 680-685.

The Relationship between Structure and Function for and the Regulation of the Enzymes of Fatty Acid Synthesis[a]

SALIH J. WAKIL

Verna and Marrs McLean Department of Biochemistry
Baylor College of Medicine
Houston, Texas 77030

Biosynthesis of long-chain fatty acids from acetyl-coenzyme A (acetyl-CoA) is carried out by two major steps.[1,2] The first is synthesis of malonyl-CoA from acetyl-CoA, a reaction catalyzed by acetyl-CoA carboxylase (reaction 1). The second is conversion of acetyl-CoA and malonyl-CoA to long-chain fatty acids by the enzyme fatty acid synthase in the presence of NADPH (reaction 2).

$$CH_3COS\text{-}CoA + ATP + CO_2 \rightleftarrows HOOCCH_2COS\text{-}CoA + ADP + P_i \qquad (1)$$

$$CH_3COS\text{-}CoA + 7HOOCCH_2COS\text{-}CoA + 14NADPH + 14H^+ \rightarrow$$
$$CH_3CH_2(CH_2CH_2)_6CH_2COOH + NADP^+ + 6H_2O + 8CoA\text{-}SH + 7CO_2 \qquad (2)$$

The two enzymes are multifunctional and carry out sequential reactions. In this review we would like to sum up our recent studies of these enzymes.

ACETYL-CoA CARBOXYLASE

Acetyl-CoA carboxylase is a biotin enzyme that functions in two stages: first, carboxylation of the the protein-bound biotin to form a carboxy-biotinyl derivative (reaction 3); second, transfer of the carboxyl group to the acceptor acetyl group to form the product malonyl-CoA (reaction 4).

$$\text{Biotin-BCCP} + ATP + HCO_3^- \xrightleftharpoons{Mn^{2+} \text{ or } Mg^{2+}} \text{carboxybiotin-BCCP} + ADP + P_i \qquad (3)$$

$$\text{Carboxybiotin-BCCP} + \text{acetyl-CoA} \rightleftarrows \text{biotin-BCCP} + \text{malonyl-CoA} \qquad (4)$$

In bacteria, the biotinyl carboxyl-carrier protein (BCCP), biotin carboxylase (reaction 3), and transcarboxylase (reaction 4) can be readily separated and purified.[3,4]

[a] This work was supported in part by grants from the National Institutes of Health (GM19091), the National Science Foundation (PCM8206562), the Robert A. Welch Foundation (Q-587), and the Clayton Foundation.

In animal tissues, however, carboxylation is performed by a single enzyme. Early disagreements over the subunit structure of animal acetyl-CoA carboxylase[5,6] were spawned by the proteolytic generation of catalytically active fragments during carboxylase isolation. The problem has been overcome with the development of rapid purification methods involving polyethylene glycol precipitation and avidin affinity chromatography. The enzyme, as isolated from chicken liver,[7,8] rat liver,[9] and rabbit mammary gland,[10] is composed of identical subunits with molecular weights of 225 000-250 000, and each subunit contains the allosteric site, one biotin prosthetic group, and two enzymes: biotin carboxylase and transcarboxylase.

Regulation of the acetyl-CoA carboxylase reaction, the committed step in fatty acid synthesis, has been actively investigated in many laboratories. Certain lines of evidence suggest the involvement of citrate, and some of these suggest a role for phosphate groups. In both cases apparently conflicting data have been obtained.

Citrate, which is a precursor of cytoplasmic acetyl-CoA, is considered to be the feed-forward activator of the carboxylase.[11] Moreover, it has been shown to rapidly activate purified carboxylase preparations, which otherwise have little activity. In this capacity, citrate has been regarded as an "obligatory allosteric activator" of animal acetyl-CoA carboxylase.[4,12] The exact underlying mechanism of this activation is not known. Recent *in vitro* evidence indicates that the enzymes undergoes conformational changes in the presence of citrate and that these changes result in the activation of the enzyme.[13] The protein then polymerizes into a filamentous form.[14] Half-maximal activation ($K_{0.5}$) of the carboxylase, however, was found between 2 and 10 mM citrate,[15] which is several times higher than the intracellular concentration of citrate (less than 0.5 mM),[16] suggesting that citrate may not play a significant role *in vivo*.

In 1972, Inoue and Lowenstein[6] suggested that phosphorylation may play a role in the regulation of acetyl-CoA carboxylase based on their finding that the carboxylase is a phosphoprotein. Support for this notion came from Carlson and Kim,[17] who observed the incorporation of labeled phosphate from [α-^{32}P]ATP into immunoprecipitated carboxylase and an ATP-dependent inhibition of acetyl CoA carboxylase in crude preparations. More recently, the phosphorylation and simultaneous inactivation of rat liver acetyl-CoA carboxylase by both a cAMP-independent protein kinase[18] and the catalytic subunit of cAMP-dependent protein kinase[19] have been reported. Consistent with these findings is the observation that dephosphorylation of rat liver carboxylase, by an "acetyl-CoA carboxylase phosphatase" isolated from the same tissue, resulted in activation.[20] It has also been shown that the carboxylases of rabbit mammary gland[10,21] and chicken liver[8] can be activated by heterologous phosphatases.

In other experiments, however, no direct relationship between phosphorylation and catalytic activity was detected, raising doubts about the role of phosphorylation and dephosphorylation in regulating the acetyl-CoA carboxylase. No change in carboxylase activity was found upon the incorporation of phosphate (2 mol of phosphate/mol of carboxylase subunit), an incorporation catalyzed by two cAMP-independent kinases of rat liver (casein kinases I and II),[22] nor was there any diminution of activity in carboxylase prepared from cultured chick liver cells under conditions in which 9-10 phosphate groups/carboxylase subunit were labeled.[23] Moreover, alkaline phosphatases produced no change in carboxylase activity upon the hydrolysis of phosphate from carboxylases from chicken[8] and rat[24] livers (2 and 3 mol of phosphate/mol of carboxylase subunit, respectively).

Recent investigations of the role of phosphorylation and dephosphorylation in the regulation of carboxylase activities and the relationship of this regulation to citrate activation have begun to produce some interesting results. It has long been known that acetyl-CoA carboxylase is isolated as a phosphoprotein (6-15 mol of phosphate/mol of carboxylase subunit).[9,15,25] The enzyme is a good substrate for additional phosphorylation. In some instances the citrate-dependent, inactive enzyme is converted, by phosphorylation by cAMP-dependent and -independent protein kinases,[14,15] to another inactive form that can no longer be activated by citrate. Yet in other cases (phosphorylation by casein kinases I and II) the incorporation of phosphate (2 mol of phosphate/mol of carboxylase subunit) has no effect on carboxylase activity.[22] Likewise, the isolated enzyme can be dephosphorylated. The action of protein phosphatases results in the conversion of citrate-dependent, inactive carboxylase to other inactive forms that have higher activities in the presence of citrate,[20] whereas the hydrolysis of phosphate (2-3 mol of phosphate/mol of carboxylase subunit) with alkaline phosphatase brings about no change in carboxylase activity.[24] Given the fact that several sites of phosphorylation exist on the carboxylase protein, it might be possible to account for these different and apparently conflicting findings by postulating that some of the sites are important in regulating catalytic activity whereas others are unimportant.

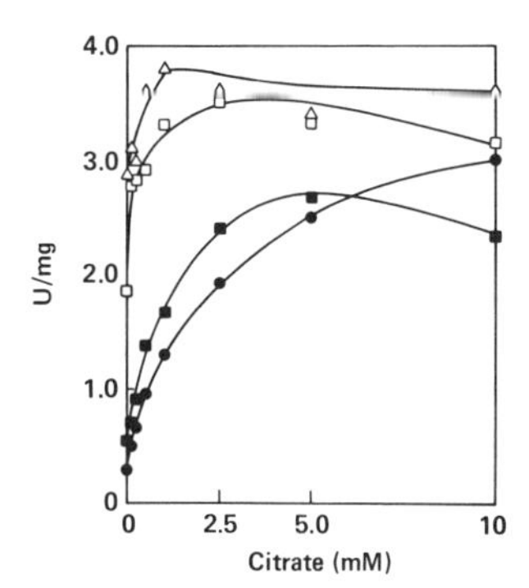

FIGURE 1. Citrate activation of different preparations of acetyl-CoA carboxylase. Acetyl-CoA carboxylase prepared by various procedures was assayed at the indicated citrate concentration as described previously.[15,26] Δ and ■: Carboxylases prepared by quick and late freeze-clamped livers, respectively;[12] ●: enzyme prepared by various methods;[15] □: carboxylase prepared by late freeze-clamping, but activated in the presence of Mn^{2+} with carboxylase D phosphatase, then assayed in presence of citrate at the concentration indicated.[15,26]

We have recently carried out systematic investigations of the role of phosphorylation in the regulation of acetyl-CoA carboxylase and have obtained findings compatible with this thesis. Initially, a rapid procedure for the isolation of highly purified preparations of acetyl-CoA carboxylase was developed. Electrophoresis of such preparations on sodium dodecyl sulfate-polyacrylamide gels showed the presence of a single protein band; the protein in this band has an estimated molecular weight of 250 000. Like other preparations,[4,12] that for the enzyme has little activity in the absence of citrate, but undergoes a rapid activation when citrate is added (FIG. 1). We have also isolated a Mn^{2+}-dependent phosphatase (M_r: 90 000) that increases the activity of the carboxylase approximately 10-fold, as shown in FIGURE 2. This phosphatase shows high affinity for acetyl-CoA carboxylase (K_m: 0.2 μM) as compared to its action on phosphorylase *a* (K_m: 5.5 μM) and phosphohistone (K_m: 20 μM).

In TABLE 1 the catalytic activity and phosphate content of acetyl-CoA carboxylase are compared before and after dephosphorylation by the phosphatase. Treatment with the phosphatase removed phosphate (2-3 mol of phosphate/mol of carboxylase subunit) and increased the specific activity 10-fold, to 2.6-3.0 U/mg, in the absence of citrate. These activities are comparable to those of the phosphorylated carboxylase in the presence of 10 mM citrate. Thus, dephosphorylation by the Mn^{2+}-dependent phosphatase renders the carboxylase more citrate independent than the phosphorylated form, which is heavily citrate dependent.

Up until this point, the activities of all animal acetyl-CoA carboxylase preparations have shown a requirement for citrate, that is, are citrate dependent (carboxylase D).

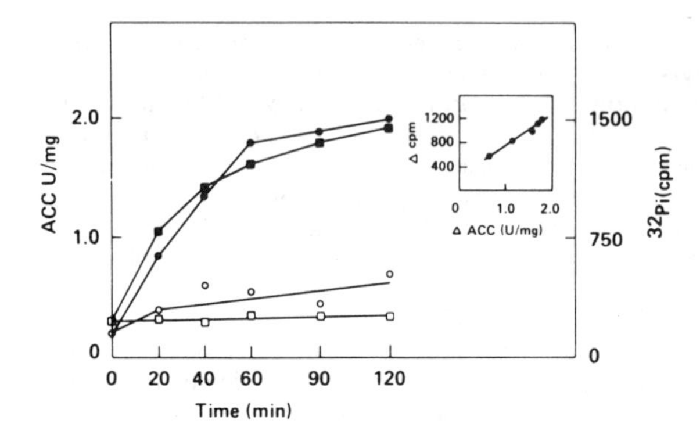

FIGURE 2. Activation and simultaneous dephosphorylation of ^{32}P-labeled acetyl-CoA carboxylase by acetyl-CoA carboxylase-activating phosphatase. The reaction mixture contained 50 mM HEPES, pH 7.5; 40 μg of ^{32}P-labeled acetyl-CoA carboxylase; 1.0 mM $MnCl_2$; 0.75 mg of BSA; and 2 mM DTT in a final volume of 1.0 ml. The mixture was incubated at 37° C in the presence (solid symbols) or absence (open symbols) of acetyl-CoA carboxylase-activating phosphatase (6 μg). At the time indicated, 5 μl was withdrawn and diluted 30-fold into acetyl-CoA carboxylase (squares). For the determination of $^{32}P_i$, 150 μl was withdrawn, added to an equal volume of 25% trichloroacetic acid, cooled over ice, and centrifuged. A sample (280 μl) was withdrawn from the supernatant fluid and counted in a liquid scintillation spectrophotometer (circles). Alternatively, the $^{32}P_i$ released was determined after extraction of the phosphomolybdate complex with 2-butanol/benzene. Both methods gave the same results. Inset: plot of the increase in $^{32}P_i$ in counts/minute versus the increase in acetyl-CoA carboxylase activity.

More recently we have succeeded in isolating an active acetyl-CoA carboxylase in the absence of citrate.[26] The enzyme was prepared rapidly from frozen powders obtained from livers that were freeze-clamped immediately ("quick" freeze-clamped) after their excision from the animal. Extraction of the enzyme from such livers and subsequent purification by avidin chromatography gave a preparation that, upon electrophoresis on sodium dodecyl sulfate gels, showed the presence of only one major band. The estimated molecular weight for the protein in this band is 250 000,[26] which is in agreement with values reported for the enzyme prepared by previous methods.[9,15]

TABLE 1. Phosphate Content and Specific Activity of Carboxylase Prepared by Various Methods

Method of Preparation	Phosphate (mol P_i/mol subunit)	Specific Activity (U/mg)	
		−Citrate	+Citrate
Freeze-clamping			
Quick	4.5 ± 0.6	2.9	3.9
Late	7.1 ± 1.0	0.5	2.6
Previous methods	8.2 ± 0.6	0.3	3.0
Dephosphorylation	5.1 ± 0.2	2.6	3.6

NOTE: The alkali-labile phosphate was determined in purified preparations of acetyl-CoA carboxylase as described elsewhere.[15] The specific activities of various preparations were taken from FIGURE 1.

The purified enzyme exhibited significant activity (2.9 U/mg) in the absence of citrate (TABLE 1), an activity similar to that of the citrate-activated rat liver acetyl-CoA carboxylase prepared by other methods (3.0 U/mg).[15] If enzyme was prepared from livers that were freeze-clamped a few minutes ("late" freeze-clamped) after their excision, however, the carboxylase activity was relatively low (0.5 U/mg), and activation then required high concentrations of citrate.

Phosphate analyses of such preparations showed that the enzyme prepared from quick freeze-clamped livers and late freeze-clamped livers contained 4.5 $\pm$ 0.8 and 7.1 $\pm$ 1.0 mol of phosphate/mol of carboxylase subunit, respectively (TABLE 1). Thus the enzyme from quick freeze-clamped liver resembles the *in vitro* dephosphorylated enzyme described above both in terms of its relatively low phosphate content and its citrate-independent activity (TABLE 1). These results demonstrate that there is a reciprocal relationship between phosphate content and catalytic activity. The correlation between phosphate content and citrate-dependent activity is less apparent because both active and inactive forms have high activity in the presence of citrate (TABLE 1). Carboxylase preparations with low phosphate content are citrate independent whereas those with high phosphate content are citrate dependent.

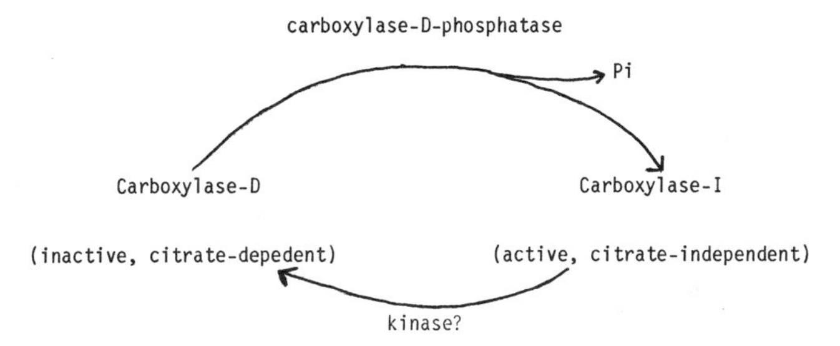

FIGURE 3. Model for the interconversion between the active and inactive forms of acetyl-CoA carboxylase.

The above studies suggest that the citrate-independent, active form of the enzyme (carboxylase I) predominates in the liver of starved and refed animals. The low activity of the enzyme prepared from late freeze-clamped livers appears to be related to its high phosphate content because dephosphorylation by carboxylase D phosphatase converts the enzyme to the active form (FIGURE 1). The enzyme prepared either by the usual methods or from late freeze-clamped livers contains about 3-4 mol of phosphate/mol of carboxylase subunit over that of the active enzyme. Dephosphorylation activates these inactive preparations, so we ascribe their initial lack of activity to phosphorylation during isolation. The phosphorylation most likely occurs after tissue excision because freeze-clamping the tissue immediately following excision arrests phosphorylation, yielding an active enzyme with lower phosphate content.

Based on these results we propose the model shown in FIGURE 3, which shows an interconversion of two forms of acetyl-CoA carboxylase, active and inactive, by phosphorylation and dephosphorylation.

According to the model, the active (I) form of acetyl-CoA carboxylase predominates in livers of starved and refed rats, where fatty acid synthesis is very active. This form may be converted to the inactive (D) form by phosphorylation. The kinase(s) that effects this change has not yet been identified. It is not known whether those kinases that phosphorylate the inactive enzyme[18,19] also phosphorylate carboxylase I, and thus convert it to carboxylase D, under controlled conditions. The inactive form, which heretofore has been the only form isolated, may very well exist *in vivo,* particularly under conditions that do not favor fatty acid synthesis. It could become reactivated either by citrate, which can override the effect of phosphorylation on catalytic activity, or by dephosphorylation. Carboxylase D phosphatase, a specific protein phosphatase, has been shown to effect this change *in vitro*.[15] The model in FIGURE 3 depicts not only the importance of both allosteric activation and covalent modification but also their interdependence in regulating the catalytic activity of acetyl-CoA carboxylase.

THE FATTY ACID SYNTHASE

Identification of Functional Domains

The synthesis of long-chain fatty acids from acetyl-CoA and malonyl-CoA involves numerous sequential reactions and acyl intermediates. The nature of these reactions and the intermediates involved became known primarily from studies of fatty acid synthesis in cell-free extracts of *Escherichia coli*.[2,3] A protein with a 4′-phosphopantetheine prosthetic group (M_r: 8847), a protein known as acyl carrier protein (ACP), was identified as the coenzyme that binds all acyl intermediates as thioester derivatives. The individual enzymes were then isolated and utilized in the reconstitution of the synthase system.[2,3] The following enzymes and reactions are involved in the synthesis of palmitate, the major product:

Acetyl transacylase

$$CH_3COS\text{-}CoA + ACP\text{-}SH \rightleftarrows CH_3COS\text{-}ACP + CoA\text{-}SH \quad (5)$$

Malonyl transacylase

$$HOOCCH_2COS\text{-}CoA + ACP\text{-}SH \rightleftarrows HOOCCH_2COS\text{-}ACP + CoA\text{-}SH \quad (6)$$

β-Ketoacyl-ACP synthase (condensing enzyme)

$$CH_3COS\text{-}ACP + \text{synthase-SH} \rightleftarrows CH_3COS\text{-synthase} + ACP\text{-}SH \quad (7a)$$

$$CH_3COS\text{-synthase} + HOOCCH_2COS\text{-}ACP \rightarrow CH_3COCH_2COS\text{-}ACP + CO_2 + \text{synthase-SH} \quad (7b)$$

β-Ketoacyl-ACP reductase

$$CH_3COCH_2COS\text{-}ACP + NADPH + H^+ \rightleftarrows \text{D-}CH_3CHOHCH_2COS\text{-}ACP + NADP^+ \quad (8)$$

β-Hydroxyacyl-ACP dehydratase

$$CH_3CHOHCH_2COS\text{-}ACP \rightleftarrows \textit{trans}\text{-}CH_3CH{=}CHCOS\text{-}ACP + H_2O \quad (9)$$

Enoyl-ACP reductase

$$CH_3CH{=}CHCOS\text{-}ACP + NADPH + H^+ \rightarrow CH_3CH_2CH_2COS\text{-}ACP + NADP^+ \quad (10)$$

Thioesterase

$$CH_3(CH_2)_{14}COS\text{-}ACP + H_2O \rightarrow CH_3(CH_2)_{14}COOH + ACP\text{-}SH \quad (11)$$

In prokaryotes and plants, the above enzymes are loosely associated with each other and can be readily separated into homogenous states by conventional procedures. In animal tissues, the enzymes are covalently linked on a single polypeptide chain (M_r: 250 000). The native synthase isolated from animal tissues consists of two such multifunctional polypeptides (M_r: 500 000). In yeast and other fungi, the enzymes distributed themselves on two different subunits: subunit α (M_r: 213 000), which contains two enzymes (the β-ketoacyl synthase and the β-ketoacyl reductase) and one active site (the 4′-phosphopantetheine site), and subunit β (M_r: 203 000), which contains the remaining five enzymes (the acyl transacylase, the enoyl reductase, the dehydratase, and the malonyl and palmitoyl transacylases). An active enzyme has a molecular weight of 2.4×10^6 or an $\alpha_6\beta_6$ complex.

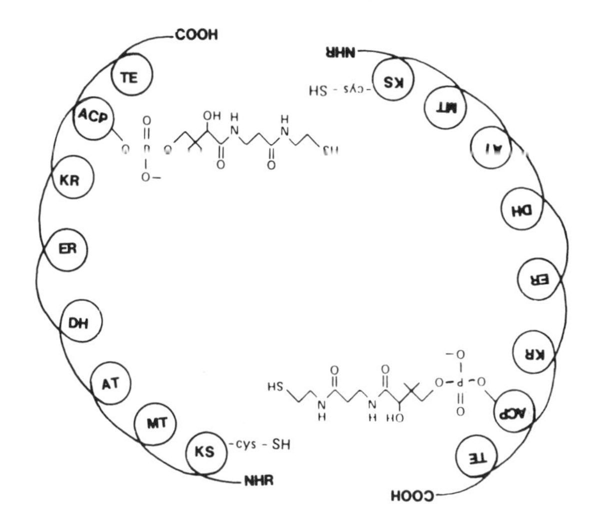

FIGURE 4. Schematic representation of the two multifunctional polypeptides and their head-to-tail association to form the enzymatically active homodimer. TE: thioesterase; ACP: acyl carrier protein; KR: ketoreductase; ER: enoyl reductase; DH: dehydratase; AT: acetyl transacylase; MT: malonyl transacylase; and KS: ketoacyl synthase.

The multifunctional nature of the fatty acid synthases of yeast and animal tissues was documented by investigations in our laboratory[27] and in the laboratory of Schweizer.[28] Before these investigations the prevailing concept was that the fatty acid synthases were complexes of noncovalently associated enzymes, but this concept proved to be untenable and has since been abandoned.

The animal fatty acid synthase was found to be a dimer of two identical subunits (FIG. 4). The identity of the subunits was based on the following observations: 1) The polypeptide subunits have the same size, charges, and shapes.[29] 2) Analyses of the 4′-phosphopantetheine content of the enzymes revealed the presence of 1.4-1.8 mol of the vitamin per mol of enzyme, or about 1 mol per subunit.[30] 3) Electron microscopic studies of negatively stained rat liver fatty acid synthase indicated that each subunit has a linear structure, 50 Å in diameter and 200 Å long, and has an estimated mass of 200 000 daltons, which is consistent with a subunit molecular mass of 250 000 daltons.[29] 4) Synthase mRNA from goose uropygial gland[31,32] and rat mammary gland[33] was isolated and characterized as a single contiguous mRNA, which can be translated in a cell-free system to a protein that has a molecular weight of 250 000, that specifically binds to the antisynthase antibodies, and that is competed for by native synthase.[33] The size of the purified mRNA is estimated to be 12-16 kilobases, which is large enough to code for the synthase subunit of 250 000 daltons, or for about 2300 amino acids.[33] 5) Finally, dissociation of the native enzyme to monomers resulted in the retention of six of the enzymic activities: the activities of acetyl transacylase, malonyl transacylase, β-ketoacyl reductase, β-hydroxyacyl dehydratase, enoyl reductase, and thioesterase.[29,34-37] The one activity absent from the monomer is the one provided by the β-ketoacyl synthase (condensing enzyme), whose active cysteine-SH is present on the subunit.[37] The presence, therefore, of eight distinct sites on the single polypeptide chain of the synthase presupposes identity of the two subunits. The requirement of the dimer for palmitate synthesis, however, was puzzling, and led to investigations into the relative location and functional organization of the partial activities along the polypeptide subunits.

There is increasing evidence that multifunctional proteins are actually arranged as a series of globular domains (constituting the sites of catalytic or regulatory activity) connected by polypeptide bridges that are more sensitive than the globular domains to proteolytic attack.[38,39] Thus it has been possible to isolate active fragments from several multifunctional proteins. All evidence indicates that the synthase is organized in a similar fashion and hence is amenable to controlled analysis by proteolytic dissection.

The proteolysis of chicken liver synthase by a variety of proteases (chymotrypsin, elastase, trypsin, *Myxobacter* protease subtilisins A and B, and kallikrein), utilized either individually or in combination, gave fragments that were analyzed with respect to both kinetics and size to establish the precursor-product relationships required for mapping.[40] Chymotrypsin showed the most restricted cleavage of the synthase by hydrolyzing its subunits into two fragments that had masses of 230 000 and 33 000 daltons, respectively (FIG. 5). The smaller fragment contained the thioesterase activity and could be readily separated from the 230 000-dalton fragment. Trypsin and elastase cleaved the synthase subunits into 230 000- and 33 000-dalton thioesterase domains, and the 230 000-dalton fragment was then centrally split into two major fragments: one fragment had a mass of 127 000 daltons and the other one had a mass of 107 000 daltons. A similar cleavage was obtained with *Myxobacter* protease. At a higher trypsin concentration, the 107 000-dalton fragment was further degraded to yield a 94 000-dalton polypeptide and a 15 000-dalton fragment. When the synthase was labeled with [^{14}C]pantethenate, the ^{14}C-labeled prosthetic group was sequentially found in 230 000-107 000-, and 15 000-dalton polypeptides, but not in the other tryptic fragments, suggesting that the ACP domain was associated with these fragments. The results of these cleavage patterns are summarized in FIGURES 6 & 7. When the [^{14}C]pantetheine-labeled synthase was cleaved with kallikrein and subtilisin, a new set of peptide fragments was obtained that contained the [^{14}C]pantetheine, as shown in FIGURE 7.

The results obtained from the individual and combined proteolytic digestions uphold the contention that the subunits of chicken liver synthase are identical, and that the polypeptides consist of domains that are linked by sequences susceptible to proteolysis. Also, a reasonably detailed map of the synthase subunit has been constructed (FIG. 7). Analyses of all the fragment patterns and summation of their masses consistently gave a value of 263 000 daltons for the mass of the intact synthase subunit. The synthase subunit can be divided into a terminal 33 000-dalton thioesterase (domain III) and a large 230 000-dalton peptide that contains all of the "core" activities of the synthase sequence (FIG. 7). α-Chymotrypsin specifically cleaves the synthase at this site and separates the thioesterase from the multifunctional complex. Other proteases attack this site, in addition to their hydrolysis of the synthase subunit at other sites (FIG. 7). The ability of α-chymotrypsin to simply cleave the synthase into two fragments made it possible to determine the H_2N-COOH orientation of the subunit polypeptide. Because the intact synthase has a blocked NH_2-terminal,[27] NH_2-terminal sequence analysis of each of the chymotryptic fragments identified which had the free

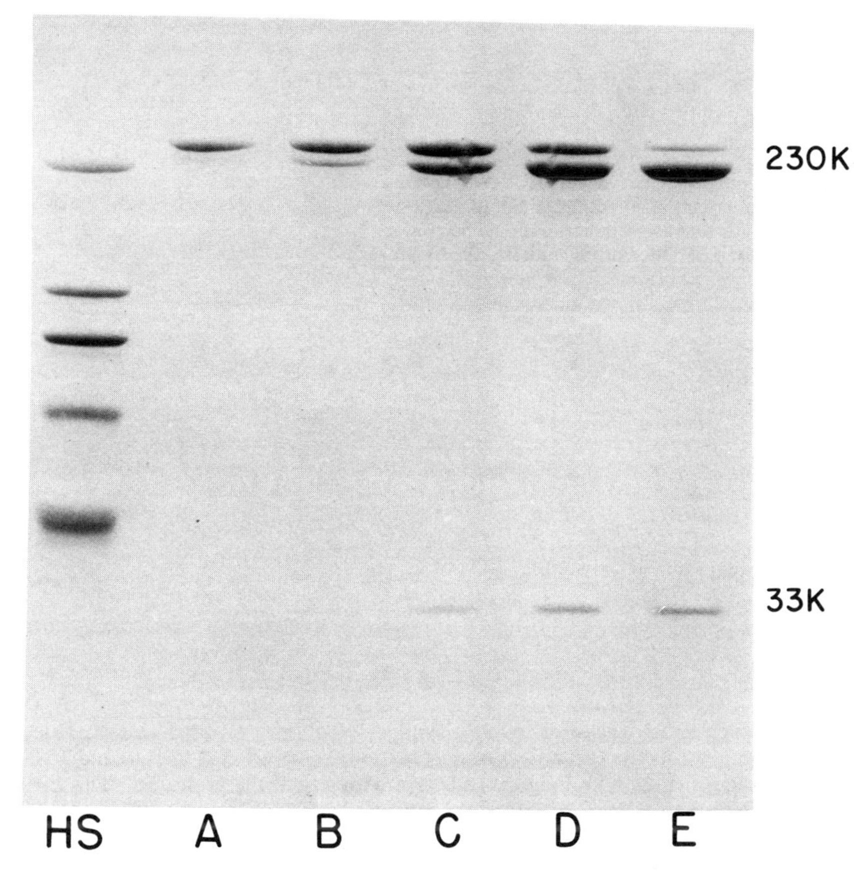

FIGURE 5. Electrophoretic display of chymotrypic digests of chicken synthase (1:1000; w:w). Lane A contains undigested synthase (2.5 μg), and lanes B to E contain hydrolysates after proteolysis for 7 min (4 μg), 15 min (7.5 μg), 30 min (10 μg), and 60 min (12.5 μg), respectively. The lane designated HS contains a mixture of high-molecular-weight standard proteins: myosin (200K), β-galactosidase (116K), phosphorylase *b* (94K), bovine serum albumin (68K), and ovalbumin (43K). The numbers suffixed by K should be multiplied by 10^3 to obtain the molecular weights.

NH_2-terminal, thereby specifying the orientation of the protein. The results showed that the thioesterase (M_r: 33 000) has a free NH_2-terminal whereas the 230 000-dalton fragment, like the intact synthase, has a blocked NH_2-terminal.[41] The sequence of the thioesterase at the NH_2-terminal is H_2N-Lys-Thr-Gly-Pro-Gly-Glu-Pro-Pro. These results place the thioesterase (domain III) at the COOH-terminal of the subunit.

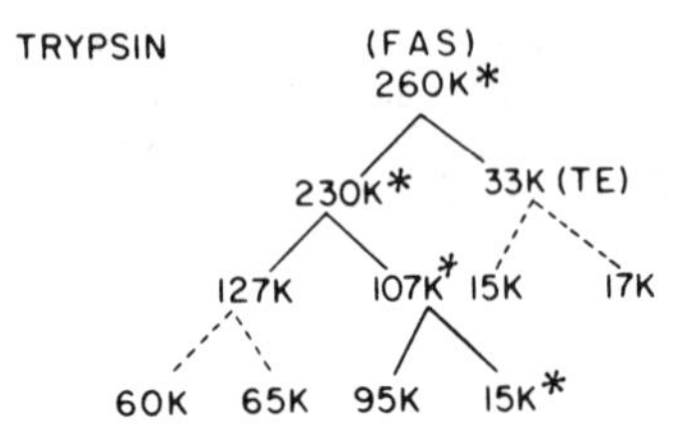

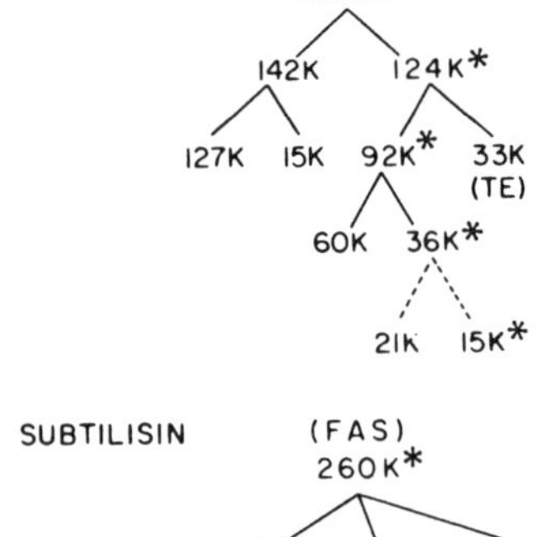

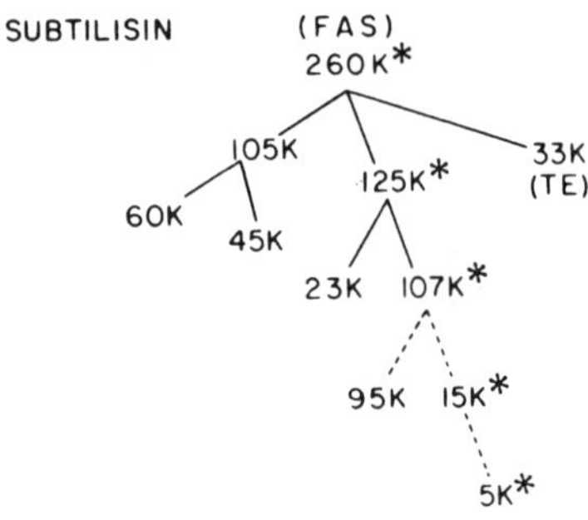

FIGURE 6. Summary of the cleavage patterns of [^{14}C]pantetheine-labeled chicken liver synthase (FAS) by proteases. Asterisks in the figure indicate the radioactive fragments that contain [^{14}C]phosphopantetheine of the ACP. Numbers are molecular weights of polypeptides, and K stands for 10^3. A continuous arrow indicates the main pathway of proteolysis; a dotted arrow indicates the hydrolysis by longer incubation. TE: thioesterase.

The thioesterase is released from the synthase complex as an intact and catalytically active fragment. It undergoes a slow degradation, however, into 18 500- and 15 000-dalton fragments upon prolonged incubation with any of the proteases.[41] This breakdown, which is accompanied by loss of catalytic activity, is evident at later stages in the cleavage patterns.

As previously stated, the 230 000-dalton core region can be segregated into two principal domains, designated I (M_r: 127 000) and II (M_r: 107 000). Domain II is located adjacent to the thioesterase (domain III) and contains the primary kallikrein site, a 92 000-dalton peptide from the thioesterase junction (see below). Domain II also contains a secondary tryptic site, which leads to the release of a terminal 15 000-dalton segment identified as the ACP (FIG. 6).

The recognition that the thioesterase is located at the COOH-terminal of the synthase polypeptide made it possible to establish the mapping of the functional centers on the synthase subunit. In these studies,[42,43] whenever possible, the known properties of the active sites of the component activities of the synthase were employed. For instance, the ACP site of the synthase was labeled with radioactive pantetheine, which was then followed throughout the course of proteolysis (FIG. 6). Similar approaches were followed using assays of the catalytic activities, labeled substrates, specific inhibitors, or antibodies, either monoclonal or polyclonal, developed against homogenous components or domains. For instance, antithioesterase antibodies were found to completely inhibit synthase activity and to bind both intact synthase and isolated thioesterase. Using the Western transfer technique and visualizing the antibody-binding sites with ^{125}I-labeled protein A, it became apparent that all of the 33 000-dalton fragments produced by the different proteases (chymotrypsin, trypsin, elastase, and subtilisin) were related, if not identical, and therefore represented the thioesterase moiety. A closer examination of the results showed that the antithioesterase antibody reacts only with the 33 000-dalton peptides, their breakdown products, and the intermediate fragments of the synthase that were predicted to include the 33 000-dalton domain on the basis of proteolytic mapping (FIG. 7).

The results of these types of studies were incorporated together into a proposed two-dimensional diagram of the synthase polypeptide (FIG. 8) (see references 46-48 for more details). This diagram illustrates the relative sizes of the domains and the associated partial activities. Two subunits are drawn in a head-to-tail arrangement so that two sites of palmitate synthesis are constructed (functional division). As can be seen from this model, the β-ketoacyl synthase (KS) and the ACP sites are located in two separated domains of the synthase subunit and are far removed from each other. Domain I contains the acetyl and malonyl transacylases as well as the condensing enzyme site, thus making this domain the substrate-entry and chain-elongation domain. Domain II contains the β-ketoacyl reductase, the dehydratase, and the enoyl reductase partial activities; thus this domain functions as the processing domain for the reduction

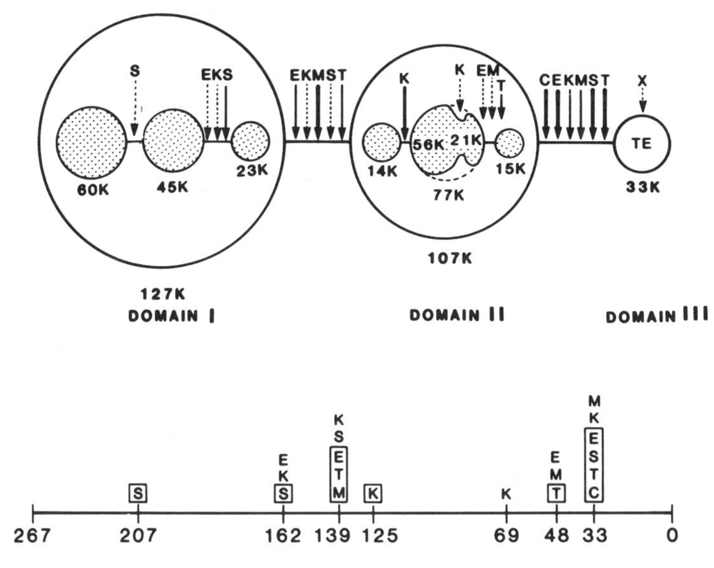

FIGURE 7. Proteolytic map of the chicken synthase. In the upper figure, thick arrows indicate primary cleavage sites by individual proteases; thin arrows, secondary cleavage sites; and dashed arrows, substantial cleavage sites at longer time courses. Abbreviations for proteases are as follows: T: trypsin; E: elastase; M: *Myxobacter* protease; S: subtilisin A or B; C: α-chymotrypsin; K: kallikrein; X: all of the above proteases. The molecular weights of the fragments are as indicated, and K stands for 10^3. The lower figure shows the actual distances (in daltons $\times$ 10^{-3}) of the protease cleavage sites from the thioesterase terminal of the monomer. These distances were derived by the summation of the molecular weights of individual peptides released by proteolysis. Proteases in boxes are capable of inflicting primary cleavage at the sites indicated.

of the carbonyl carbon to the methylene analogue by NADPH. The ACP and its 4′-phosphopantetheine arm is located next to the reduction domain (domain II) and connects it to the thioesterase of the chain-termination or palmitate-release domain (domain III).

Subunit Arrangement and Mechanism of Action

Palmitate synthesis is a cyclical process that requires an orderly involvement of seven different enzyme domains, five of which participate sequentially eight times

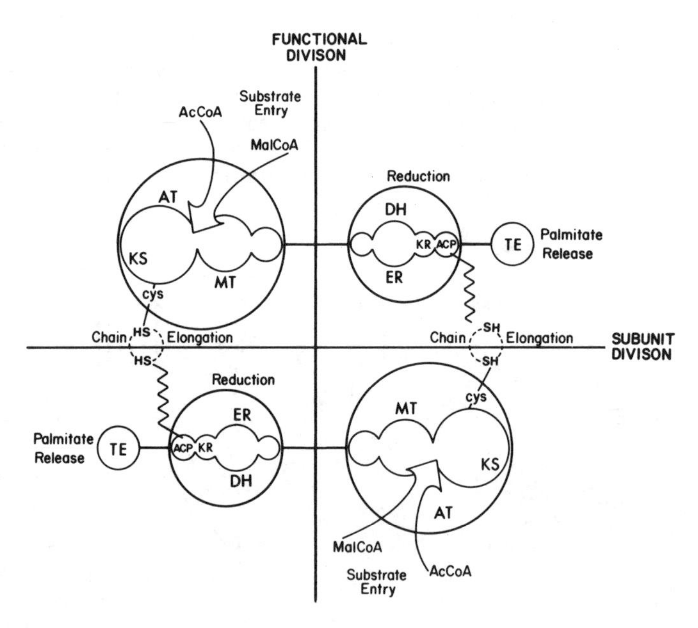

FIGURE 8. Functional map of chicken fatty acid synthase. Two subunits are drawn in a head-to-tail arrangement (subunit division) so that two sites for palmitate synthesis are constructed (functional division). Abbreviations are same as in FIGURE 4. The wavy line represents the 4′-phosphopantetheine prosthetic group.

during the synthesis of one molecule of palmitate. Thus, the synthesis of palmitate from acetyl-CoA and malonyl-CoA involves at least 30 acyl intermediates that are covalently bound to the enzyme. Thus the organization of the component enzymes on the polypeptide subunits and the special arrangement of the two subunits in the native animal synthase become important in the understanding of the mechanism of action of this complex enzyme.

Initially, the acetyl and malonyl transacylases prime the synthase with the carbon atoms required for fatty acid synthesis (reactions 5 and 6). The malonyl transacylase is specific for the malonyl group. The acetyl transacylase, however, may be active on other acyl groups with short chains (proprionyl and butyryl groups) in addition to being active on its usual substrate, acetyl-CoA. All transacylases have an active serine residue where the acyl groups are bound as an O-ester linkage. The acetyl group is transferred from the seryl residue to the active cysteine-SH of the β-ketoacyl synthase. The malonyl group, on the other hand, is transferred from the seryl residue of the malonyl transacylase to the cysteamine-SH of the ACP moiety where it forms a thioester ready for condensation with an acetyl group to form a β-ketoacyl derivative.

The coupling of the acyl and malonyl groups to form β-ketoacyl derivatives is catalyzed by the β-ketoacyl synthase component of the synthase system. Information concerning this reaction was obtained initially from studies of the condensation of acetyl-ACP and malonyl-ACP (reactions 7A and 7B) by the *E. coli* enzyme.[47,48] In these studies it was shown that the β-ketoacyl synthase contains an active cysteine-SH, which forms an acyl-S-enzyme intermediate prior to its coupling with the malonyl-S-ACP to yield β-ketoacyl-ACP and CO_2. This same general mechanism was shown to be operative in the animal synthase.[37] An active cysteine-SH was identified, which on alkylation with iodoacetamide caused inhibition of the condensing activity only. Acetyl-CoA but not malonyl-CoA protected the enzyme against inhibition, suggesting the formation of an acetyl-S-enzyme.

Available evidence indicates that in the eukaryotic synthases the acetyl and other acyl groups form primarily the acyl derivative of the cysteine-SH of the condensing enzyme, and the malonyl group forms the acyl derivative of the pantetheine-SH prior to their condensation to form the β-ketoacyl derivative. We studied the structure and mechanism of the β-ketoacyl synthase of the animal enzyme[37,44,45] and identified the residues to which the acetyl and malonyl groups are bound at the β-ketoacyl synthase site.

Our studies showed that the dimer is the active form of the animal synthase.[29] This conclusion was based on the values of the sedimentation coefficients ($s_{20,w}$:15.0-16.5 S) measured in the presence and then in the absence of substrates using the technique of active enzyme centrifugation. These values were shown by sedimentation equilibrium experiments to correspond to the dimer form of the enzyme. The reasons for the requirement of the dimer form of the enzyme became evident from our studies of the role of active thiols in fatty acid synthesis.[38,45] The synthase of chicken liver, for instance, was completely inhibited in the presence of 0.5 mM iodoacetamide in a pseudo-first-order process. When iodo[^{14}C]acetamide was reacted with the synthase, over 80% of the ^{14}C label was recovered as ^{14}C-labeled S-carboxymethylcysteine, but none as ^{14}C-labeled S-carboxymethylcysteamine after HCl hydrolysis, indicating that the inhibition of the enzyme was due to the alkylation of the active cysteine-SH, but not the pantetheine-SH.[37] The cysteine residue was identified as an essential component of the β-ketoacyl synthase site because this was the only partial activity lost. Preincubation of the synthase with acetyl-CoA protects the enzyme from inhibition by iodoacetamide, suggesting that this thiol is the site of binding of the acetyl group to the β-ketoacyl synthase site. Preincubation of the enzyme with malonyl-CoA prior to treatment with iodoacetamide, however, does not protect the synthase against iodoacetamide inhibition, suggesting that the site of binding of the malonyl group is not the cysteine.

In contrast to the slow inhibition of the chicken synthase by iodoacetamide, the bifunctional reagent 1,3-dibromo-2-propanone inhibits the enzyme rapidly (within 30 sec) and completely.[37] The loss of synthase activity is due to inhibition of the β-ketoacyl synthase activity only. Preincubation of the synthase with acetyl-CoA protects the enzyme against inhibition by dibromopropanone whereas malonyl-CoA does not. These results are similar to those found for iodoacetamide inhibition and clearly show that, like iodoacetamide, the dibromopropanone competes with acetyl-CoA for the same thiol in the β-ketoacyl synthase site.

When the dibromopropanone-inhibited synthase was analyzed by sodium dodecyl sulfate-polyacrylamide gel electrophoresis, the 250 000-dalton synthase subunit was nearly absent, with a concomitant appearance of 500 000-dalton oligomers. These observations suggested that the synthase subunits were cross-linked by the bifunctional reagent dibromopropanone. Preincubation of the synthase with acetyl-CoA or malonyl-CoA prevented the cross-linking. The stoichiometry of inhibition was determined by binding studies, which indicated that the binding of about 1.8 mol of dibromopropanone per mole of enzyme was required for complete inactivation of the synthase.[37] Altogether, these results indicated that the dibromopropanone reacts as a bifunctional reagent, cross-linking the two subunits that constitute the enzymically active synthase dimer.

When [^{14}C]dibromopropanone was used as the cross-linking reagent, the cross-linked oligomers separated by denaturing gel electrophoresis contained over 85% of the protein-bound radioactivity. Hydrolysis of the ^{14}C-labeled oligomers with HCl after Baeyer-Villiger oxidation with performic acid yielded ^{14}C-labeled sulfones of S-carboxymethylcysteine and S-carboxymethylcysteamine sulfones in equal amounts.[37] These results indicated that the dibromopropanone cross-links the two synthase subunits by reacting with a cysteine-SH of one subunit and the cysteamine-SH of the adjacent subunit (FIG. 4).

The formation of vicinal sulfhydryl groups, which resulted from the cysteine residue of one subunit being juxtapositioned with the pantetheine residue of the adjacent subunit, suggested that the Ellman's reagent, 5,5′-dithiobis(2-nitrobenzoic acid) (DTNB), would oxidize the two SH groups and thus form the mixed disulfide. Indeed, this reagent at a concentration of 10^{-5} M rapidly inhibited the synthase and resulted in cross-linking of the two subunits as shown by electrophoresis in denaturing polyacrylamide gels. Acetyl-CoA, but not malonyl-CoA, prevented this inhibition; however, both acyl groups prevented cross-linking. Inhibition by DTNB resulted in inactivation of the β-ketoacyl synthase reaction only: other partial reactions were not affected. Analyses of the DTNB reaction with the synthase indicated that 2 mol of DTNB per mol of enzyme was required for complete inactivation of the enzyme.[45] DTNB inhibition of synthase can be reversed by mercaptoethanol, indicating that the cross-linking of the synthase subunits is via disulfide bridges between the cysteine-SH and the cysteamine-SH (FIG. 4). The results of these studies are consistent with those obtained with dibromopropanone and further emphasize the close proximity of the SH groups of the two residues. The fact that the DTNB cross-linking reaction occurs places these SH groups of the two subunits within bonding distance, which is about 2 Å. The requirement for the vicinal arrangement of the pantetheine and cysteine residues on separate subunits for the β-ketoacyl synthase reaction explains, therefore, the loss of this partial activity and the synthase activity on dissociation of the homodimer.

In conclusion, since the subunits of the animal synthase are identical (with each subunit containing the same catalytic domains, including the active cysteine-SH of the β-ketoacyl synthase and 4′-phosphopantetheine-SH of the ACP domain), it was proposed from the cross-linking studies with dibromopropanone or Ellman's reagent that in the dimer state the two subunits are arranged in head-to-tail fashion,[37] as shown in FIGURES 4 & 8. The head-to-tail arrangement of the two subunits of the synthase of animal tissues suggests the presence of two centers for β-ketoacyl synthase,[37] and therefore two centers for palmitate synthesis; this is consistent with the stoichiometry of binding of dibromopropanone and DTNB. Studies using the core complex of the 230,000-dalton peptide allowed the estimation of the stoichiometry of NADPH oxidation and fatty acids synthesized relative to the pantetheine content of the 230 000-dalton core dimer. The results show that in the absence of thioesterase

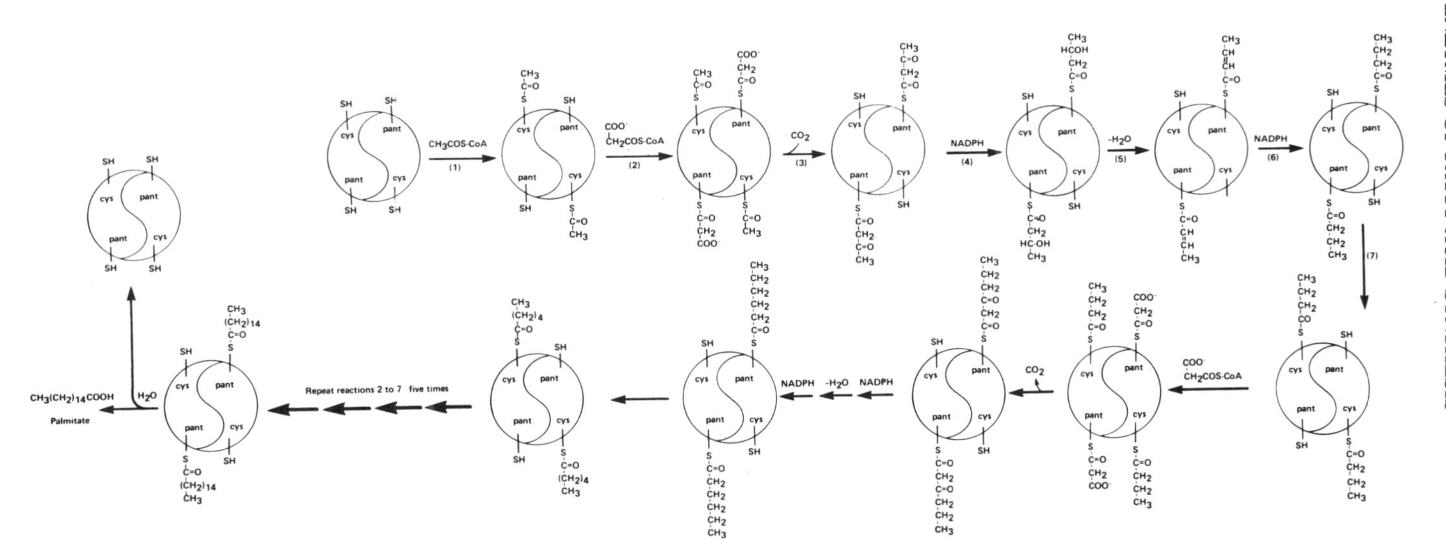

FIGURE 9. Proposed mechanism for palmitate synthesis. The circles represent the multifunctional subunits of the synthase in its dimer form. The cys-SH represents the active cysteine thiol of the β-ketoacyl synthase, and the pant-SH represents the pantetheine thiol of the ACP. Other catalytic domains are not shown and are presumed to be present in both subunits (see reactions 5-11 and FIG. 8).

the core enzyme continues the chain elongation and reduction processes until fatty acids of C_{20} and C_{22} are synthesized as a limit.[46] Little or no palmitate or stearate, normally synthesized, is formed under these conditions, indicating that these fatty acyl groups are still attached to the pantetheine-SH and are further elongated to the C_{20} and C_{22} acids. The chain-terminating process in the native synthase, therefore, is dictated by the thioesterase, which has the highest activity for palmitoyl and stearoyl thioesters. In any case, quantitation of the fatty acids bound to the enzyme and the stoichiometry of the NADPH oxidation show that 1 mol of long-chain fatty acids is synthesized per mole of pantetheine associated with the core dimer; therefore, the two centers for palmitate synthesis are active simultaneously.[46]

Our studies suggest that two centers may function independently of each other and engage catalytic domains on the two subunits. In this arrangement (FIG. 8), each center has the entire complement of enzymes. Based on these results, the following mechanism for palmitate synthesis (FIG. 9) was proposed:[46] The active fatty acid synthase in the dimer form interacts with the substrates acetyl-CoA and malonyl-CoA. The active cysteine-SH of the condensing site forms a thioester linkage with the acetyl group, and the cysteamine-SH of the acyl carrier site forms a similar linkage with the malonyl group via their respective transacylases (steps 1 and 2, FIG. 9). The acetyl group of one subunit is coupled to the β-carbon of the malonyl group of the second subunit with a simultaneous release of CO_2 and the formation of acetoacetyl product. The cysteine-SH of the condensing enzyme is reset in the free thiol form. The acetoacetyl-S-pantetheine derivative is then reduced as outlined in FIG. 9. The butyryl group is then transferred to the cysteine-SH of the condensing enzyme, freeing the cysteamine-SH to accept a malonyl group from malonyl-CoA. The butyryl and malonyl groups condense to form a β-ketohexanoyl derivative, which is then processed as outlined in FIGURE 9, ultimately yielding palmitic acid. The essence of this mechanism is 1) the involvement of the two subunits in the condensation reaction, where the acyl group "seesaws" between the cysteine-SH and cysteamine-SH of the two subunits with each cycle adding C_2 units, and 2) the occurrence of the two centers for the synthesis of palmitate within each active synthase dimer, where each has its own complement of enzymes and perhaps functions independently of the other. The multifunctional nature of the subunit and its organization and structural arrangement into the dimer form produces a highly efficient enzyme complex capable of carrying out sequentially and repetitively a total of 37 reactions in the synthesis of a molecule of palmitate from acetyl-CoA and malonyl-CoA.

REFERENCES

1. WAKIL, S. J. 1958. J. Am. Chem. Soc. **80:** 6465.
2. WAKIL, S. J. 1970. *In* Lipid Metabolism. S. J. Wakil, Ed.: 1-48. Academic Press. New York, NY.
3. VOLPE, J. J. & P. R. VAGELOS. 1977. Physiol. Rev. **56:** 339-417.
4. LANE, M. D., J. MOSS & E. POLAKIS. 1974. Curr. Top. Cell. Regul. **8:** 129-195.
5. GUCHHAIT, R. B., E. E. ZWERGEL & M. D. LANE. 1974. J. Biol. Chem. **249:** 4776-4780.
6. INOUE, H. D. & J. M. LOWENSTEIN. 1972. J. Biol. Chem. **247:** 4825-4832.
7. BEATY, N. B. & M. D. LANE. 1982. J. Biol. Chem. **257:** 924-929.
8. WADA, K. & T. TANABE. 1983. Eur. J. Biochem. **135:** 17-23.
9. WITTERS, L. A. & B. VOGT. 1981. J. Lipid Res. **22:** 364-369.
10. HARDIE, D. G. & P. COHEN. 1978. FEBS Lett. **91:** 1-7.
11. LOWENSTEIN, J. M. 1968. Biochem. Soc. Symp. **27:** 61-86.

12. KIM, K.-H. 1983. Curr. Top. Cell. Regul. **22:** 143-176.
13. BEATY, N. B. & M. D. LANE. 1983. J. Biol. Chem. **258:** 13043-13050.
14. BEATY, N. B. & M. D. LANE. 1983. J. Biol. Chem. **258:** 13051-13055.
15. THAMPY, K. G. & S. J. WAKIL. 1985. J. Biol. Chem. **260:** 6318-6323.
16. SIESS, E. A., D. G. BROCKS & O. H. WIELAND. 1978. Hoppe-Seyler's Z. Physiol. Chem. **359:** 785-798.
17. CARLSON, C. A. & K.-H. KIM. 1973. J. Biol. Chem. **248:** 378-380.
18. LENT, B. & K.-H. KIM. 1982. J. Biol. Chem. **257:** 1897-1901.
19. TIPPER, J. P. & L. A. WITTERS. 1982. Biochim. Biophys. Acta **715:** 162-169.
20. KRAKOWER, G. R. & K.-H. KIM. 1981. J. Biol. Chem. **256:** 2408-2413.
21. HARDIE, D. G. & P. COHEN. 1979. FEBS Lett. **103:** 333-338.
22. TIPPER, J. P., G. W. BACON & L. A. WITTERS. 1983. Arch. Biochem. Biophys. **227:** 386-396.
23. PEKALA, P. H., M. J. MEREDITH, D. M. TARLOW & M. D. LANE. 1978. J. Biol. Chem. **253:** 5267-5269.
24. SONG, C. S. & K.-H. HIM. 1981. J. Biol. Chem. **256:** 7786-7788.
25. AHMAD, F. & P. M. AHMAD. 1981. Methods Enzymol. **71:** 16-26.
26. THAMPY, K. G. & S. J. WAKIL. 1986. J. Biol. Chem. In press.
27. STOOPS, J. K., M. J. ARSLANIAN, Y.-H. OH, K. C. AUNE, T. C. VANAMAN & S. J. WAKIL. 1975. Proc. Natl. Acad. Sci. USA **72:** 1940-1944.
28. SCHWEIZER, E., G. DIETLEIN, G. GIMMLER, A. KNOBLING, H. W. TAHEDL & M. SCHWEIZER. 1975. FEBS Proc. Meet. **40:** 85.
29. STOOPS, J. K., P. R. ROSS, M. J. ARSLANIAN, K. C. AUNE, S. J. WAKIL & R. M. OLIVER. 1979. J. Biol. Chem. **254:** 7418-7426.
30. ARSLANIAN, M. J., J. K. STOOPS, Y.-H. OH & S. J. WAKIL. 1976. J. Biol. Chem. **251:** 3194-3196.
31. ZEHNER, Z. E., J. S. MATTICK, R. STUART & S. J. WAKIL. 1980. J. Biol. Chem. **255:** 9519-9522.
32. MORRIS, S. M., JR., J. H. NILSON, R. A. JENIK, L. K. WINBERRY, M. A. MCDEVITT & A. G. GOODRIDGE. 1982. J. Biol. Chem. **257:** 3225-3229.
33. MATTICK, J. S., Z. E. ZEHNER, M. A. CALABRO & S. J. WAKIL. 1981. Eur. J. Biochem. **114:** 643-651.
34. BUTTERWORTH, P. H. W., P. C. YANG, R. M. BOCK & J. W. PORTER. 1967. J. Biol. Chem. **232:** 3508-3516.
35. MUESING, R. A., F. A. LORNITZO, S. KUMAR & J. W. PORTER. 1975. J. Biol. Chem. **250:** 1814-1823.
36. YUNG, S. & R. Y. HSU. 1972. J. Biol. Chem. **247:** 2689-2698.
37. STOOPS, J. K. & S. J. WAKIL. 1981. J. Biol. Chem. **256:** 5128-5133.
38. KIRSCHNER, K. & H. BISSWANGER. 1976. Annu. Rev. Biochem. **45:** 143-166.
39. WETLAUTER, D. B. 1973. Proc. Natl. Acad. Sci. USA **70:** 697-701.
40. MATTICK, J. W., Y. TSUKAMOTO, J. NICKLESS & S. J. WAKIL. 1983. J. Biol. Chem. **258:** 15291-15299.
41. MATTICK, J. S., J. NICKLESS, M. MIZUGAKI, C. Y. YANG, S. UCHIYAMA & S. J. WAKIL. 1983. J. Biol. Chem. **258:** 15300-15304.
42. TSUKAMOTO, Y., J. S. MATTICK, H. WONG & S. J. WAKIL. 1983. J. Biol. Chem. **258:** 15312-15322.
43. WONG, H., J. S. MATTICK & S. J. WAKIL. 1983. J. Biol. Chem. **258:** 15305-15311
44. STOOPS, J. K. & S. J. WAKIL. 1982. J. Biol. Chem. **257:** 3230-3235.
45. STOOPS, J. K. & S. J. WAKIL. 1982. Biochem. Biophys. Res. Commun. **104:** 1018-1024.
46. SINGH, N., S. J. WAKIL & J. K. STOOPS. 1984. J. Biol. Chem. **259:** 3605-3611.
47. TOOMEY, R. E. & S. J. WAKIL. 1966. J. Biol. Chem. **241:** 1159-1165.
48. D'AGNOLO, G., I. S. ROSENFELD & P. R. VAGELOS. 1975. J. Biol. Chem. **250:** 5283-5288.

Structural Relationships in Glycogen Phosphorylases[a]

R. J. FLETTERICK, J. A. BURKE, P. K. HWANG,
K. NAKANO, AND C. B. NEWGARD

*Department of Biochemistry and Biophysics
School of Medicine
University of California
San Francisco, California 94143*

INTRODUCTION

The energy that cells require for activity and growth is maintained in part by regulated synthesis and breakdown of stored saccharide. The most prevalent form of stored saccharide is glycogen, which is found in bacterial, plant, and animal cells. The breakdown of glycogen to form glucose 1-phosphate in response to energy demands of the cell is carried out by glycogen phosphorylase. This enzyme has been carefully studied for more than 40 years since its discovery by Carl Cori. The enzyme from skeletal muscle tissue is the best characterized; amino acid sequence determination and three-dimensional crystallography have helped to define this complex molecular structure at near-atomic resolution.[1–3]

The wealth of structural information for (rabbit) muscle phosphorylase has been invaluable in interpreting kinetic studies, *in vitro* and *in vivo,* of the regulation of this enzyme. The enzyme functions as a dimer with five allosteric regulator sites. The biochemical and physical characteristics of the rabbit muscle enzyme are shown in TABLE 1. TABLE 2 lists the allosteric effectors and provides descriptions of their binding sites on the phosphorylase dimer. The phosphorylated enzyme, phosphorylase *a,* is active without additional effectors but can be inhibited by glucose or purines in a synergistic fashion. The phosphorylase *b* molecule is dependent on AMP for its activation, though it can be weakly activated by larger concentrations of inorganic phosphate or glucose 1-phosphate. The role of the pyridoxal phosphate is to participate directly in the catalytic phosphorolysis at the active site. The coenzyme is buried inside the COOH-terminal domain of the glycogen phosphorylase monomer with only its phosphoryl group penetrating the active site. The coenzyme group interacts with the phosphoryl substrate, in a manner yet undetermined, in order to carry out the phosphorolysis reaction.[4,5]

Phosphorylase activity can be altered by ligand binding, as shown in TABLE 2, or by covalent interconversion through kinase and phosphatase enzymes. The process of interconversion remains coupled to the regulation of activity by the allosteric effectors. Thus, for example, phosphorylase *a,* which is normally active, can be inhibited by

[a]This research was supported by Grant R01 AM32822 from the National Institutes of Health and a postdoctoral grant to C. Newgard from the Juvenile Diabetes Foundation.

purines or glucose. The linkage between covalent interconversion and allosteric regulation is important physiologically because inactivation of phosphorylase *a* by glucose or purines leads to its dephosphorylation by phosphorylase phosphatase. This occurs by roughly the following mechanism: Tight binding exists between glycogen phosphorylase *a* in its active conformation and the phosphatase, causing the inhibition of the phosphatase. The inhibition is then released through binding of glucose, which induces a conformational change, and renders phosphorylase *a* a substrate for the phosphatase. Phosphorylase *b,* in contrast, does not bind the phosphatase very tightly. In addition, there are interactions among the allosteric effectors that are of physiological significance. For example, AMP can overcome glucose inhibition if the AMP: glucose ratio is high enough. Analogously, glucose and purines interact cooperatively and can overcome AMP activation. The convoluted and interacting controls on glycogen phosphorylase attest to the importance of its regulation inside the cell.

Different organisms and different cell types have altered requirements for regulation through allosteric effectors and covalent interconversion. In the potato, for example, phosphorylase is not a phosphorylated molecule, shows none of the allosteric effector sites, and has no binding site for glycogen.[6] In *Escherichia coli,* the enzyme is also unregulated. The primary sequences of the potato[6] and bacterial enzymes[7] have been previously determined and compared with the muscle isozyme. The use of recombinant DNA methods to obtain sequence information for other phosphorylase isozymes, such as the yeast and liver forms, allows the study of the evolution and de-evolution of the regulatory sites on glycogen phosphorylase.

TABLE 1. Biochemical and Physical Properties of Glycogen Phosphorylase from Rabbit Muscle

Active species	Dimer
M_r per subunit	97 500
Number of amino acids	842
Covalent interconversion	Ser 14 (phosphorylation)
Coenzyme	Pyridoxal phosphate (at Lys 680)
Reaction catalyzed	$(\text{Glucose})_n + P_i \rightarrow (\text{glucose})_{n-1} +$ glucose 1-phosphate

MATERIALS AND METHODS

A previously isolated rabbit muscle phosphorylase cDNA fragment[8] was used to probe cDNA libraries prepared from rabbit muscle, human muscle, and human liver. The rabbit muscle cDNA library was provided by David MacLennan; the human muscle library, by Peter Gunning; and the human liver cDNA library, by Savio Woo. Clones were identified by colony or plaque hybridization, restriction mapped, and subsequently sequenced by the dideoxy method. The yeast genomic DNA for glycogen phosphorylase was isolated from a library constructed by Rose *et al.*[9] An oligonucleotide corresponding to the published amino acid sequence of the pyridoxal phosphate-binding site was used to probe this library. A 14-kb yeast genomic DNA fragment was isolated and restriction mapped; appropriate restriction fragments were sequenced by the dideoxy method. The gene for muscle glycogen phosphorylase was isolated from a cosmid library provided by Roger Lebo.

RESULTS

Muscle Phosphorylase cDNAs

The nucleotide sequences for human, rat, and rabbit muscle phosphorylases were shown to be over 90% homologous.[8] There are relatively few sequence variations between the human and rabbit enzymes, and none of them significantly affect the tertiary structure as judged from a visual inspection of computer graphics models. We have completed sequencing a full-length cDNA clone for rabbit muscle glycogen phosphorylase.[10] The cDNA sequence begins 70 bases upstream of the initiation AUG codon and continues through to the poly(A) site. At eight positions in the amino acid sequence, the cDNA sequence is different from the published protein sequence.[11] These differences have been resolved in favor of the cDNA sequence (K. Walsh, personal communication) and are confined to the 100 amino acids at the NH_2-terminal end.

TABLE 2. Regulators of Muscle Phosphorylase Activity

Ligand	Binding Site	Effects
Ser 14-P_i	Between subunits	Opens active site, breaks down glycogen
P_i, Glucose 1-phosphate	Active site, between NH_2- and COOH-terminal domains	Activates weakly
Glycogen	Storage site, NH_2-terminal	Opens active site
ATP	AMP site	Blocks AMP activation of phosphorylase *b*
Glucose 6-phosphate	AMP site	Blocks AMP activation of phosphorylase *b*
Glucose	Active site	Inhibits enzyme, disorders NH_2-terminal
Purine	Inhibitor site, near active site	Inhibits enzyme, disorders NH_2-terminal

The comparison of the 3′ untranslated regions of the cDNAs from rat, rabbit, and human muscles reveals a conserved 11-nucleotide segment.[8] The conservation of this segment in an otherwise divergent region of DNA suggests that it may have a regulatory role.

The Human Muscle Phosphorylase Gene

A single gene for the human muscle isozyme of phosphorylase has been identified and is being characterized in our laboratory. FIGURE 1 shows the partial physical map of this gene. The full gene is made up of at least 11 exons and 10 introns. Other introns are anticipated in the unsequenced regions. Surprisingly, most introns are relatively short: none are larger than 2 kb. The intron positions of the human gene correspond to positions on the surface of the three-dimensional structure of the rabbit

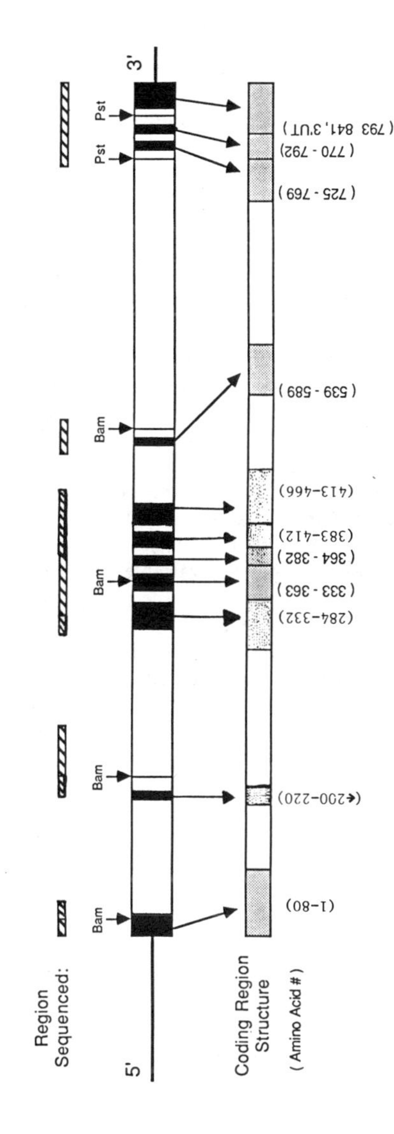

FIGURE 1. The human muscle glycogen phosphorylase gene. The short arrows in the figure show where the restriction sites are, and the bars marked with diagonal lines extend over the regions of the gene that have been sequenced. The darkened regions on the bar for the gene represent the exon sequences, and the long arrows, which point to the shaded regions on the lower bar, show the amino acid sequences of glycogen phosphorylase that correspond to the exon sequences. The vertical lines on the lower bar correspond to the intron sequences of the gene.

muscle phosphorylase monomer (data not shown). This is in agreement with data from numerous other genes for which corresponding three-dimensional protein structures are known.[12] The gene has been mapped by dot blot hybridization to human chromosome 11 (q13q TER).[13] Attempts to find restriction-fragment-length polymorphisms using cloned DNA probes from the characterized region of the glycogen phosphorylase gene have so far proved unsuccessful. It is hoped that the further characterization of the human gene will provide probes that can be effectively used to study the dysfunction in McArdle's disease, a relatively rare inherited disorder of glycogen metabolism in which muscle glycogen phosphorylase is either not functioning or not present.

Human Liver Phosphorylase cDNA

Several clones corresponding to human liver glycogen phosphorylase were identified from a cDNA library. Attempts to isolate clones using COOH-terminal cDNA from rabbit or human muscle proved to be unsuccessful. It was necessary to use a probe that corresponded to the highly conserved coenzyme-binding site in order to produce sufficiently strong hybridization for liver phosphorylase cDNA identification. The longest of the isolated liver cDNAs was shown to begin at amino acid 70 and to continue through the 3′ untranslated region to the poly(A) site. This clone has been fully sequenced by the dideoxy method. Comparison of the 3′ untranslated regions of the liver and muscle cDNAs revealed no homology. The sequence homology at the DNA level in the coding region is somewhat variable with the overall homology at 70%. The two isozymes are 78% homologous at the protein level.

Most of the changes in amino acid side chains between the muscle and liver isozymes occur at positions on the surface of the phosphorylase molecule, as shown in FIGURE 2. These changes would not be expected to affect activity or regulatory properties in general. The region around the active site of the molecule is remarkably conserved (FIG. 3). It is especially interesting that the three residues involved in purine's allosteric binding—Asn 284, Phe 285, and Tyr 613—are conserved between muscle and liver phosphorylases. This implies that the kinetic and structural analyses carried out on the muscle enzyme's purine site are applicable to the liver enzyme. The observed discrepancy in ligand specificity at this site between rabbit liver and rabbit muscle enzymes[1] is therefore unexplained. Rabbit liver enzyme could have a sequence different from that of the human liver enzyme, and structural changes at the purine site of the liver enzyme could be caused by amino acid substitutions, which could occur at locations far from the purine site. For example, substitutions of hydrophobic amino acids that form part of the core of the protein could cause these structural changes. Examples of buried positions that may be occupied by different residues in the rabbit muscle and liver enzymes are as follows: positions 561 and 562 are occupied by Leu and Phe in muscle but by Met and Val in human liver (FIG. 3), and position 765 is occupied by Met in muscle but by Phe in liver (data not shown).

Amino acid substitutions that are more likely to directly influence binding of the effector ligand to the phosphorylase molecule are found at the glycogen-storage site and at the AMP-binding site. Some of the changes involve residues that are not directly involved in binding the effector but are of possible importance in allosteric transmission. Substitutions of this type may explain why liver phosphorylase exhibits weaker activation on binding AMP than its muscle counterpart. Alternatively, the observed functional changes might be due to damaged enzyme. Work from our laboratory has

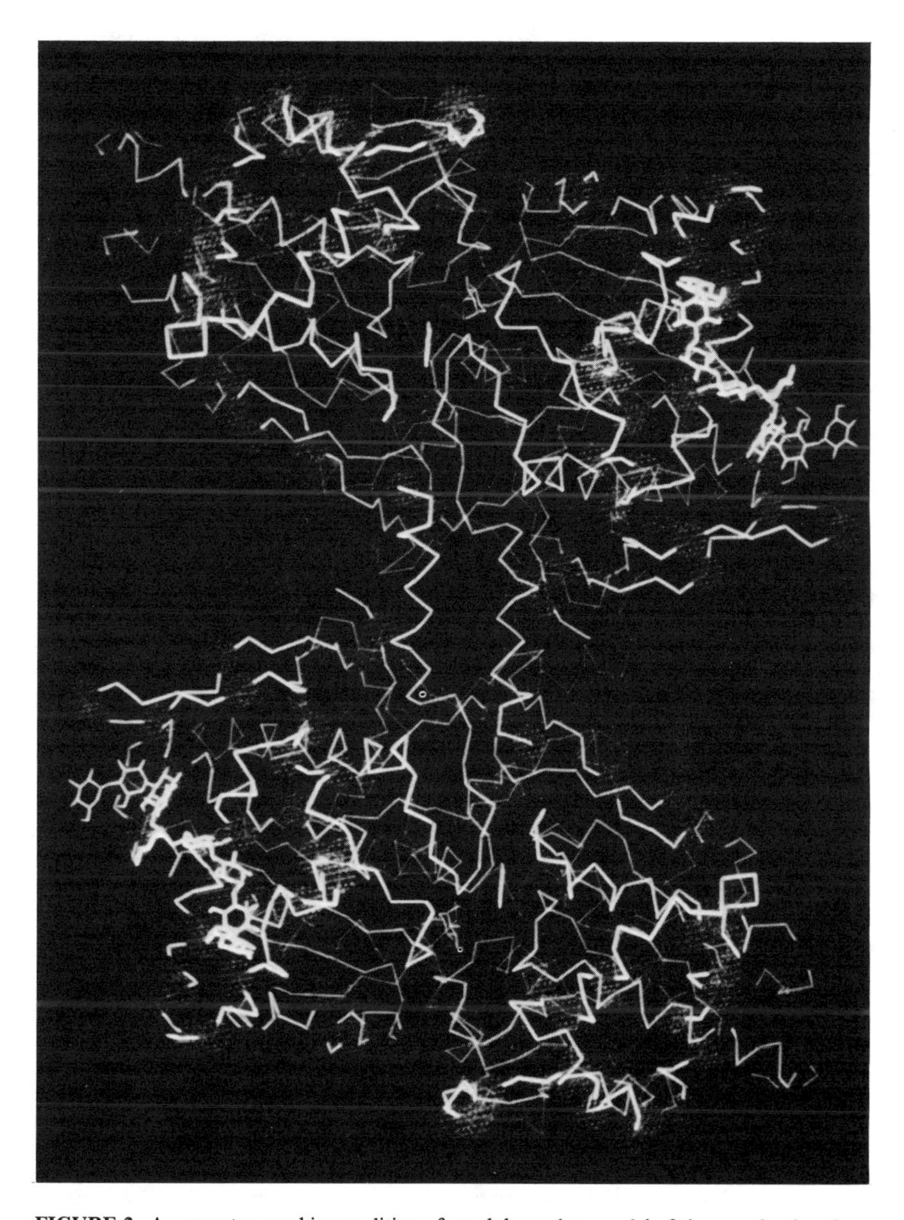

FIGURE 2. A computer graphics rendition of an alpha carbon model of the muscle phosphorylase dimer. Glucose and a glycogen fragment are shown bound to the dimer. The glucose is found in the cleft separating the NH_2- and COOH-terminal domains (near the lower center of the bottom subunit). The glycogen fragment is bound to the NH_2-terminal (left side of bottom subunit). Part of the glycogen fragment is seen to be jutting away from the protein. The positions where side chains are variant between human liver and rabbit muscle are shown as dotted spheres. These spheres are seen to be, in general, at surface locations.

```
             120               140
             --------        @@@@@@@
                               **    *
RABBIT       MEELEEIEED AGLGNGGLGR LAACFLDSMA
LIVER        IEELEEIEED AGLGNGGLGR LAACFLDSMA
POTATO       LENVASQEPD AALGNGGLGR LASCFLDSLA
YEAST        LEDVLDQEPD AGLGNGGLGR LAACFVDSMA
E-COLI       LTDLLEEEID PALGNGGLGR LAACFLDSMA

               280
             ----- ---
                  **
RABBIT       RVLYPNDNFF
LIVER        RVLYPNDNFF
POTATO       YILYPGDESE
YEAST        AVLYPNDNFA
E-COLI       KVLYPNDNHT

                           380
                -----         ----
                     * *
RABBIT       TVKTCAYTNH TVLPEALERW
LIVER        NQKTFAYTNH TVLPEALERW
POTATO       TQRTVAYTNH TVLPEALEKW
YEAST        VTKTFAYTNH TVMQEALEKW
E-COLI       TSKTFAYTNH TLMPEALERW

             450
             -----   @@
                  *
RABBIT       HAVNGVARIH
LIVER        HAVNGVAKIH
POTATO       HAVNGVAEIH
YEAST        HKVNGVVELH
E-COLI       FAVNGVARLH

             480
              -------
                 *
RABBIT       -QNKTNGITPR
LIVER        -QNKTNGITPR
POTATO       -QNKTNGVTPR
YEAST        FVNVTNGITPR
E-COLI       -HNVTNGITPR

                                  570
                             ------     @@
                               * * * **
RABBIT       ------PNSLFDVQVK RIHEYKRQLL
LIVER        ------PSSMVDVQVK RIHEYKRQLL
POTATO       ------PDAMFDIQVK RIHEYKRQLL
YEAST        INREYLDDTLFDMQVK RIHEYKRQQL
E-COLI       ------PQAIFDIQIK RLHEYKRQHL

                               680
             ----            @@@@@@@    -----
                  ***    **    *
RABBIT       ISTAGTEASG TGNMKFMLNG ALTIGTMDGA
LIVER        ISTAGTEASG TGNMKFMLNG ALTIGTMDGA
POTATO       ISTAGMEASG TSNMKFAMNG CIQIGTLDGA
YEAST        ISTAGTEASG TSNMKFVMNG GLIIGTVDGA
E-COLI       ISTAGKEASG TGNMKLALDG ALTVGTLD--
```

FIGURE 3. A comparison of the active-site regions for several glycogen phosphorylases. Each side chain identified as being crucial to binding is marked (*). Also marked are the portions of the secondary structure of the protein in the α conformation (@) and those in the β conformation (-). The binding residues are all conserved (except for E284 in potato). Asn 284 binds glucose, which does not inhibit potato phosphorylase.

shown that even crystals of liver phosphorylase are partially degraded.[14] The effect of proteolysis on the allosteric regulation of liver glycogen phosphorylase has not yet been evaluated. Further analysis of the three-dimensional structure of phosphorylase with the superimposed liver enzyme sequence will perhaps resolve these issues.

The alignment of the amino acid sequences for glycogen phosphorylases from muscle, liver, potato, yeast, and *E. coli* is presented in FIGURE 3 for the active-site loops. The high conservation in this area makes the alignment of the amino acid sequences quite reliable. Residues that are believed to be involved in binding ligands at the active site (but not the complete list) are shown with an asterisk above them. The secondary structure in the region of the binding residues is also indicated.

Variation in G + C

The cDNAs for muscle and liver phosphorylases show a striking difference in their G + C contents. The G + C content of the muscle cDNA is 60%, compared to only 48% in the liver homologue. FIGURE 4 is a plot of the G + C content by codon position over the segments of the liver and muscle phosphorylase cDNAs encoding the COOH-terminal domain (amino acids 480-845). It is apparent that most of the increases in G + C content that are found in the muscle cDNA are a result of a strong bias for G or C at the third position of the codon. Analysis of other liver and muscle proteins from rat, rabbit, and human show that this increased frequency is a general feature; that is, genes that are expressed in muscle tissue tend to be richer in G + C.[15]

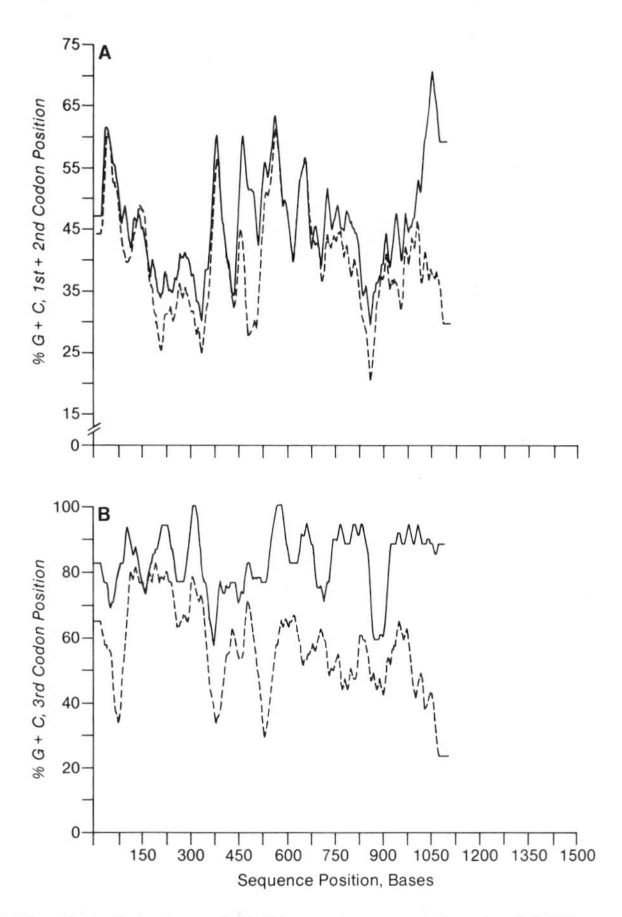

FIGURE 4. The G + C content of cDNA sequences encoding the COOH-terminal domains of human liver (----) and rabbit muscle (——) phosphorylases are plotted for the first plus second position of the codon (**A**) and for the third position of the codon (**B**). The increase in G + C content of the gene for the muscle phosphorylase results from the frequency with which G or C occupies the third position.

Yeast Phosphorylase Gene

A single gene for the yeast enzyme has been cloned and sequenced. It codes for a protein that is 891 amino acids long and is 45% homologous with the muscle protein. The sequence of the yeast enzyme in the active-site region is given in FIGURE 3. Comparisons of specific residues show that the substrate- and inhibitor-binding residues are conserved between yeast and muscle phosphorylases. The purine-binding residues, Phe 285 and Tyr 613, are also conserved, suggesting a regulatory role for purine or analogous ligands in yeast. Residues that are involved in binding glycogen at the storage site are mostly conserved, though some replacements have occurred. These have not yet been analyzed. There are 14 length variations (insertions or deletions) between the yeast and rabbit muscle enzymes, located at amino acid positions 1, 114, 214, 317, 438, 475, 508, 557, 591, 722, 735, 767, 794, and 830. These positions correspond to surface locations on the three-dimensional structure for rabbit muscle phosphorylase. It is interesting that the positions determined for four out of the seven introns in the human muscle gene correspond to positions on the yeast enzyme where insertions and deletions may be found. This may be a common relationship between gene and protein structure.[16]

Unlike the changes just discussed, there are two effector regions where the substitutions of amino acid residues in yeast are likely to have significant effects. These are the Ser 14-phosphorylation site and the AMP-binding site. Both of these sites are formed at the subunit interface of the muscle phosphorylase dimer. It is expected that the subunit contacts in the yeast enzyme are significantly altered from the subunit contacts in the muscle dimer. This alteration should affect the allosteric properties of the yeast phosphorylase dimer. The AMP-binding site is disrupted by a 14-residue insertion at position 114, and most of the AMP-binding residues are not conserved. AMP-binding residues are marked with an A in FIGURE 5. It is not surprising that these residues are altered because the yeast enzyme is shown to be unaffected by AMP.[17]

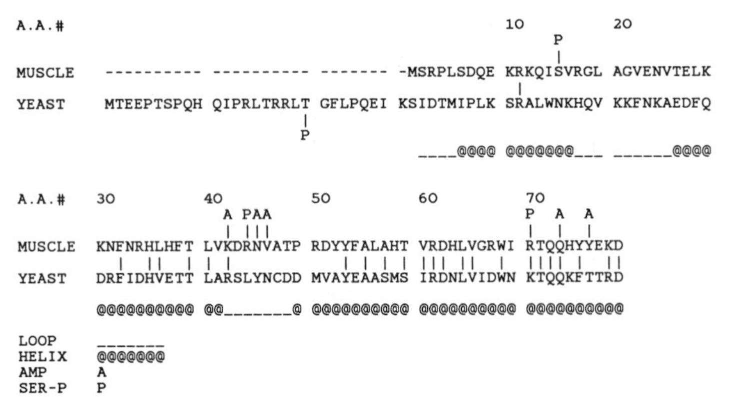

FIGURE 5. A comparison of the NH_2-terminal regions of the yeast and muscle phosphorylases. The portions of the secondary structure in the α conformation (@) and the portions in the β conformation (-) are marked. Also marked are the AMP-binding residues (A) and the phosphoserine-binding residues (P).

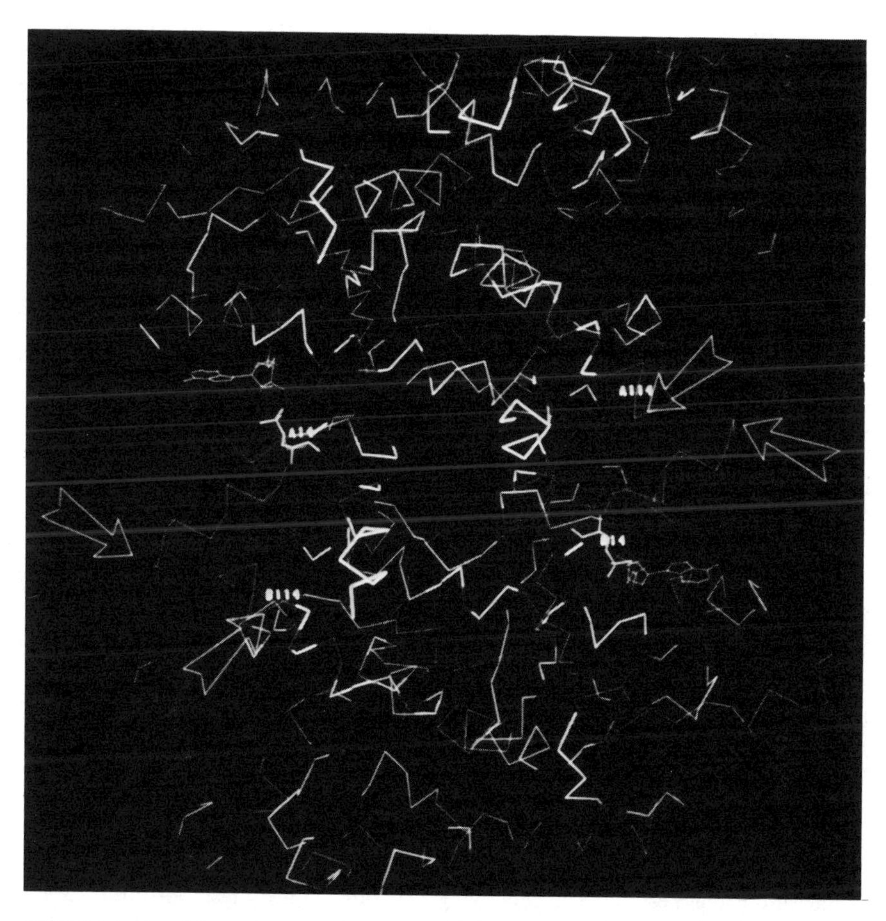

FIGURE 6. An alpha carbon drawing of the interface between the phosphorylase dimer's two monomers. Each monomer's Ser 14-phosphate residue is marked (A14 for the upper monomer, B14 for the lower), and both of these monomers are shown with AMP. Each monomer's residue at position 114 is marked (A114 and B114). The regions around these residues, which are marked with arrows, are partly responsible for binding the NH_2-terminal and are necessary for the activation by phosphorylation. Yeast phosphorylase has an insertion at each of these residues. Other arrows mark the 30-residue extensions that harbor the yeast phosphothreonine.

Muscle phosphorylase is activated by a specific muscle phosphorylase kinase. In the case of yeast phosphorylase, a cAMP-dependent protein kinase phosphorylates a threonine residue. There is no conservation of protein sequence in the vicinity of phosphorylation sites between yeast and muscle enzymes. The secondary structure known to be present in the muscle enzyme is indicated in FIGURE 6, which shows an alpha carbon drawing of the subunit contact region. It can be seen that the homology between yeast and muscle enzymes vanishes around the AMP-binding residues, positions 40 through 50, and that there is also no sequence homology between the first 33 residues of muscle phosphorylase and the first 61 residues of yeast phosphorylase. The 27-residue extension found in the yeast enzyme probably has a function similar to that of 20 residues at the NH_2-terminal end of the muscle enzyme because both segments contain phosphorylation sites.[18]

DISCUSSION

Complementary DNAs and two genes for glycogen phosphorylases have been cloned and characterized by DNA sequencing. Glycogen phosphorylase shows greater homology within a tissue type between two species than it does within a single species between tissue types. Rat, rabbit, and human muscle glycogen phosphorylases are 90% homologous. Human liver glycogen phosphorylase is 78% identical with the rabbit muscle enzyme. There is little length variation between the tissue isozymes, but there are regions where hydrophobic residues have been altered between the two molecules. It is expected that this may change some of the allosteric or regulatory properties. Most of the changes between the liver and muscle enzymes are confined to the surface of the protein, including a few found at the AMP- and glycogen-binding sites. The effects of substitutions that might make liver glycogen phosphorylase less responsive to AMP are not yet completely analyzed. Possible functional differences might be ascribed to sequence variations in the glycogen-binding site as well. The strong homology between the active sites of the tissue-specific isozymes suggests that modeling the three-dimensional structure of the liver enzyme based on its muscle counterpart is rigorously valid. It remains to be determined how the amino acid substitutions at the two effector sites account for the difference in properties between the muscle and liver enzymes. This analysis is being carried out using computer graphics, computer modeling, and biochemical approaches. The exact effects of specific residues identified as having functional roles by these types of studies can then be addressed by site-directed mutagenesis studies.

The difference in G + C composition between muscle and liver phosphorylases appears to apply to other muscle- and liver-specific genes as well. The higher G + C content found in muscle cDNA sequences is primarily due to G + C enrichment at the third codon position. The underlying evolutionary mechanism for this discrepancy is unclear, but may reflect the need for stabilizing the processes of replication, transcription, and translation in the more physically stressful muscle environment.

In contrast to the conservation that has been observed among species and different tissue types, a marked divergence between the yeast glycogen phosphorylase molecule and the vertebrate enzymes may be seen. The potato, yeast, and *E. coli* enzymes are all approximately 45% identical with the rabbit muscle enzyme at the amino acid level. In comparing the DNA sequences, the identity between the muscle and liver enzymes is 70%; between the muscle and yeast enzymes, 51%; and between the liver and yeast enzymes, 56%. The better agreement in the DNA comparison between the liver and yeast enzymes is due to the similar G + C content and the relatively unbiased distribution of nucleotides at the third position of the yeast enzyme and liver enzyme codons.

Though the sequence homology is only 45%, it is reasonable to assume that in many places the tertiary structure for the yeast phosphorylase is similar to that for the muscle phosphorylase. Most of the amino acid substitutions in yeast enzyme relative to the muscle enzyme are conservative. Only surface residues are markedly different between the yeast and muscle enzymes. Particularly striking are the number of insertions and deletions between the yeast and muscle sequences. All of the length variations correspond to variations in the surface of the three-dimensional structure for the muscle phosphorylase; four of these, in fact, occur at positions where there are intron-exon junctions in the muscle gene.

The conservation of the residues associated with the regulatory binding site for the synergistic inhibitors glucose and purine was surprising. No evidence of glucose or purine inhibition has been found for the yeast enzyme. It is possible that these binding sites are nonfunctional, though they will certainly be worth analyzing biochemically in view of the sequence data. The long NH_2-terminal extension (FIG. 6) and radical divergence of the phosphorylation site in the yeast enzyme suggests that phosphorylation of glycogen phosphorylase as a control mechanism developed independently in the yeast and muscle enzymes. None of the phosphorylation-site-binding residues are common between these two sequences. FIGURE 6 shows that with the 14-residue insertion at position 114 it is likely that the yeast phosphorylation site, which presumably regulates the yeast phosphorylase activity, is structurally and functionally different. The yeast enzyme appears simpler in that the activation process would not be affected by AMP. It will be interesting to determine whether the inhibitors, glucose or purine, affect the yeast enzyme, and whether the phosphorylation and dephosphorylation events are regulated by allosteric ligand interactions. If they are, it will be important to identify the communication linkage between the phosphorylation activation of yeast and its active site and to compare it to the analogous linkage in the muscle enzyme.

ACKNOWLEDGMENTS

The authors would like to thank Prudence S. Bothen for preparing this manuscript.

REFERENCES

1. FLETTERICK, R. J. & N. B. MADSEN. 1980. The structures and related functions of phosphorylase *a*. Annu. Rev. Biochem. **49:** 31-62.
2. FLETTERICK, R. J. & S. R. SPRANG. 1982. Glycogen phosphorylase structures and function. Acc. Chem. Res. **15:** 361-369.
3. DOMBRADI, V. 1981. Structural aspects of the catalytic and regulatory function of glycogen phosphorylase. Int. J. Biochem. **13:** 125-140.
4. TAGAYA, M. & T. FUKUI. 1984. Catalytic reaction of glycogen phosphorylase reconstituted with a coenzyme substrate conjugate. J. Biol. Chem. **259:** 4860-4865.
5. KLEIN, H. W., M. J. IM, D. PALM & E. J. M. HELMREICH. 1984. Does pyridoxal 5′-phosphate function in glycogen phosphorylase as an electrophilic or a general acid catalyst? Biochemistry **23:** 5853-5861.
6. NAKANO, K. & T. FUKUI. The primary structure of potato phosphorylase. J. Biol. Chem. In press.
7. PALM, D., R. GOERL & K. BURGER. 1985. Evolution of catalytic and regulatory sites in phosphorylases. Nature **313:** 500-502.
8. HWANG, P. K., Y. P. SEE, A. M. VINCENTINI, M. A. POWERS, R. J. FLETTERICK & M. M. CRERAR. 1985. Comparative sequence analysis of rat, rabbit, and human muscle glycogen phosphorylase cDNAs. Eur. J. Biochem. **152:** 267-274.
9. ROSE, M., P. NOVICK & J. THOMAS. Cell. In press.
10. NAKANO, K., P. HWANG & R. J. FLETTERICK. The cDNA sequence of rabbit muscle glycogen phosphorylase. FEBS Lett. In press.

11. TITANI, K., A. KOIDE, J. HERMANN, L. H. ERICSSON, S. KUMAR, R. D. WADE, K. A. WALSH, H. NEURATH & E. H. FISCHER. 1977. Complete amino acid sequence of rabbit muscle glycogen phosphorylase. Proc. Natl. Acad. Sci. USA **74:** 4762-4766.
12. CRAIK, C. S., S. SPRANG, R. J. FLETTERICK & W. J. RUTTER. 1982. Intron-exon splice junctions map at protein surfaces. Nature **299:** 180-182.
13. LEBO, R. V., F. GORIN, R. J. FLETTERICK, F. T. KAO, M. C. CHEUNG, B. D. BRUCE & Y. W. KAN. 1984. High-resolution chromosome sorting and DNA spot blot analysis assign McArdle's syndrome to chromosome 11. Science **225:** 57-59.
14. HWANG, P., M. STERN & R. J. FLETTERICK. 1984. Purification and crystallization of bovine liver phosphorylase. **791:** 252-258.
15. NEWGARD, C. B., K. NAKANO, P. W. HWANG & R. J. FLETTERICK. Sequence analysis of the cDNA encoding human liver glycogen phosphorylase reveals tissue-specific codon usage. Proc. Natl. Acad. Sci. USA. In press.
16. CRAIK, C. S., W. J. RUTTER & R. J. FLETTERICK. 1983. Splice junction's association with variation in protein structure. Science **220:** 1125-1129.
17. LERCH, K. & E. H. FISCHER. 1975. Amino acid sequence of two functional sites in yeast glycogen phosphorylase. Biochemistry **14:** 2009-2014.
18. HWANG, P. K. & R. J. FLETTERICK. Convergent and divergent evolution of regulatory sites in eukaryotic phosphorylases. Submitted for publication.

Efficient Expression of Heterologous Genes in *Escherichia coli*

The pAS Vector System and Its Applications

ALLAN R. SHATZMAN AND MARTIN ROSENBERG

Department of Molecular Genetics
Smith Kline and French Laboratories
Swedeland, Pennsylvania 19479

INTRODUCTION

There are numerous gene products of biological interest that cannot be obtained from natural sources in quantities sufficient for detailed biochemical and physical analysis. Moreover, the limited bioavailability of these molecules has made it impossible to consider their potential utilization as either pharmacological agents or targets. One solution to this problem has been the development of recombinant vector systems that are designed to achieve efficient expression of cloned genes in bacteria.

In general, the rationale used in the design of these systems involves insertion of the gene of interest into a multicopy vector system (usually a plasmid) such that the gene is efficiently transcribed and translated. Because most genes do not naturally contain the proper signals for ribosome recognition and translation initiation in *Escherichia coli,* special procedures must be devised to supply this information. This is done either by fusing the gene to a bacterial ribosome binding site or to the NH_2-terminal coding region of a bacterial gene. In the former case, some difficulties in obtaining efficient translation have been encountered owing to sequence alterations made in the ribosome recognition region prior to or as a result of gene insertion; in the latter case, the gene product is a fusion protein carrying additional peptide information at its NH_2-terminal. The fusion products may have physical and functional properties that differ from the normal protein, thereby limiting their value for biological study. In addition to factors such as promoter strength, gene copy number, and translational efficiency, several other factors may influence the expression of a cloned gene in bacteria. These include 1) the reduction of transcription resulting from polarity effects, 2) the stability of the mRNA, 3) the stability of the gene product, and 4) the potential lethality of the product to the growth of the host.

This chapter describes the development and application of a particular set of vectors that have been designed to potentially achieve efficient regulated expression in *E. coli* of any gene coding sequence. A variety of host strains has also been developed in order to help control, stabilize, and thereby maximize expression of various cloned genes. The ability to carefully regulate, achieve rapid production of, and minimize the proteolytic degradation of gene products has proven particularly useful in expressing potentially lethal or unstable gene products. The system has been used successfully to express more than 75 different prokaryotic and eukaryotic gene products. In each case, the protein of interest has been obtained at levels varying from 1 to 40% of total cellular protein. Identification of the product usually requires no functional assay because the protein is visualized directly from total cellular protein by resolution on a one-dimensional polyacrylamide gel. Moreover, the coding sequence of interest, either natural or synthetic, is fused directly and precisely to the translation initiation signal provided on these vectors. This allows the expression of authentic gene products, rather than the more commonly achieved expression of gene fusion products. Of course, the expression of gene fusions and gene deletions can also be obtained readily. We will demonstrate the application of this system to the expression and characterization of several gene products of biological and biomedical interest.

THE pAS EXPRESSION VECTOR SYSTEM

The family of vectors we have constructed are pBR322 derivatives carrying regulatory signals derived primarily from the bacteriophage λ (FIG. 1). Phage regulatory

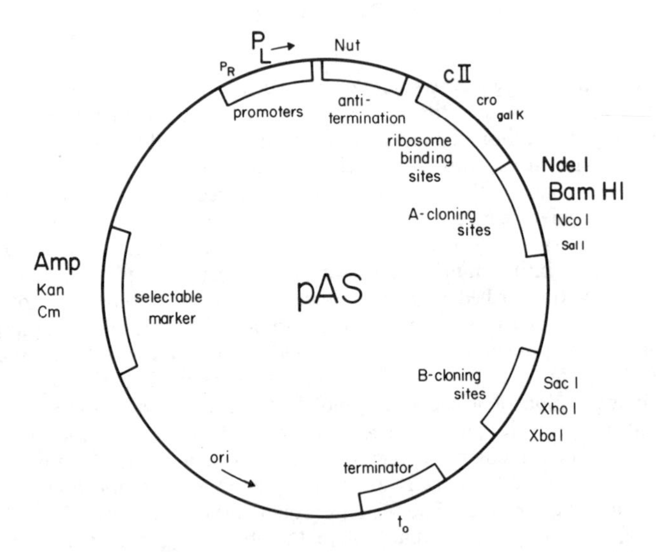

FIGURE 1. Schematic diagram of the pAS plasmid vector for the overproduction of heterologous gene products. The phage λ P_L or P_R promoters are provided to direct efficient transcription. N utilization site(s) and a transcription terminator (t_o) are provided to maximize transcription and plasmid stability. The λ *c*II, *cro,* or *E. coli gal*K gene ribosome binding sites are provided to direct efficient translation initiation signals. Unique *Nde*I, *Bam*HI, *Nco*I, and *Sal*I restriction sites have been engineered adjacent to the ribosome binding sites (A-cloning sites) to permit easy insertion of genes adjacent to the translational regulatory information. Several other unique restriction sites (B-cloning sites) are present to further facilitate gene insertion and manipulation. Ampicillin (Amp), kanamycin (Kan), and chloramphenicol (Cm) resistance markers have been inserted in multiple positions and orientations. The vector contains a colE1 origin of replication. The diversity of lettering sizes used to represent the functional elements is a relative measure of the degree of use and of data that have been generated with each of these elements (that is, P_L is present and has been used far more often than P_R).

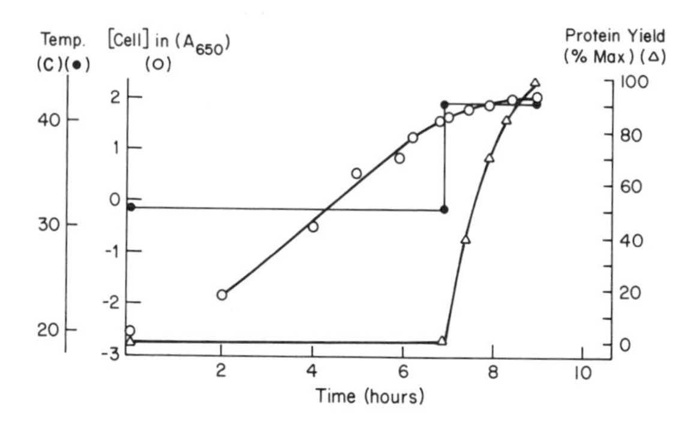

FIGURE 2. Typical production fermentation run of an *E. coli* K12-defective lysogen carrying a pAS expression vector. Cells are grown to high optical density in log phase at 32° C (open circles). Just prior to stationary phase the temperature of the culture is shifted to 42° C (closed circles). Immediately upon temperature shift the gene expressed into the pAS expression vector is expressed rapidly and efficiently into protein (triangles).

information was chosen because of its high efficiency and ability to be regulated. The system provides a promoter that can be fully controlled, an antitermination mechanism to ensure efficient transcription across any gene insert, high vector stability, several choices of antibiotic selection, and easy and flexible insertion of the gene of interest adjacent to efficient translation regulatory information.

Efficient Regulated Transcription

The phage λ P_L promoter was been chosen for inclusion in these vectors as it has been shown by gene fusion experiments[1] to be more efficient than many bacterial promoters. Plasmids carrying P_L are often unstable, presumably owing to the high level of P_L-directed transcription. This problem of instability was overcome by repressing P_L transcription, using bacterial hosts that contain an integrated copy of the λ genome (bacterial lysogens). In these cells, P_L transcription is controlled by the phage λ repressor protein (cI), a product synthesized continuously and regulated autogenously in the lysogen.[2] The high level of transcription obtained from P_L is apparently sufficient to maximize transcription of any desired gene product, as the addition of multiple promoters in tandem with P_L does not lead to further increases in gene expression (A. R. Shatzman, unpublished results).

The *E. coli* lysogens used with P_L-containing vectors have typically carried a temperature-sensitive mutation in the phage λ cI gene (cI^{857}), a mutation that allows P_L-directed transcription to be activated by temperature shift. Induction is accomplished by simply raising the temperature of the cell from 32° C to 42° C. Thus, cells carrying the vector can be grown initially to high density without expression of the cloned gene (at 32° C) and subsequently induced to rapidly synthesize the product at 42° C (FIG. 2).

It has been shown recently that the pAS system can also be induced chemically in lysogens carrying a wild-type repressor gene. This is accomplished by inducing the SOS response[3] of the bacterial host with a DNA-damaging agent such as nalidixic acid.[4] In this case, the wild-type repressor protein is cleaved by the *recA* protease, which is induced by the SOS response. In contrast to the very rapid heat inactivation of the cI^{ts} protein (maximum levels of gene product accumulate within 45-90 min), naladixic acid mediated inactivation of cI^{+} is comparatively slow (maximum product yield obtained 5-6 hr post induction).

Termination-Antitermination

In addition to providing a strong, regulatable promoter, the system also ensures that P_L-directed transcription efficiently traverses any gene insert. This is accomplished by providing both the phage λ antitermination function (N) and sites on the P_L transcription unit necessary for N utilization (Nut site). N expression from the host lysogen reduces transcriptional polarity, thereby helping to ensure that transcription traverses the entire P_L transcription unit.[5] For example, a termination site, tR1, is present within the P_L transcription unit just upstream of the *c*II ribosome binding information. We have demonstrated that N expression from the host lysogen is sufficient to completely antiterminate transcription through this site on the vector.[6]

In an effort to further stabilize these vector systems, a highly efficient transcription termination signal (t_o) was inserted downstream from the site of gene insertion (FIG. 1). It was thought that insertion of this termination signal would prevent P_L transcription from proceeding through regions of the plasmid necessary for replication, and thereby stabilize these vectors in nonlysogens. Surprisingly, these vectors remained unstable in nonlysogens, suggesting that the t_o terminator may not be as efficient in the context of this plasmid as it had been in other systems.[7] Inclusion of this transcriptional signal did, however, increase the stability of these vectors in various lysogenic hosts and resulted in a two- to threefold increase in plasmid copy number.

Use of the Phage λ cII *Ribosome Binding Site*

In order to use the pAS system to express genes lacking *E. coli* translational regulatory information, several efficient ribosome recognition and translation initiation sites (rbs) were engineered into the P_L transcription unit. The primary site chosen was that of the efficiently translated λ phage gene, *c*II. The entire coding region of this gene was removed, leaving only its ATG initiation codon and translational regulatory sequences upstream. Neither the sequence nor the position of any nucleotides in the rbs were altered. Instead, a unique *Bam*HI restriction site was introduced immediately downstream from the ATG initiation codon (ATGGATCC) (FIG. 1). This system allows direct fusion of any coding sequence to the translation initiation signal and, furthermore, allows any coding information to be adapted for insertion into the vector. Those genes which contain a *Bam*HI, *Bcl*I, *Bgl*II, *Xho*II, or *Sau*3AI restriction site at or near their own initiation codon may be inserted directly into these vectors. As most genes do not contain such conveniently located restriction sites, it is often necessary to convert the *Bam*HI cloning site into a blunt-ended cloning site by treatment with a single-strand-specific nuclease.[8] Any gene that has been modified such that it has a blunt end at or near its 5′ end (in the appropriate reading frame) can be inserted adjacent to the ATG of the *c*II rbs. In addition to the *Bam*HI site immediately downstream of the ATG, there is also a naturally occurring *Nde*I site immediately preceding and overlapping the ATG (CATATG). In order to make this *Nde*I site unique on the vector, a pAS derivative was constructed from which all other *Nde*I sites were removed.[9] The resulting vector, pMG27N$^-$, permits the insertion of a gene containing its own ATG initiation codon without altering the sequence of the *c*II rbs. This vector has proven particularly useful for cloning foreign genes such as the human oncogene H*ras*,[9] various metallothioneins,[10] and human interleukin-1 (P. Young, unpublished results), in which the 5′ ends were reconstructed using synthetic oligonucleotides such that an authentic gene product was obtained.

While the authentic *c*II rbs site has been used successfully in the expression of many genes, several altered *c*II rbs sites have been constructed in order to provide alternative restriction sites for gene insertion. For example, an *Nco*I restriction site was engineered into the *c*II rbs (CCATGG). This site permits cloning of a gene in-frame immediately adjacent to the ATG following *Nco*I cleavage and a polymerase-mediated fill-in reaction to create a blunt end. This method of inserting information adjacent to the ATG was more efficient than that described above for the *Bam*HI site. In this vector, however, both the sequence and spacing of the *c*II rbs are altered slightly. Several genes have been inserted into this vector, and expression levels were similar to those found using the authentic *c*II rbs. Apparently there exists some flexibility in the information needed for efficient ribosome recognition and translation initiation at the *c*II rbs.

Several pAS derivatives have been constructed that use ribosome binding sites other than *c*II for expression. For example, both the λ *cro* gene rbs and the *E. coli gal*K gene rbs have been adapted with unique restriction sites for the insertion and expression of heterologous genes. In the case of the *cro* rbs, expression levels similar to those observed with the *c*II rbs were obtained with most genes. In contrast, genes linked to the *gal*K rbs have exhibited significantly decreased expression as compared to the *c*II or *cro* rbs fusions.

Insertion Sites, Resistance Markers, and Replication Origins

In addition to providing transcriptional and translational regulatory sequences, several other readily exchangeable elements have been engineered into the pAS vector. Synthetic oligonucleotides have been inserted at the *Sal*I site to add unique *Sac*I, *Xho*I, *Xba*I, and *Stu*I restriction sites. These sites facilitate both the insertion of genes into the vector as well as their removal and movement to other vector systems. Three different antibiotic selection markers are available for use in these systems. Chloramphenicol and kanamycin resistance markers have been inserted such that they occur in addition to or in place of the ampicillin resistance marker. The origin of replication for the single-stranded DNA phage, F1, has also been added to the vector to create a system, PFCE4, that generates single-stranded plasmid DNA in cells infected with F1 phage.[11] This feature is similar to that found in the commonly used pEMBL vectors;[12] however, the pAS derivatives are unique in that they combine the utility of generating single-stranded DNA with the elements of an efficient gene expression vector. Thus, the pAS vector can be used to generate single-stranded DNA for rapid DNA sequencing of gene inserts as well as for providing templates for site-directed mutagenesis. This technology may therefore be useful in generating either mutational alterations that lead to higher levels of gene expression or gene analogues in which point mutations may be introduced to study the relationships between structure and function in proteins. Once isolated, the effect of the mutation on expression levels or gene product function may be examined directly using the PFCE4 derivative.

Bacterial Hosts

The pAS expression vector system has been used in combination with a variety of *E. coli* K12 lysogenic hosts. In each case the host lysogen carries a replication-defective λ phage so that induction of the expression system does not lead to either phage growth or replication. Gene expression from various pAS derivatives has been examined in these K12 backgrounds, and the results indicate that the host background can play an important role in the final yield of accumulated gene product. For the most part, we do not understand the reasons for the rather dramatic differences seen in product yield from different host strains. Product stability, however, often appears to be a determining factor. Recently, host lysogenic strains have been developed that are defective in certain proteolytic functions (J. Auerbach and M. Rosenberg, unpublished results). These proteolytic-deficient strains are obtained as temperature-sensitive conditional lethals and thus are compatible with our temperature-induction system. Cells are grown to high optical density at low temperature and are then shifted to high temperature for expression. Concomitant with the induction of gene expression is the turning off of the synthesis of these particular host proteolytic systems. These specialized host strains have had a dramatic impact on the expression of several (but not all) gene products in *E. coli* (see FIG. 3 and the following section) (C. Debouck and M. Rosenberg, unpublished results).[13,14]

In addition to the genotype of the host strain, the mode of induction can also affect the overall accumulation of gene product.[4] For example, temperature induction is known to result in a generalized heat shock response in the bacteria. At least some of the proteins induced by this response are thought to be proteases involved in the turnover of certain cellular gene products.[15] In contrast, chemical induction leads to a typical SOS response by the host.[3] Again, a set of host proteins are induced, only some of which overlap with the heat shock-induced proteins. Thus, the two modes of induction lead to distinctly different cellular states, and these variations can, in turn, lead to significant differences in pAS-derived product accumulation. Indeed, expression of the human α_1-antitrypsin gene has been shown to increase threefold when chemical means, rather than thermal induction, are employed (H. Johansen, J. Sutiphong, A. R. Shatzman, and M. Rosenberg, in preparation).

In conclusion, it has become increasingly clear that the variety of K12 strains now available for pAS expression, in combination with the two different modes of induction for the system, has contributed significantly to the overall success of this system in producing foreign gene products.

EXPRESSION OF ONCOGENE PRODUCTS USING THE pAS SYSTEM

The pAS expression vector system has been used successfully to obtain the protein products of a variety of mammalian cellular and viral oncogenes.[9,13,16–21] Previously, these gene products have proven extremely difficult, if not impossible, to obtain from

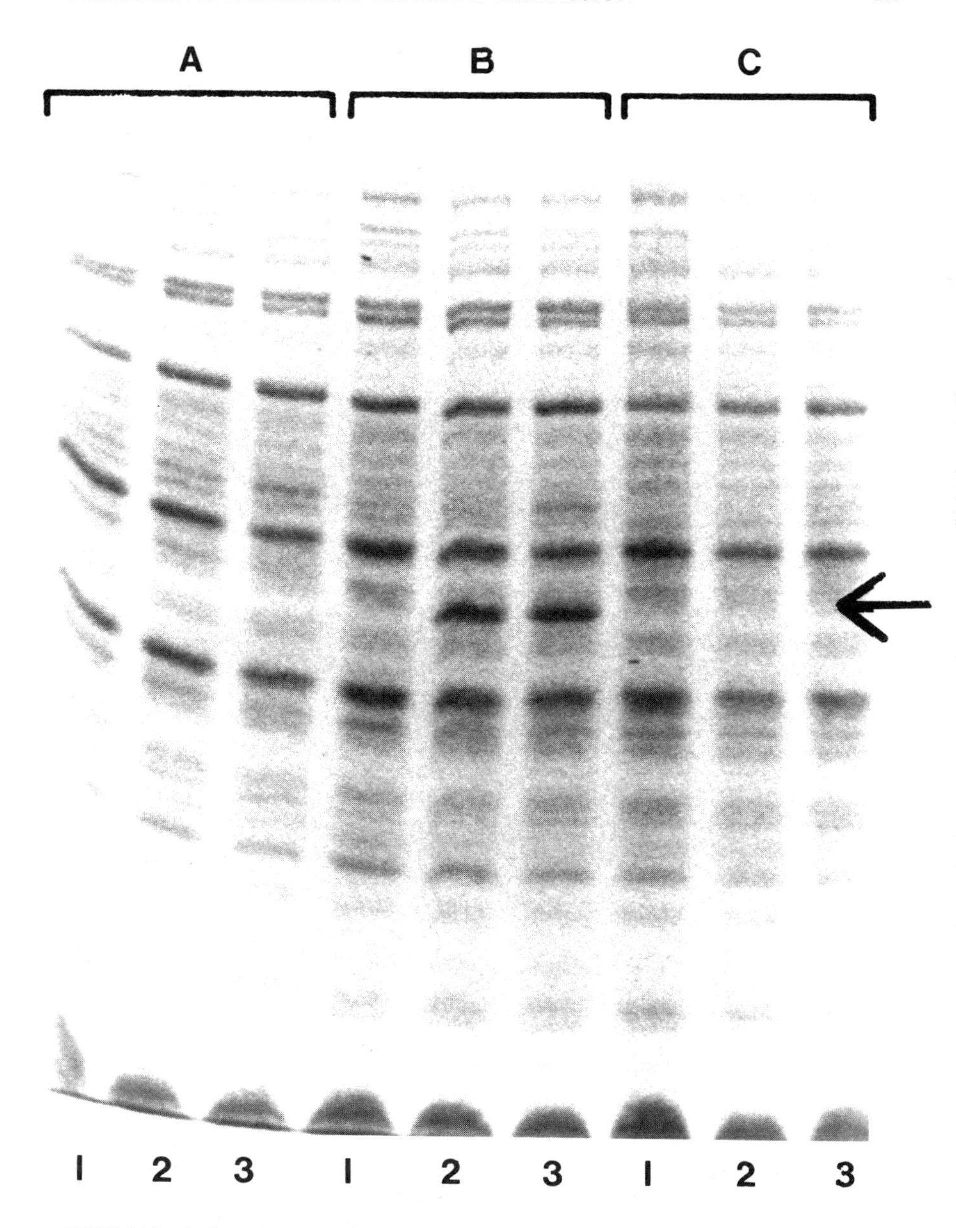

FIGURE 3. Sodium dodecyl sulfate-polycrylamide gel electrophoresis (SDS-PAGE) analysis of the expression of the human c-*myc* gene product in three different *E. coli* K12 lysogenic strains (A, B, and C) using the pAS vector system. The strains shown in A and B are isogenic except for the temperature-sensitive mutation, which inactivates certain host proteolytic activities (see text for details). The strain shown in C is a standard K12 wild-type strain, N99. Lanes 1, 2, and 3 represent total cellular protein prior to, 20 min after, and 40 min after temperature induction, respectively. The human c-*myc* protein (arrow) is seen to accumulate only after induction in the *E. coli* strain carrying the proteolytic deficiency (B).

natural sources in quantities sufficient for detailed study. Our approach has been to express these proteins at high levels in bacteria using the pAS vector system and thereby obtain amounts of these materials that allow both their biochemical and biophysical analysis. To date, efficient expression has been achieved for several genes from this class, including both the oncogenic and proto-oncogenic forms of the human H-*ras* gene,[9,18] the viral forms of the *abl*[21] and *myb* gene products (A. S. Ferguson and M. Rosenberg, unpublished results), the human c-*myc* gene product,[13] and the two forms of the human adenovirus E1A gene product.[16,17,19,22] In each case, the protein products have been produced and purified in quantities sufficient to allow both structural and functional characterization. The yields of these products have varied from between 2 to 10% of total bacterial cell protein. One of the immediate advantages of obtaining these purified proteins in quantity has been that they provide a ready source of material for producing high-titre-specific polyclonal and monoclonal antibodies that recognize these gene products. These antisera have been shown to be superior diagnostic reagents for detecting and monitoring these oncogene proteins in both normal and transformed mammalian cells. In this section we will describe the utility of the pAS expression system as it has been applied to the study of oncogene products and, in particular, to the nuclear-localized transcription activators derived from the adenovirus ElA early gene region.

The adenovirus ElA gene encodes a nuclear-localized function that stimulates transcription of the other viral early transcription units.[23] In addition, ElA stimulates transcription of certain endogenous cellular genes[24,25] and activates several genes introduced into mammalian cells by DNA transfection.[26] Evidence now suggests that ElA is also capable of repressing transcription: in particular, transcription mediated by certain enhancer elements.[27] ElA function is also required for the oncogenic properties exhibited by adenovirus. ElA alone has been shown to be sufficient for immortalization of certain mammalian cells[28] and, in combination with ElB or the human H-*ras* oncogene, results in full morphologic transformation of primary cells.[28,29]

Our approach to understanding the structure and function of ElA has been to use the pAS expression vector system to efficiently produce, purify, and characterize the full-length protein products encoded by the human subgroup C adenovirus ElA 12S and 13S mRNAs.[17,21] The vector system also lends itself readily to the construction and expression of various truncated derivatives of a gene. NH_2-terminal, COOH-terminal, or internal deletion mutants can be constructed and readily inserted in-frame with the ATG initiation codon on the vector. This approach has allowed us to rapidly construct, express, and purify a set of deletion mutant protein derivatives of ElA (FIGS. 4A & 4B).[19,30] These mutant derivatives have been used successfully to map distinct functional domains in the ElA protein.

Microinjection techniques were used to introduce these proteins into either frog oocytes or mammalian cells in order to examine their ability to nuclear localize, activate gene expression, and induce DNA synthesis and morphologic transformation.[17,19,30,31] Our results demonstrate that both the 12S and 13S mRNA products localize rapidly to the cell nucleus and both activate adenoviral gene expression. Analysis of the ElA variant proteins indicates that sequences essential for rapid and efficient nuclear localization are located at the COOH-terminal of the protein. In contrast, information essential for efficient gene activation is located in the NH_2-terminal domain of the protein. In fact, the first exon coding region of ElA will induce low-level gene expression when introduced into cells at high concentrations. Thus, information for nuclear localization and gene activation are encoded by distinct domains of the protein. Surprisingly, although ElA localizes to the nucleus and activates gene expression, the purified protein does not bind to DNA under conditions where the human c-*myc* and viral *myb* proteins exhibit strong DNA interaction.[21]

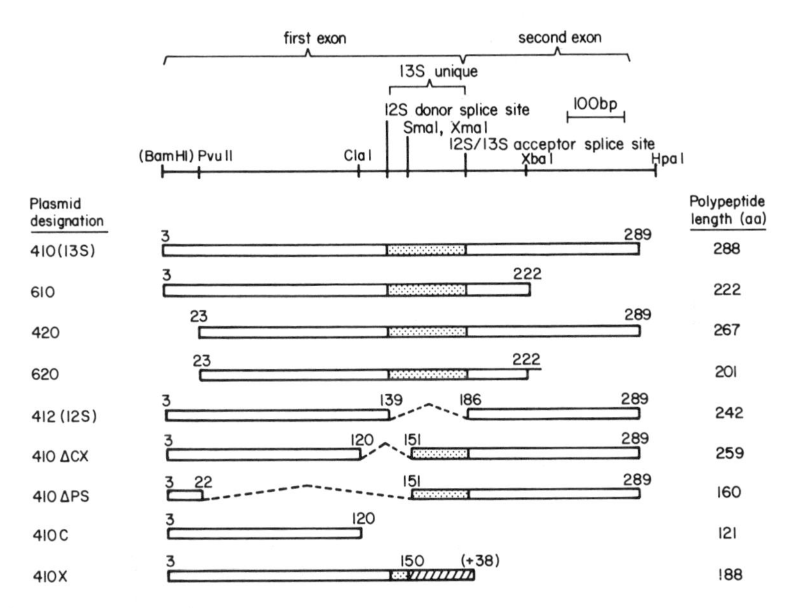

FIGURE 4A. Schematic representation of various deletion and insertion mutant ElA gene products expressed in the pAS vector system. The numbers of the first and last amino acids encoded by segments of the E1A coding sequence (blank area) that were fused to the pAS translation initiation codon are indicated. Amino acids unique to the ElA 13S mRNA product (stippled area) as well as mis-sence (hatched area) and deleted ElA-specific (dashed line) amino acids are indicated. Also shown are the regions of the ElA protein encoded by the first and second exons, the region unique to the 13S mRNA product, the restriction sites used in the mutant construction, and the amino acid lengths of the various ElA derivatives.

FIGURE 4B. Immunoblot analysis of the full-length ElA 13S (410), E1A 12S (412), and various deletion-mutant proteins expressed in *E. coli* that were purified and analyzed by SDS-PAGE analysis (12.5% acrylamide). The size in kilodaltons of several marker proteins are indicated.

The c-*myc* and *myb* gene products used in these experiments were also produced in *E. coli* using the pAS vector system. It is of interest to note that the full-length *myc* protein accumulated only in a bacterial cell line that was deficient in certain proteolytic functions (see the previous section and FIG. 3).[13] It was found that the c-*myc* protein was highly unstable in all other thermally induced *E. coli* strains tested. Clearly, the proteolytic-deficient strains have proven to be extremely important in efficiently producing certain gene products.

It is known that during an adenovirus infection ElA protein undergoes some post-transcriptional modification(s). We have demonstrated that the same modification(s) occur when the product expressed by *E. coli* is introduced either into mammalian cells, frog oocytes, or *in vitro* cytoplasmic lysates.[31] Analysis of the variant ElA proteins localizes the region of modification to the same first exon domain required for gene activation. Preliminary evidence suggests that the human c-*myc* protein may undergo very similar modifications. Most importantly, the large amounts of these proteins made available by the bacterial expression system now make it possible to use the purified protein as a substrate for these modifications. This should lead to the precise characterization of the nature of these modifications.

Finally, the availability of these materials has also allowed us to develop a rapid assay to assess the oncogenic properties of both the ElA and c-*myc* gene products.[32] The assay involves the coinjection of ElA or c-*myc* in combination with the oncogenic H-*ras* protein (also made in *E. coli* using the pAS vector system) into early passaged primary fibroblasts. H-*ras* alone has little effect on primary cells, although H-*ras* will induce morphologic transformation and cell proliferation in many established cell lines.[33] ElA or c-*myc* alone also do not affect these primary cells. In contrast, cointroduction of H-*ras* plus ElA or H-*ras* plus c-*myc* proteins into rat embryo fibroblasts results in a rapid increase in DNA synthesis and the induction of cell division. This assay is now being used to analyze the transforming properties of the variant ElA proteins and similarly constructed variant c-*myc* proteins with the intention of achieving a domain analysis of the oncogenic properties of these molecules.

In conclusion, the pAS vector system has now made this most interesting class of oncogene products available to us in unprecedented quantities. These materials are being used as affinity probes for those cellular components with which they naturally interact in cells. They have already helped to define new biochemical properties of some of these oncogene products.[13,18,21] In addition, the large amounts of these materials in highly pure form should allow determination of the tertiary structure of these molecules by crystallographic analysis. Clearly, the ability to efficiently synthesize and obtain a gene product either in native or mutant form provides us with an entirely new perspective on how to approach the study of the structure and function of proteins that nature did not provide in adequate amounts for study.

USE OF THE pAS SYSTEM TO INCREASE GENE EXPRESSION VIA GENE FUSIONS

The insertion of a gene into an expression vector does not always guarantee high-level expression of the desired gene product. Often, levels of accumulation of the desired gene product are lower than expected. In such cases, gene expression may be optimized via the use of gene fusion technology.[34] Fusing a heterologous gene to the end of a well-expressed *E. coli* gene (such as *lac*Z) is a commonly used technique.[34–36] Such *lac*Z fusions, however, result in the synthesis of very large fusion proteins (> 100 kDa), and such syntheses limit the further manipulation and utility of the fusions. Now that many genes have been expressed efficiently in *E. coli,* there are clearly alternatives to *lac*Z for generating gene fusions. We have adapted the pAS vector system for optimizing expression using gene fusion technology.

The influenza NS1 gene has been cloned and efficiently expressed using the pAS system.[37] In addition to expressing at high levels (~30% of total cellular protein), this gene product exhibits unusual stability. We reasoned that fusion of heterologous gene products that were normally found to be expressed poorly in bacteria, to the 80 NH_2-terminal amino acids of the NS1 gene product, might result in stable high-level expression. This, indeed, was the case. For example, the human α_1-antitrypsin gene, which normally expresses very poorly in bacteria (see below), accumulates to at least 20% of total cellular protein when produced as a fusion with the first 80 amino acids of NS1 (A. Shatzman, unpublished results). The fact that our constructs only add 80 amino acids to the desired gene product makes manipulation of these fusion proteins far easier than similar fusions with *lac*Z. The availability of antisera that recognize this portion of the NS1 protein also facilitates the identification and purification of such protein fusions.

In contrast to the NS1 fusion vectors in which sequences are fused downstream of a well-expressed gene, coding information can also be fused upstream of a well-expressed product. For example, the pASK fusion vector system has been constructed by inserting the *E. coli gal*K gene into the pAS vectors adjacent to the *c*II rbs, such that the *Bam*HI site is present in all three reading frames (pASK 1-3) (FIG. 5A). Any coding sequence may now be inserted between the *c*II rbs and the *gal*K gene to form a translational fusion in which *gal*K coding information will be fused downstream of the inserted information. There are two principal methodologies in which the pASK system should be useful. Cells that contain pASK2 or pASK3 do not express the *gal*K gene because *gal*K is out of frame with the *c*II rbs. The cells appear as white colonies on indicator plates. Insertion of DNA fragments (at the *Bam*HI site) that contain open reading frames (ORFs) and shift *gal*K into the appropriate reading frame can be selected as *gal*K$^+$ red colonies on indicator plates in the appropriate *gal*E$^+$T$^+$K$^-$ host. This approach is similar to that used for detecting open reading frames by creating *β-gal* fusions.[34]

Another application of translational fusion vectors is to obtain increased levels of expression of those genes that are not translated efficiently. Alterations in the translation of these sequences may be phenotypically detected using the pASK system and assayed by monitoring changes in the expression of the *gal*K gene fusion product. This approach will be discussed below.

Although the expression levels of certain full-length gene products may be quite low, deletion or alteration of a small portion of the coding sequence (usually near the initiation signal) has in some cases resulted in dramatic increases in gene expression. Such sequences are commonly thought to form secondary structures that are inhibitory to translation. Computer analysis of potential RNA structure is now often used to help predict problems associated with secondary structure. Such analysis and subsequent sequence alteration (via oligonucleotide replacement or site-directed mutagenesis) has resulted in increased gene expression in several cases.[38,39] This approach is not always successful, however, and many predicted secondary structures, when "disrupted" by sequence changes, have shown little, if any, increase in gene expression. Because predictions of secondary structures and their effects are not completely reliable, a system such as pASK, which allows direct monitoring of changes in translational efficiency of any *gal*K fusion, can, and has, proven to be extremely useful. Such a system has the potential to tell us precisely what sequence changes will increase gene expression.

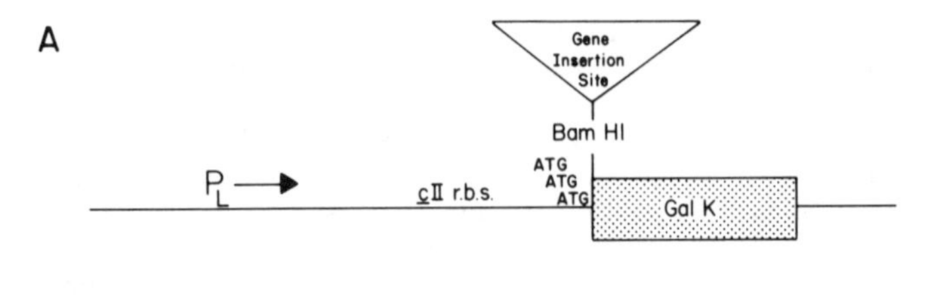

B

Asp Pro Gln Gly Asp Ala Ala Gln Lys Thr Asp Thr Ser His His
5' GAT CCC CAG GGA GAT GCT GCC CAG AAG ACA GAT ACA TCC CCC CAT 3'
GG GTC CCT CTA CGA CGG GTC TTC TGT CTA TGT AGG GGG GTA CTA G

FIGURE 5. **(A)** Transcriptional unit of the pASK translational fusion vectors for selection of "up expression" mutants and open reading frames. The λ P_L promoter is used to direct efficient transcription of the *E. coli gal*K gene (stippled area). This gene has been inserted adjacent to the *c*II ribosome binding site with an ATG initiation codon available in all three reading frames (pASK 1-3). A unique *Bam*HI site has been engineered adjacent to the ATG initiation codon to permit insertion of any gene coding sequence as a translational fusion with the *gal*K gene. **(B)** The sequence of the α_1-antitrypsin gene (and the protein sequence it codes for) that has been inserted into pASKl to form the translational fusion vector pαK.

α_1-Antitrypsin is the major human antiprotease that inhibits many serine proteases including elastase. The human gene encodes a protein of 394 amino acids. The coding region for α_1-antitrypsin, when cloned into our expression systems, expresses very poorly (<0.1% of total cellular protein). A deletion of the first 15 codons (45 bp) of their gene, however, results in expression of a truncated α_1-antitrypsin (αBcl) protein as 20-30% of total cellular protein (H. Johansen, unpublished results). Analysis of secondary structure and introduction of mutations to disrupt putative inhibitory structures had only minor effects on expression levels. In order to study the apparent inhibitory effects of this 45-bp region, we inserted a fragment encoding this region into the *Bam*HI site of pASK1 (FIG. 5B). The resulting vector, pαK, produced an in-frame fusion of the first 15 codons of α_1-antitrypsin with the *gal*K gene. In contrast to the control vector, pASKl, which expresses *gal*K as 10-20% of total cellular protein, the pαK vector expresses the *gal*K fusion product as only <0.1% of total cellular protein (FIG. 6A). Thus, the 45-bp fragment from the α_1-antitrypsin gene reduces *gal*K expression some 200-300-fold, just as it does to its original authentic gene product. Clones that express this fusion gene product are phenotypcially *gal*K$^-$, and therefore appear on indicator plates as white colonies. Mutations that result in increased levels of expression are readily selected as red colonies.

The pαK plasmid DNA was mutagenized *in vitro* by treatment with hydroxylamine or *in vivo* by passage through Mut T or Mut D strains of *E. coli* and transformed into a *gal*K$^-$ strain of *E. coli.* A spectrum of mutants was obtained, and some mutants were shown to produce levels of the αK fusion protein 5-20 times greater than those of the parent vector (FIG. 6A). The three most productive clones were found to have the same mutation: a G-to-A change in the first nucleotide following the initiator ATG codon. Clones containing this mutation express the fusion protein at the level of about 1% of total cellular protein (pαKA) (that is, at a level 20 times that of the wild-type αK protein). Other mutants have been characterized within this 45-bp α_1-antitrypsin-coding fragment that increase expression only 3-5-fold over that of the wild type. The fact that no single mutation has given greater than a 20-fold increase in αK expression indicates that multiple mutations may be necessary to obtain optimum levels of expression equivalent to those observed for the unfused *gal*K gene product.

Each of the mutations selected in the gene fusion system has been engineered back into the original α_1-antitrypsin-coding sequence. These changes increase α_1-antitrypsin expression commensurate to their effects on the *gal*K fusion product (that is, pαA that contains the G-to-A change described above expresses 20 times greater than authentic α_1-antitrypsin (FIG. 6B).

In order to see if combinations of these mutations would further increase α_1-antitrypsin expression, multiple mutations were introduced into the NH_2-terminal of the α_1-antitrypsin-coding sequence. One particular combination of four changes (pα4), which included the G-to-A change adjacent to the initiation codon, increased expression 150-200-fold over that of the wild type (FIG. 6B). This is about 60% of the level obtained for the deletion mutant, pαBcl, missing the entire NH_2-terminal end of the protein (FIG. 6B).

The ability to enhance α_1-antitrypsin expression 150-200-fold is a dramatic demonstration of the utility of the pASK translational fusion vectors. The selection of the G-to-A change in the second codon was crucial to achieving this level of expression and could not have been predicted from secondary structure analysis. The 20-fold increase in α_1-antitrypsin expression produced by this mutation is, in fact, probably independent of secondary structure and is instead the result of a stabilized translation initiation complex.[40]

The pASK gene fusion technology should be applicable to the study of other poorly expressed genes, and provides a reliable methodology for increasing heterologous gene expression in *E. coli.* Furthermore, the identification of specific mutations

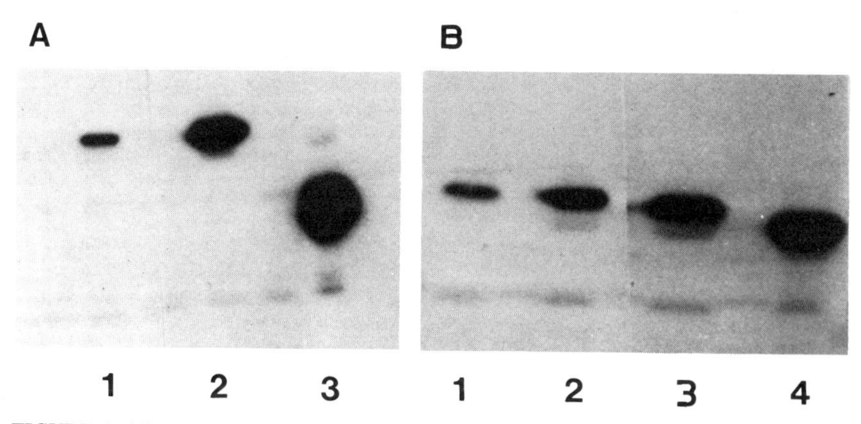

FIGURE 6. (**A**) Detection of authentic and fusion galactokinase proteins synthesized by *E. coli* by immunoblot analysis. The *gal*K protein synthesized in *E. coli* was detected by Western blot analysis with an *E. coli galK*-specific antiserum. Total cell extracts were prepared from *E. coli* strain AR120 transformed with either pαK (lane 1), pαKA (lane 2), or pASK1 (lane 3) after a 5-hr naladixic acid induction. Equal volumes of total cell extracts were resolved by SDS-PAGE analysis (10% acrylamide). The ratio of synthesis of αK:αKA:pASK1 was 1:20:200 as measured by laser scanning densitometry. (**B**) Detection of α_1-antitrypsin protein and variants synthesized by *E. coli* by immunoblot analysis (as above) using goat anti-α_1-antitrypsin (Cooper Biomedical). The ratio of expression of authentic α_1-antitrypsin (lane 1): αA (lane 2): α4 (lane 3): αBcl (lane 4) was 1:20:200:300 as determined by laser scanning densitometry.

that affect translational regulation may allow us to discern the parameters important for obtaining efficient translation of cloned genes in the pAS system.

ACKNOWLEDGMENTS

We thank C. Debouck, H. Johansen, J. Sutiphong, J. Young, B. Ferguson, and J. Auerbach for supplying unpublished information to this manuscript. We also thank Mary McCullough for typing and editing the manuscript.

REFERENCES

1. McKenney, K., H. Shimatake, D. Court, U. Schmeissner, C. Brady & M. Rosenberg. 1981. A system to study promoter and terminator signals recognized by *E. coli* RNA polymerase. *In* Gene Amplification and Analysis. Vol. 2. Analysis of Nucleic Acids by Enzymatic Methods. J.G. Chirikjian & T. Papas, Eds.: 383-415. Elsevier/North-Holland. New York, NY.
2. Ptashne, M., K. Backman, M. Z. Humayun, A. Jeffrey, R. Mauer, B. Meyer & R. Sauer 1976. Autoregulation and function of a repressor in bacteriophage lambda. Science **194:** 156-160.
3. Little, J. W. & D. W. Mount. 1982. The SOS regulatory system of *E. coli.* Cell **29:** 11-18.
4. Mott, J., R. Grant, Y. S. Ho & T. Platt. 1985. Maximizing gene expression from plasmid vectors containing the λ P_L promoter: Strategies for overproducing transcription termination factor. Proc. Natl. Acad. Sci. USA **82:** 88-92.
5. Rosenberg, M., D. Court, D. L. Wulff, H. Shimatake & C. Brady. 1978. Structure and function of an intercistronic regulatory region in bacteriophage lambda. *In* The Operon. J. Miller, Ed.: 345-371. Cold Spring Harbor Laboratory. Cold Spring Harbor, NY.
6. Shatzman, A. R., Y. S. Ho & M. Rosenberg. 1983. Use of phage λ regulatory signals to obtain efficient expression of genes in *E. coli. In* Experimental Manipulation of Gene Expression. M. Inouye, Ed.: 1-14. Academic Press. New York, NY.
7. Riccio, A., C. B. Bruni, M. Rosenberg M. Gottesman, K. McKenney & F. Blasi. 1985. Regulation of single and multicopy *his* operons in *Escherichia coli.* J. Bacteriol. **163:** 1172-1179.
8. Rosenberg, M., Y. S. Ho & A. R. Shatzman. 1983. The use of pKC30 and its derivatives for controlled expression of genes. *In* Methods in Enzymology. R. Wu, Ed. **101:** 123-138. Academic Press. New York, NY.
9. Gross, M., R. W. Sweet, G. Sathe, S. Yokoyama, O. Fasano, M. Goldfarb, M. Wigler & M. Rosenberg. 1985. Purification and characterization of human H-*ras* proteins expressed in *Escherichia coli.* Mol. Cell. Biol. **5:** 1015-1024.
10. Berka, T., J. Zimmerman, M. Rosenberg & A. Shatzman. 1986. Expression and function of yeast metallothionein in *E. coli.* In preparation.
11. Lorenzetti, R., L. Dani, D. Lappi, D. Martineau, M. Casati, L. Monaco, A. Shatzman, M. Rosenberg & M. Soria. 1985. pFCE4: A new system of *E. coli* expression-modification vectors. Gene **39:** 85-87.
12. Dente, L., G. Cesareni & R. Cortese. 1983. pEMBL: A new family of single-stranded plasmids. Nucleic Acids Res. **11:** 1648-1655.
13. Watt, R., A. Shatzman & M. Rosenberg. 1985. Expression and characterization of the human c-*myc* DNA-binding protein. Mol. Cell. Biol. **5:** 448-456.

14. Ho, Y.-S., D. Wulff & M. Rosenberg. 1986. Protein-nucleic acid interactions involved in transcription activation by the phage lambda regulatory protein *c*II. *In* Regulation of Gene Expression: 25 Years On. I. R. Booth & C. F. Higgins, Eds. Cambridge University Press. In press.
15. Baker, T., A. Grossman & C. Gross. 1984. A gene regulating the heat shock response in *E. coli* also affects proteolysis. Proc. Natl. Acad. Sci. USA **81:** 6779-6783.
16. Devare, S., A. Shatzman, K. Robbins, M. Rosenberg & S. Aaronson. 1984. Expression of the PDGF-related transforming protein of simian sarcoma virus in *E. coli.* Cell **36:** 43-49.
17. Ferguson, B., N. Jones, J. Richter & M. Rosenberg. 1984. Adenovirus Ela gene product expressed at high levels in *E. coli* is functional. Science **224:** 1343-1346.
18. Sweet, R., S. Yokoyama, T. Kamata, J. Feramisco, M. Rosenberg & M. Gross. 1984. The product of *RAS* is a GTPase and the T24 oncogenic mutant is deficient in this activity. Nature **311:** 273-275.
19. Ferguson, B., B. Krippl, H. Westphal, N. Jones, J. Richter & M. Rosenberg. 1985. Functional characterization of purified adenovirus ElA proteins expressed in *Escherichia coli. In* Cancer Cells: Growth Factors and Transformation. Vol. **3:** 265-274. Cold Spring Harbor Laboratory. Cold Spring Harbor, NY.
20. Ferguson, B., M. L. Pritchard, J. Feild, D. Rieman, R. G. Greig, G. Poste & M. Rosenberg. 1985. Isolation and analysis of an Abelson murine leukemia virus-encoded tyrosine-specific kinase produced in *E. coli.* J. Biol. Chem. **260:** 3652-3657.
21. Ferguson, B., B. Krippl, O. Andrisani, N. Jones, H. Westphal & M. Rosenberg. ElA 13S and 12S mRNA products made in *E. coli* both function as nuclear-localized transcription activators but do not directly bind DNA. Mol. Cell. Biol. **5:** 2653-2661.
22. Krippl, B., B. Ferguson, M. Rosenberg & H. Westphal. 1984. Functions of purified ElA protein microinjected into mammalian cells. Proc. Natl. Acad. Sci. USA **81:** 6988-6992.
23. Berk, A. F., F. Lee, T. Harrison, J. Williams & P. A. Sharp. 1979. Pre-early adenovirus 5 gene product regulates synthesis of early viral messenger RNAs. Cell **17:** 935-944.
24. Kao, H. T. & J. R. Nevins. 1983. Transcriptional activation and subsequent control of the human heat shock gene during adenovirus infection. Mol. Cell. Biol. **3:** 2058-2063.
25. Stein, R. & E. B. Ziff. 1984. HeLa cell β-tubulin gene transcription is stimulated by adenovirus 5 in parallel with viral early genes by an Ela-dependent mechanism. Mol. Cell. Biol. **4:** 2792-2799.
26. Gaynor, R. B., D. Hillman & A. J. Berk. 1984. Adenovirus early region lA protein activates transcription of a nonviral gene introduced into mammalian cells by infection or transfection. Proc. Natl. Acad. Sci. USA **81:** 1193-1197.
27. Schrier, P. I., R. Bernards, R. T. M. J. Vaessen, A. Houweling & A. J. van der Eb. 1983. Expression of class I major histocompatibility antigens switched off by highly oncogenic adenovirus 12 in transformed rat cells. Nature **305:** 771-774.
28. van den Elsen, P. J., S. de Peter, A. Houweling, J. van der Veer & A. J. van der Eb. 1982. The relationship between region Ela and Elb of human adenoviruses in cell transformation. Gene **18:** 175-188.
29. Ruley, H. E. 1983. Adenovirus early region lA enables viral and cellular transforming genes to transform primary cells in culture. Nature **304:** 602-606.
30. Richter, J. D., P. Young, N. C. Jones, B. Krippl, M. Rosenberg & B. Ferguson. 1985. A first exon-encoded domain of ElA sufficient for posttranslational modification, nuclear localization and induction of adenovirus E3 promoter expression in *Xenopus* oocytes. Proc. Natl Acad. Sci. USA **82:** 8434-8438.
31. Krippl, B., B. Ferguson, N. Jones, M. Rosenberg & H. Westphal. 1985. Mapping of functional domains in adenovirus type C ElA proteins. Proc. Natl Acad. Sci. USA **82:** 7480-7484.
32. Sullivan, N. F., D. Bar-Sagi, T. L. Chao, B. Q. Ferguson, M. S. Gross, T. Kamata, R. W. Sweet, R. A. Watt, W. J. Welch, M. Rosenberg & J. R. Feramisco. 1985. Microinjection of oncogene proteins causes DNA synthesis and cell division in nonestablished cells. *In* Cancer Cells.: Growth Factors and Transformation. Vol. **3:** 243-249. Cold Spring Harbor Laboratory. Cold Spring Harbor, NY.

33. FERAMISCO, J., M. GROSS, T. KAMATA, M. ROSENBERG & R. SWEET. 1984. Microinjection of the oncogene form of the human H-*Ras* (T-24) protein results in rapid proliferation of quiescent cells. Cell **38:** 109-117.
34. CASADABAN, M. J., A. MARTINEZ-ARIAS, S. K. SHAPIRA & J. CHOU. 1983. β-Galactosidase gene fusions for analyzing gene expression in *Escherichia coli* and yeast. Methods Enzymol. **100:** 293-308.
35. ROSE, M. & D. BOTSTEIN. 1983. Construction and use of gene fusions to *lac*Z (β-galactosidase) that are expressed in yeast. Methods Enzymol. **101:** 167-180.
36. GUARENTE, L. 1983. Yeast promoters and *lac*Z fusions designed to study expression of cloned genes in yeast. Methods Enzymol. **101:** 181-191.
37. YOUNG, J. F., V. DESSELBERG, P. PALESE, B. FERGUSON, A. SHATZMAN & M. ROSENBERG. 1983. Efficient expression of influenza virus NS1 nonstructural proteins in *E. coli.* Proc. Natl. Acad. Sci. USA **80:** 6105-6109.
38. SCHONER, B., H. HSIUNG, R. BELAGAJE, N. MAYNE & R. SCHONER. 1984. Role of mRNA translational efficiency in bovine growth hormone expression in *E. coli.* Proc. Natl. Acad. Sci. USA **81:** 5403-5407.
39. WOOD, C., M. BOSS, T. PATEL & J. EMTAGE. 1984. The influence of mRNA secondary structure on expression of an immunoglobulin heavy chain in *E. coli.* Nucleic Acids Res. **12:** 3937-3950.
40. GOLD, L., D. PRIBNOW, T. SCHNEIDER, S. SHINEDLING, S. SWEBILIUS & G. STORMO. 1981. Translation initiation in prokaryotes. Annu. Rev. Microbiol. **35:** 364-403.

Structure and Expression of 3-Hydroxy-3-methylglutaryl Coenzyme A Reductase[a]

KENNETH L. LUSKEY[b]

Departments of Molecular Genetics and Internal Medicine
University of Texas Health Science Center
Dallas, Texas 75235

Cholesterol balance in cells is maintained by the coordinate regulation of the endogenous pathway of cholesterol synthesis and the exogenous pathway of cholesterol delivery.[1] Endogenous synthesis is accomplished by a series of over 20 enzymatic reactions that catalyze the formation of cholesterol from acetyl CoA. The rate of cholesterol synthesis is regulated primarily by regulating the expression of 3-hydroxy-3-methylglutaryl coenzyme A (HMG CoA) reductase, the enzyme that mediates the conversion of HMG CoA to mevalonate. This is one of the early steps in the pathway of cholesterol synthesis. Cells also obtain cholesterol from the receptor-mediated endocytosis of cholesterol-carrying low-density lipoprotein (LDL) in the extracellular environment. This process is mediated by the LDL receptor, a cell surface molecule that specifically binds LDL. This LDL is then internalized and delivered to lysosomes where it is degraded. Hydrolysis of the cholesterol esters yields cholesterol for the cell. The expression of both HMG CoA reductase and the LDL receptor are regulated such that adequate cholesterol is available for the synthesis of plasma membranes, steroid hormones, bile acids, and lipoproteins. When sufficient cholesterol is delivered via the LDL receptor, cholesterol synthesis is shut off by suppressing the activity of HMG CoA reductase. In this manner, the overproduction and excess accumulation of cholesterol is avoided.

The ability to suppress HMG CoA reductase is critical to the control of cholesterol homeostasis; however, the mechanisms responsible for such control were not clear. It has recently become possible to study the regulation of HMG CoA reductase using the techniques of molecular biology. The isolation and characterization of both the HMG CoA reductase gene and cDNA has provided the tools necessary to analyze the mechanisms responsible for the regulation of this enzyme at a molecular level. These studies have shown that the regulation of this enzyme is complex, with actions occurring at the levels of the gene, mRNA, and protein that can ultimately affect the expression of enzyme activity.

In most cells HMG CoA reductase is present in very low amounts, even under conditions where the enzyme is maximally induced. This made characterization of the enzyme quite difficult. High-level expression of HMG CoA reductase was accom-

[a] This research was supported by Grant HL 20948 from the National Institutes of Health.

[b] An Established Investigator of the American Heart Association.

plished by adapting Chinese hamster ovary cells to growth in high levels of compactin,[2] a competitive inhibitor of HMG CoA reductase isolated by Endo *et al.*[3] These cells, designated UT-1 cells, have amplified the gene for HMG CoA reductase and can express 500- to 1000-fold higher reductase activity than the parental cell. Approximately 1% of the cell protein in UT-1 cells is HMG CoA reductase. Despite the high levels of expression of reductase in these cells, the enzyme still is regulated by sterols.[2,4] Using UT-1 cells, the structures of the gene, mRNA, and protein for HMG CoA reductase have been characterized in detail. A summary of these findings is shown in FIGURE 1. The gene is 25 kb long and contains 20 exons.[5] Following transcription and processing to generate the mature mRNA, mRNA species 5.0-5.5 kb long are generated. When translated, these mRNAs yield a protein product with a molecular mass of 97 kDa.

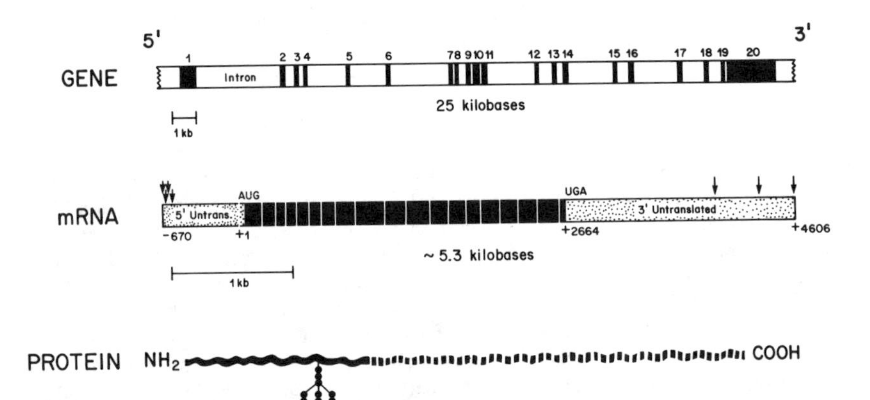

FIGURE 1. Schematic representation of the HMG CoA reductase gene, mRNA, and protein structures. The gene is shown with exons in black that are separated by 19 introns. The mRNA structure indicates the 5′ and 3′ untranslated regions and the coding region. The arrows show the multiple sites of transcription initiation and polyadenylation. Variation in the splicing of the 5′ untranslated region is not shown. The protein is shown with the hydrophobic and catalytic domains. The N-linked oligosaccharide is shown. Reproduced with permission from the *Journal of Lipid Research*.[1]

The first level at which HMG CoA reductase is regulated is at the level of the gene. The rate of transcription of the reductase gene declines after the addition of sterols.[4] Experiments in UT-1 cells have shown that following the addition of LDL as a source of cholesterol or the addition of oxysterols dissolved in ethanol, reductase mRNA synthesis falls to 7-8% of the induced rate.[4] Similar changes in reductase mRNA levels have been seen in other cultured cells and animal tissues associated with suppression of HMG CoA reductase activity.[6,7] This ability to regulate transcription is a property of the promoter of the HMG CoA reductase gene. Following the isolation of the gene, Osborne *et al.* studied the ability of the 5′ flanking region of the gene to promote and regulate transcription while it was linked to an independent marker.[8] When a 508-bp fragment that included 300 bp of DNA 5′ of the major site of transcription initiation was fused with the chloramphenicol acetyltransferase (CAT) gene, CAT activity was generated in cells in which this DNA had been introduced by DNA transfection. Most importantly, CAT activity and CAT RNA was suppressed in these cells following the addition of sterols. Constructs that contained smaller amounts of the 5′ flanking sequence were less active in promoting transcription and were less susceptible to regulation by sterols.

The structure of the region of the HMG CoA reductase gene that is involved in the sterol-mediated regulation of transcription is different from many other eukaryotic promoters.[5] In the region upstream of transcription initiation, sequences commonly found in other promoters, such as a TATA box or a CAAT box, are absent. Instead this region is very GC rich, containing five repeats of the hexanucleotide 5′-GGGCGG-3′ or its complement, 5′-CCGCCC-3′. This same hexanucleotide repeat is found repeated six times in the SV40 promoter where it is an important sequence for promoter activity. Recently other eukaryotic promoters have been described which are GC rich and contain several copies of this hexanulceotide. The role of these sequences in promoting transcription in the HMG CoA reductase promoter is not certain. In the SV40 promoter these hexanucleotides have been shown to be an important part of the binding site for a specific transcription factor, SP1.[9]

Heterogeneity of both transcription initiation and splicing of the first intron from the reductase mRNA result in a family of mature transcripts that differ in their 5′ untranslated regions.[5,10] Reynolds *et al.* have shown that in UT-1 cells there are two classes of reductase mRNAs differing at their 5′ ends.[10] The most abundant class accounts for about 70% of the reductase transcripts in the cell. These molecules have short 5′ untranslated regions that contain no AUG codons upstream of the AUG used to initiate translation of the reductase protein. In contrast, the second class of reductase transcripts initiates 50 to 100 nucleotides downstream of the major transcription initiation site, and a variety of 5′ donor splice sites are used to remove the first intron. This results in a group of reductase mRNAs that have 5′ untranslated regions from 360 to 670 nucleotides in length. In these transcripts, from three to eight AUG codons are present upstream of the AUG used to initiate translation of the reductase protein.

Peffley and Sinensky have recently shown that mevalonate is able to suppress HMG CoA reductase activity by decreasing the translational efficiency of the reductase mRNA.[11] After sterols suppressed reductase mRNA levels and synthesis of the reductase protein, mevalonate was able to further decrease the rate of synthesis of the reductase protein without any additional suppression of the mRNA levels. This observation is consistent with the multivalent regulation of HMG CoA reductase first observed by Brown and Goldstein.[12] The relationship of translational efficiency to the variable structure of the 5′ untranslated region is unknown, although it is attractive to think that such control might be mediated through this region of the mRNA.

The third level of control of HMG CoA reductase activity is at the level of the protein. Normally HMG CoA reductase is bound to the membrane of the endoplasmic reticulum. Of interest is the fact that the membrane-bound HMG CoA reductase is degraded much more rapidly than most other cellular proteins, even those located in the membrane of the endoplasmic reticulum.[13,14] In addition to the rapid rate of degradation of the reductase enzyme, this rate can vary depending on the metabolic conditions of the cell. In the presence of sterols, the enzyme is degraded two or three times faster, and thus enhanced degradation contributes to the suppression of enzyme activity under these conditions.[13,14] Insight into the mechanism responsible for regulation of degradation came from the characterization of a full-length cDNA for the hamster enzyme.[15] The nucleotide sequence of this clone revealed that the reductase is 887 amino acid residues in length and contains several distinct domains in the protein structure. Computer analysis of the predicted protein sequence of the hamster enzyme allowed Liscum *et al.* to define a model for the secondary structure of the enzyme.[16] This model is shown in FIGURE 2. The NH_2-terminal end of the protein contains seven stretches of hydrophobic residues of sufficient length to span the membrane of the endoplasmic reticulum. The COOH-terminal contains two regions predicted to have a repeated β-pleated-sheet structure. By a combination of biochemical and immunologic studies, Liscum *et al.* were able to show that the NH_2-terminal third of the molecule was bound to the membrane of the endoplasmic reticulum and contained the carbohydrate moiety of the enzyme. A separate 53-kDa fragment that could be generated by proteolysis retained all the catalytic activity of the enzyme and was derived from the COOH-terminal portion of the enzyme.[16]

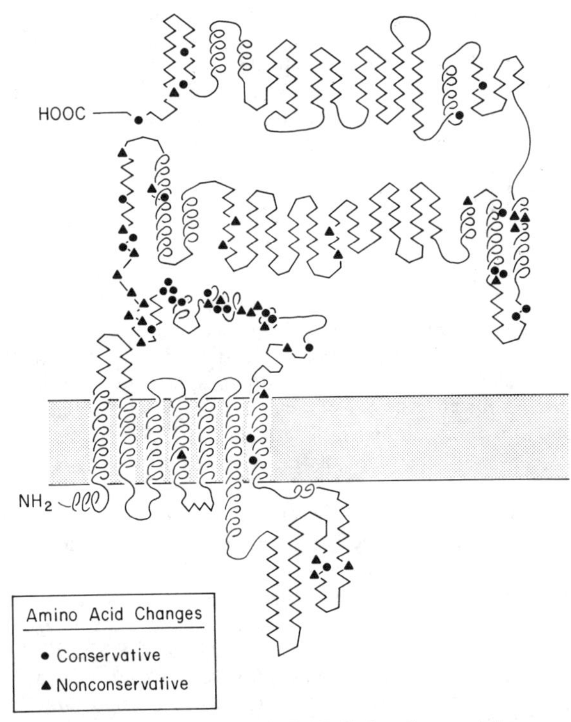

FIGURE 2. Predicted secondary structure of HMG CoA reductase. The proposed structure of the hamster enzyme is drawn as the backbone of the structure. The membrane of the endoplasmic reticulum is shaded. The sites of amino acid substitutions in the human enzyme compared to the hamster enzyme are indicated. Conservative substitutions are marked with a circle (●) and nonconservative substitutions with a triangle (▲). Reproduced with permission from the *Journal of Biological Chemistry.*[17]

To define the function of the different domains of the reductase molecule, the technique of DNA transfection was utilized to express the reductase cDNA. Plasmid DNA that contained the SV40 promoter to drive transcription of the cDNA-encoding HMG CoA reductase was introduced into reductase-deficient Chinese hamster ovary cells by calcium phosphate-mediated transfection. Reductase RNA derived from the plasmid was produced, and reductase protein was translated according to the sequence of the transfected cDNA. As the parent cell lacked its own endogenous reductase,[18] all reductase activity present in these cells was derived from the transfected reductase cDNA. By modifying the structure of the transfected reductase cDNA, the effect of a change in the protein structure of the enzyme could be studied.

Introduction of the full-length reductase cDNA resulted in the expression of a 97-kDa enzyme that was anchored to the membrane of the endoplasmic reticulum.[19] Just as in a normal cell, the rate of degradation of this enzyme was still enhanced by the addition of sterols. Gil *et al.* then studied the expression of an altered HMG CoA reductase cDNA in which the DNA corresponding to the membrane-bound domain had been specifically deleted by oligonucleotide-directed mutagenesis.[20] As the active site of this enzyme was still intact, it was expected to be catalytically active within the cell. Following transfection of this altered cDNA into the reductase-deficient cells, cells were selected that expressed HMG CoA reductase activity derived from the transfected cDNA. In this case the enzyme that was produced was no longer membrane bound, but was a 60-kDa soluble enzyme. The interesting observation was that when the degradation of this soluble reductase was studied it was no longer affected by the addition of sterols. The enzyme was also degraded at a much slower rate than the membrane-bound form of the enzyme. These experiments indicate that the membrane-bound domain of the enzyme is responsible for the rapid turnover of the enzyme and the ability of sterols to affect the degradation process.

The relative importance of this form of regulation was indicated by comparing the structure of the enzyme in hamster and man. If the membrane-bound domain is not critical to the function and regulation of the enzyme but is only present to anchor the protein to the membrane, this region of the protein should not be highly conserved. Replacement of one hydrophobic amino acid with another hydrophobic residue should not inhibit the ability of this part of the protein to transverse the membrane. By characterizing a full-length cDNA for the human HMG CoA reductase, the amino acid sequence was deduced and compared to that of the hamster enzyme.[20] Those sites at which amino acid substitutions were observed are indicated in FIGURE 2. Within the membrane-bound domain of the enzyme only 6 amino acid substitutions were found out of 339 residues. Three of these occurred within those stretches of the protein that span the membrane. This region of the protein was the most highly conserved region of the protein. The catalytic domain was also highly conserved with 22 amino acid substitutions out of 439 residues. In contrast to these highly conserved regions, the 110-residue segment that links these regions together had 32 substitutions. This degree of conservation indicates that the functional role of the membrane-bound domain is just as important and just as constrained by evolution as the active site of the enzyme.

These studies indicate how molecular biology has helped approach the complex problem of regulation of HMG CoA reductase, the rate-limiting enzyme of cholesterol synthesis. By isolating and characterizing the gene, mRNA, and protein, several different mechanisms of regulation of this enzyme can be studied independently. The use of techniques to modify the structure of the promoter, 5′ untranslated region, and protein sequence of HMG CoA reductase should allow the mechanisms by which sterols and mevalonate modulate the expression of this enzyme to be understood.

REFERENCES

1. GOLDSTEIN, J. L. & M. S. BROWN. 1984. Progress in understanding the LDL receptor and HMG CoA reductase, two membrane proteins that regulate the plasma cholesterol. J. Lipid Res. **25:** 1450-1461.
2. CHIN, D. J., K. L. LUSKEY, R. G. W. ANDERSON, J. R. FAUST, J. L. GOLDSTEIN & M. S. BROWN. 1982. Appearance of crystalloid endoplasmic reticulum in compactin-resistant

Chinese hamster cells with a 500-fold elevation in 3-hydroxy-3-methylglutaryl CoA reductase. Proc. Natl. Acad. Sci. USA **79:** 1185-1189.

3. ENDO, A., M. KURODA & K. TANZAWA. 1976. Competitive inhibition of 3-hydroxy-3-methylglutaryl coenzyme A reductase by ML-236A and ML-236B fungal metabolites having hypocholesterolemic activity. FEBS Lett. **72:** 323-326.
4. LUSKEY, K. L., J. R. FAUST, D. J. CHIN, M. S. BROWN & J. L. GOLDSTEIN. 1983. Amplification of the gene for 3-hydroxy-3-methylglutaryl coenzyme A reductase, but not for the 53-kDa protein, in UT-1 cells. J. Biol. Chem. **258:** 8462-8469.
5. REYNOLDS, G. A., S. K. BASU, T. F. OSBORNE, D. J. CHIN, G. GIL, M. S. BROWN, J.L. GOLDSTEIN & K. L. LUSKEY. 1984. HMG CoA reductase: A negatively regulated gene with unusual promoter and 5′ untranslated regions. Cell **38:** 275-286.
6. LISCUM, L., K. L. LUSKEY, D. J. CHIN, Y. K. HO, J. L. GOLDSTEIN & M. S. BROWN. 1983. Regulation of 3-hydroxy-3-methylglutaryl coenzyme A reductase and its mRNA in rat liver as studied with a monoclonal antibody and a cDNA probe. J. Biol. Chem. **258:** 8450-8455.
7. EDWARDS, P. A., S.-F. LAN & A. M. FOGELMAN. 1983. Alterations in the rates of synthesis and degradation of rat liver 3-hydroxy-3-methylglutaryl coenzyme A reductase produced by cholestyramine and mevinolin. J. Biol. Chem. **258:** 10219-10222.
8. OSBORNE, T. F., J. L. GOLDSTEIN & M. S. BROWN. 1985. 5′ End of HMG CoA reductase gene contains sequences responsible for cholesterol-mediated inhibition of transcription. Cell **42:** 203-212.
9. DYNAN, W. S. & R. TJIAN. 1985. Control of eukaryotic messenger RNA synthesis by sequence-specific DNA-binding proteins. Nature **316:** 774-777.
10. REYNOLDS, G. A., J. L. GOLDSTEIN & M. S. BROWN. 1985. Multiple mRNAs for 3-hydroxy-3-methylglutaryl coenzyme A reductase determined by multiple transcription initiation sites and intron splicing sites in the 5′ untranslated region. J. Biol. Chem. **260:** 10369-10377.
11. PEFFLEY, D. & M. SINENSKY. 1985. Regulation of 3-hydroxy-3-methylglutaryl coenzyme A reductase synthesis by a nonsterol mevalonate-derived product in Mev-1 cells: Apparent translational control. J. Biol. Chem. **260:** 9949-9952.
12. BROWN, M. S. & J. L. GOLDSTEIN. 1980. Multivalent feedback regulation of HMG CoA reductase, a control mechanism coordinating isoprenoid synthesis and cell growth. J. Lipid Res. **21:** 505-517.
13. EDWARDS, P. A. & R. G. GOULD. 1972. Turnover rate of hepatic 3-hydroxy-3-methylglutaryl coenzyme A reductase as determined by use of cycloheximide. J. Biol. Chem. **247:** 1520-1524.
14. EDWARDS, P. A., S.-F. LAN, R. D. TANAKA & A. M. FOGELMAN. 1983. Mevalonolactone inhibits the rate of synthesis and enhances the rate of degradation of 3-hydroxy-3-methylglutaryl coenzyme A reductase in rat hepatocytes. J. Biol. Chem. **258:** 7272-7275.
15. CHIN, D. J., G. GIL, D. W. RUSSELL, L. LISCUM, K. L. LUSKEY, S. K. BASU, H. OKAYAMA, P. BERG, J. L. GOLDSTEIN & M. S. BROWN. 1984. Nucleotide sequence of HMG CoA reductase, a glycoprotein of the endoplasmic reticulum. Nature **308:** 613-617.
16. LISCUM, L., J. FINER-MOORE, R. M. STROUD, K. L. LUSKEY, M. S. BROWN & J. L. GOLDSTEIN. 1985. Domain structure of 3-hydroxy-3-methylglutaryl coenzyme A reductase, a glycoprotein of the endoplasmic reticulum. J. Biol. Chem. **260:** 522-530.
17. LUSKEY, K. L. & B. STEVENS. 1985. Human 3-hydroxy-3-methylglutaryl coenzyme A reductase: Conserved domains responsible for catalytic activity and sterol-regulated degradation. J. Biol. Chem. **260:** 10271-10277.
18. MOSLEY, S. T., M. S. BROWN, R. G. W. ANDERSON & J. L. GOLDSTEIN. 1983. Mutant clone of Chinese hamster ovary cells lacking 3-hydroxy-3-methylglutaryl coenzyme A reductase. J. Biol. Chem. **258:** 13875-13881.
19. CHIN, D. J., G. GIL, J. R. FAUST, J. L. GOLDSTEIN, M. S. BROWN & K. L. LUSKEY. 1985. Sterols accelerate degradation of HMG CoA reductase encoded by a constitutively expressed cDNA. Mol. Cell. Biol. **5:** 634-641.
20. GIL, G., J. R. FAUST, D. J. CHIN, J. L. GOLDSTEIN & M. S. BROWN. 1985. Membrane-bound domain of HMG CoA reductase is required for sterol-enhanced degradation of the enzyme. Cell **41:** 249-258.

Recombinant Retroviruses in Transgenic Mice[a]

FLORENCE M. BOTTERI,[b] HERMAN VAN DER PUTTEN,[b] A. DUSTY MILLER,[c,d] HUNG FAN,[e] AND INDER M. VERMA[b]

[b] *Molecular Biology and Virology Laboratory*
Salk Institute
San Diego, California 92138

[d] *Hutchinson Cancer Research Center*
Seattle, Washington 98104

[e] *Department of Molecular Biology and Biochemistry*
University of California
Irvine, California 92717

INTRODUCTION

Gene transfer into the mouse germ line is a crucial test to monitor cell-specific gene expression because the newly introduced sequences are present in every mouse cell. Germ-line integration has primarily been achieved using direct microinjection of DNA into the pronucleus of the fertilized egg.[1] It can also be achieved, however, via incorporation into morulae or blastocysts of embryonal carcinoma (EC) and embryonic stem (ES) cells that harbor foreign genes (FIG. 1).[2–8] Furthermore, genetic information can be inserted into preimplantation and postimplantation mouse embryos via retroviral infection.[9–14]

Transgenic animals generated via DNA microinjection into the pronucleus of the fertilized egg have revealed abundant information about sequences that direct tissue-specific expression of a variety of genes. Such animals, however, harbor mostly tandem integrations (from a few up to hundreds of copies of the injected DNA molecule).[15–18] Such an aberrant organization of genes might affect neighboring chromatin structures (pleiotropic effects) or disrupt transcriptional units.[17,19–21] Tandemly arranged molecules can further undergo major rearrangements resulting, for example, in deletions.[18] Also, quantitative studies on gene expression are difficult to interpret when large copy numbers of the gene are present.

The use of EC and ES cells offers the advantage of selecting cells *in vitro* for insertion and expression of a foreign gene before these cells are incorporated into early mouse embryos to generate transgenic mice. Karyotype instability of these pluripotential cells in culture and the low contribution of transplanted EC and ES cells to the germ line of chimeric mice are some major disadvantages.[22]

[a] This work was supported in part by grants from the National Institutes of Health.

[c] A fellow of the Leukemia Society of America.

The development of techniques to insert into the germ line a single copy of a gene could have some major advantages, as illustrated by results obtained using P-element-mediated gene transfer into a *Drosophila* germ line.[23,24] Retroviruses offer such unique advantages for gene transfer into mammals. First, they mediate insertion of a single copy of a nonpermuted provirus per chromosomal site and can infect a wide variety of cells from different species. Second, retroviral vectors can harbor foreign genes expressed of internal promoters as well as genes expressed of the LTR promoter.

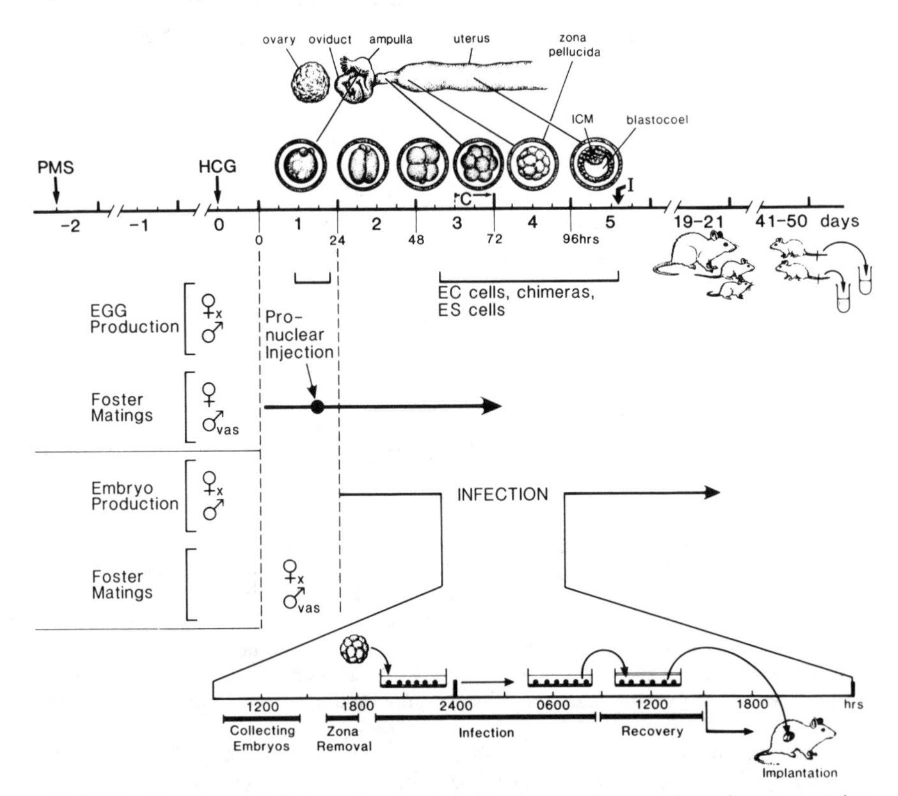

FIGURE 1. Genetic manipulations of preimplantation mouse embryos. Schematic representation of preimplantation mouse embryo stages at various times post coitum and their location in the female reproductive tract. Also, mating schedules as required for embryo and egg production and foster matings are illustrated. The various time points at which the different genetic manipulations can be carried out are indicated. C: compaction (60-72 hr post coitum); I: implantation of blastocysts into the uterine wall; ICM: inner cell mass; PMS: pregnant mares serum; HCG: human chorionic gonadotropin; vas: vasectomized.

Third, suicide vectors offer the possibility to insert single copy genes via a retroviral vector without the presence of the viral LTR-enhancer elements. Such vectors will, it is hoped, show minimal effects of the viral LTR sequences upon expression of a foreign gene's promoter within the provirus. Fourth, combinatorial specificities could easily be determined of two or more enhancer sequences on the expression of a single gene within a provirus. Fifth, and finally, it also seems possible to target and modulate the expression of a gene within a retrovirus via the enhancer sequences in the LTR.

Until recently, only the highly leukemogenic replication-competent retrovirus Moloney murine leukemia virus (M-MuLV) was successfully inserted into the mouse germ line.[9–13] Ecotropic MuLVs and a dual tropic MuLV have been found to reintegrate spontaneously into the germ line of certain laboratory strains of mice like the high-virus-titer AKR strain.[25] Although no such evidence is available for amphotropic viruses, both amphotropic and ecotropic viruses can infect EC cells,[26–30] which resemble early mouse embryo cells. These notions led us to develop a general strategy for the insertion of both replication-competent and -defective retroviruses into the mouse germ line.[13] Recombinant retroviruses can therefore be used to insert a single copy of a given gene into the mouse germ line. Also, we present evidence suggesting that recombinant retroviruses inserted in preimplantation mouse embryos can be expressed in tissues of adult mice.

RESULTS AND DISCUSSION

Mouse Embryo Infection

FIGURES 1 & 2 outline the general approach of infecting mouse embryos by recombinant retroviruses.[13] Briefly, female mice were superovulated by injecting 5 international units of pregnant mares serum (PMS) and 2.5 international units of human chorionic gonadotropin (hCG) given 42-52 hr apart. After mating, embryos were recovered in the morning of day 3 post coitum by flushing oviduct-uterus junctions and uteri with modified Whitten's medium. The zona pellucida was removed either with Pronase (22 U/ml) or by brief exposure to acidified Tyrode's solution. Denuded embryos were then layered on monolayers of virus and/or vector-producing cells in the presence of 4 μg/ml Polybrene. Infections (5-16 hr) were at 37° C in Dulbecco-modified Eagle's medium (DMEM) + 10% fetal calf serum (FCS) in 5% CO_2 in air. After infection, morulae were transferred into modified Whitten's medium under a layer of equilibrated paraffin oil. After 2-4 hr (for a 12-16-hr infection) or 12 hr (for a 5-8-hr infection) embryos were transferred into uterine horns of pseudopregnant ICR females (anesthetized with Avertin), which were at day 3 post coitum because embryos develop slower *in vitro* as compared to *in vivo*.

The infection of mouse embryos by retroviruses might be influenced by the technique used to remove the zona pellucida, the conditions affecting the affinity of the viral glycoprotein (gp70) to the cell-surface receptor, the host range of the viral coat (gp70) protein, the virus titers, the infection time, and the particular mouse embryos to be infected. Some variations in these parameters are described below.

Retroviral infection follows the pathway of receptor-mediated endocytosis.[31–33] Ecotropic and amphotropic viruses are known to bind to different cell-surface receptors, the genes of which have been mapped to different chromosomal loci.[34,35] Hence, ecotropic and amphotropic viruses show no interference during infection.[36] The interaction of the viral gp70 with its cell-surface receptor can be abolished by trypsinization.[37] Pronase, which can be used to remove the zona pellucida, would probably have the same effect. Although early mouse embryos do synthesize some RNA and proteins, the reestablishment of a cell-surface receptor might occur very slowly or not at all if the receptor was coded for by maternal mRNA because most maternal mRNA is degraded in the two-cell embryo.[38,39] Because the virus receptor has been shown to

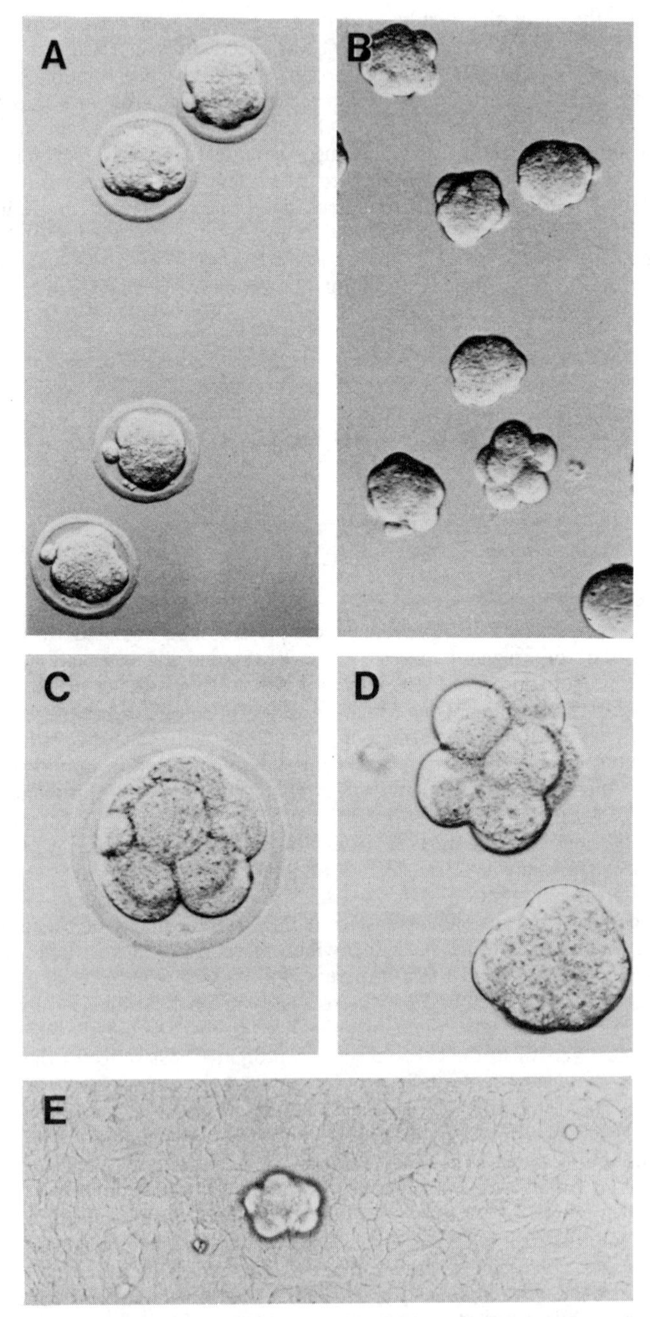

FIGURE 2. Infection of preimplantation mouse embryos. (**A**) Four compacted mouse embryos in the 8-12-cell stage, each surrounded by the zona pellucida. (**B**) Various embryos after removal of the zona pellucida. (**C**) Noncompacted embryos in the eight-cell stage, before zona removal. (**D**) Two embryos in the eight-cell stage, after zona removal: one before, and one after compaction. (**E**) An embryo in the eight-cell stage on top of a monolayer of virus-producing cells.

be rather insensitive to acid, we chose to remove the zona pellucida by treating it with acidified Tyrode's solution in most experiments.[40]

During most infections, 8 mM $CaCl_2$ was added to the culture medium. The optimal concentration for Ca^{2+} has been shown to be 10 mM for viral gp70 receptor interaction *in vitro.*[37] Because cleavage-stage mouse embryos have the same Ca^{2+}-dependent cell-cell adhesion system as teratocarcinoma cells,[41] no interference with survival of embryos was observed, as expected (TABLE 2). Also, 10 mM Ca^{2+} did not affect virus production (unpublished observation). Furthermore, infections are carried out in the presence of 4 μ/ml of Polybrene.[42] Alternatively, embryos are pretreated for 2 hr with 8 μ/ml of Polybrene, and infection is carried out in medium containing 0.4 μ/ml of Polybrene.

Retroviruses, Proviral DNA, and Probes

The proviral structures of various recombinant retroviruses and M-MuLV as well as the location of the probes used to identify each provirus are shown in FIGURE 3.

General Strategy of Analysis

To monitor infections, tail DNA was analyzed for the presence of proviruses. Also, in some cases the blood was probed for the presence of infectious viruses like M-MuLV and ΔMo+Py M-MuLV. The infected eight-cell-stage mouse embryos could be mosaic because all cells of the morula may not be infected. Furthermore, if some cells are multiply infected, they would carry more than one provirus, but each in a different chromosomal site. Integration of proviral DNA can be monitored by digesting tail DNA with the restriction endonuclease *Sst*I and hybridization to the U3LTR probe (FIG. 3). Copies of helper and/or recombinant provirus in different chromosomal positions should each contribute in the detection of the respective internal *Sst*I fragments. M-MuLV proviral DNA yields a specific fragment of 5.6 kb whereas other recombinant proviruses yield fragments that are different in size as compared to M-MuLV (FIG. 3). Furthermore, if many cells in the tail carry a provirus, another *Sst*I fragment should be detected, one which represents a 5′ LTR-host DNA fusion fragment.

Insertion of Replication-competent Retroviruses

Fifty progeny were derived from embryos exposed (FIG. 1) to cells producing a mixture of 10^7 plaque-forming units (pfu)/ml M-MuLV and 10^6 colony-forming units (cfu)/ml of the recombinant LPHGL retrovirus (FIG. 3). Sixteen animals out of 50 analyzed carried M-MuLV proviral DNA,[13] of which 13 showed one or more fragments representing a 5′ LTR-host DNA junction (FIG. 4A). Eleven out of 16 animals were viremic at 6 weeks of age as detected by XC-plaque assay (TABLE 1). None of the mice analyzed were positive for DNA from the recombinant virus LPHGL.

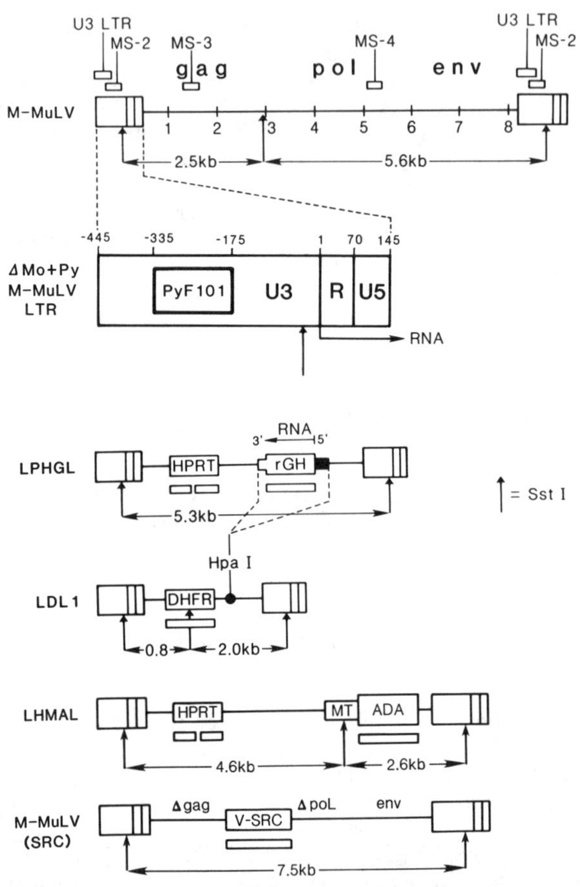

FIGURE 3. Structural maps of viruses and vectors. The probes (open bars) were used to characterize proviral DNAs in transgenic mice. MS-2, MS-3, MS-4, U3LTR, hypoxathine phosphoribosyl transferase (HPRT), rat growth hormone (rGH), dihydrofolate reductase (DHFR), and the PyF101 probes have all been described before.[13] The 2-kb *Eco*RI-*Nco* I fragment of pAMG1 was used as an human adenosine deaminase (ADA)-specific probe.[43] The 3-kb *Eco*RI fragment of psrc (*Eco*RIB) was used as an src-specific probe.[44]

Eleven out of 16 mice were bred, and the tail DNA of their offspring was tested for proviral DNA integrations (TABLE 1). Seven out of 11 yielded offspring that carried M-MuLV proviral DNA. Again, no Fl mice were found to carry an integrated LPHGL provirus. The transmission frequencies varied (5–80%) (TABLE 1), as expected from variations in mosaicism in the germ line of the parental mice. In general, all offspring had the 5′ LTR-host DNA junctions in their tail DNA that were also found in the tail DNA of their founder.[13] No selection of cells carrying proviral copies seems to occur during embryonic development because proviral DNA integrations as detected in tail DNA (largely of mesodermal origin) were equally represented in tissues of ectodermal (for example, brain) and endodermal (for example, liver) origin (FIG. 4). Therefore, genetically transmittable proviruses in transgenic mosaic animals can reliably be identified in the tail DNA of parental mice.

The ΔMo+Py M-MuLV virus is a replication-competent and ecotropic recombinant virus that lacks the M-MuLV-type enhancer in its LTR (FIG. 3).[45] Instead, it contains one of the two distinct enhancer elements of the polyoma mutant PyF101. This element confers replication competence to PyF101 in undifferentiated F9 EC cells[46] and allows gene expression in F9 EC cells of the ΔMo+Py M-MuLV LTR.[47] In contrast, M-MuLV expression does not occur in EC cells, nor does it occur after the insertion of M-MuLV into preimplantation mouse embryos because of methylation of DNA,[48,49] lack of M-MuLV LTR enhancer function,[47] splicing defects, and possible *cis*-acting factors.[27-29,50] Furthermore, ΔMo+Py M-MuLV, in contrast to M-MuLV, is nonleukemogenic upon injection into neonatal mice.[45] These features make this type of recombinant retrovirus a more useful vehicle for gene transfer into animals. Therefore, initially we addressed the question of whether such a recombinant retrovirus could be stably inserted into the mouse germ line. Also, it would show whether the PyF101 enhancer modifies the expression pattern of a ΔMo+Py M-MuLV in transgenic mice as compared to mice harboring a M-MuLV provirus. Furthermore, we chose this virus because the ΔMo+Py M-MuLV LTR is efficiently expressed in EC

TABLE 1. Embryo Infection and Germ Line Transmission

Infection (hr)	Number Born/ Number Transferred	Animals Analyzed	XC Test	Tail DNA	Number FI Positive/ Number FI Analysed
5	10/25	7	−	−	NT
		T3	+	+	3/8
		T5	+	+	5/27
		T17	−	+	NT
6	12/52	6	−	−	NT
		T1	+	+	0/8
		T2	+	+	NT
		T15	+	+	4/6
		T22	−	+	1/23
		T23	−	+	0/10
		T26	+	+	0/11
7	9/63	6	−	−	NT
		T28	+	+	4/13
		T29	−	+	NT
		T32	+	+	4/18
12	20/133	15	−	−	NT
		T40	+	+	0/7
		T48	+	+	7/9
		T50	+	+	NT
16	5/19	4	−	−	NT
		T8	−	+	NT
Totals	56/292 (19%)	50	11	16 (32%)	7/11[a]

NOTE: Infections were carried out for the indicated periods of time in the presence of 4 μg/ml of Polybrene. A virus-producing cell line was used that produces 10^7 pfu/ml M-MuLV and 10^6 cfu/ml LPHGL (TABLE 2). The number of positive offspring (FI) among the total number of FI mice tested is given in the last column. Embryos were C57BL/6 × SJL/J. XC tests were performed on 20 μl of blood (+ 100 μg/ml (174 U/mg) Heparine).[13] NT: not tested.

[a]Seven out of 11 parental animals yielded positive offspring among the number of FI mice tested.

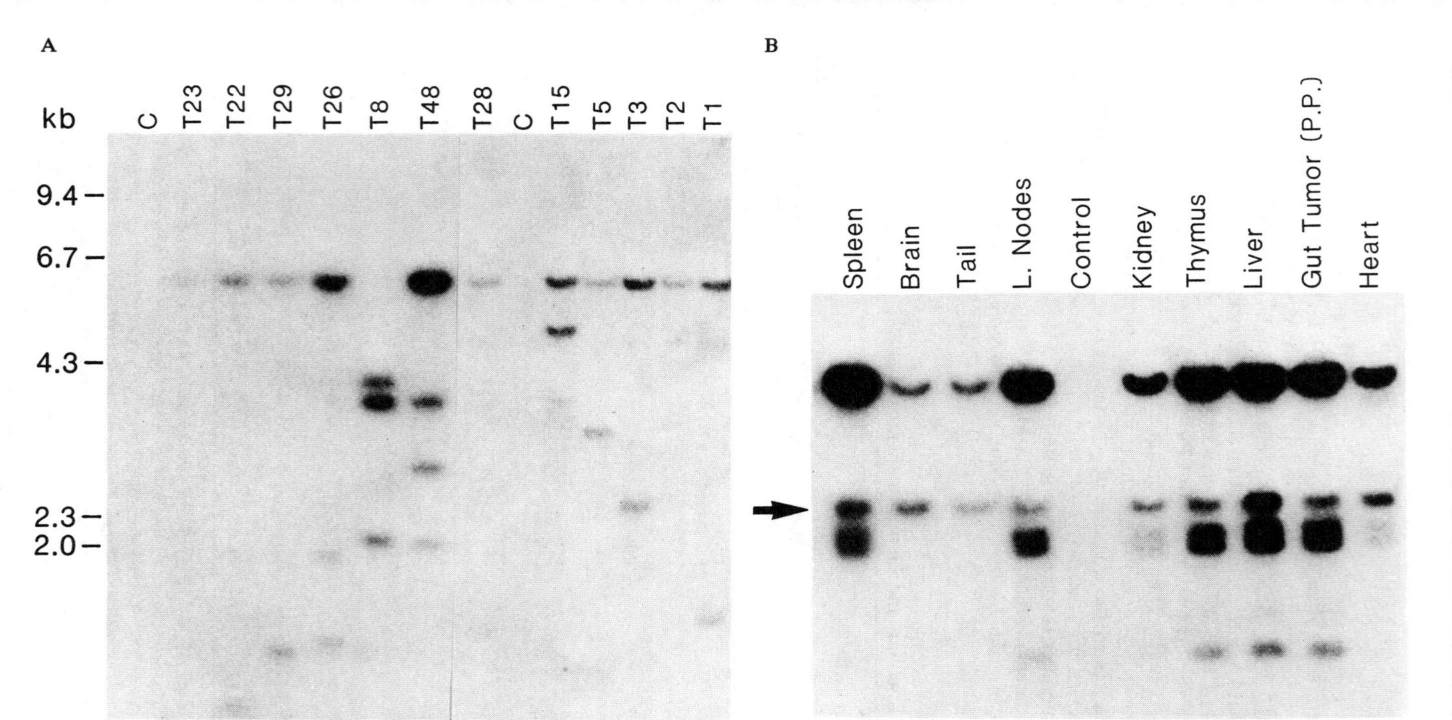

FIGURE 4. M-MuLV proviral DNA in mosaic animals. (**A**) Each lane contains 10 μg of tail DNA, restricted with *Sst*I, and is shown as it appeared after Southern blotting hybridized with the U3 LTR probe.[13] Control (C) DNA was from a tail of an unmanipulated animal. Animal T8 DNA contains two copies of an M-MuLV-derived provirus that carries a deletion. (**B**) Developmental contribution of infected embryonic cells to ecto- (for example, brain), meso- (for example, spleen), and endodermal (for example, liver) lineages. DNAs from various tissues of a leukemic mouse (T5) all show the genetically transmitted copy with the specific 5′ LTR-host DNA junction (arrow). The other three additional fragments represent newly integrated, somatically acquired MuLV copies in the clonally expanded lymphoid tumor cell population of the leukemic mouse. These copies are not represented in the germ line. P.P.: Peyer's patches (enlarged lymph nodes associated with the gut).

cells and thus offers the possibility of expression in preimplantation embryos. Therefore, the PyF101 enhancer could promote reinfection of ΔMo+Py M-MuLV during early embryogenesis.

As reported previously[13], three mouse lines were established carrying a ΔMo+Py M-MuLV provirus (TABLE 2). Expression of the two characteristic viral RNAs of 35S and 21S was detected in thymus and spleen (mesodermal) tissue of all three mouse lineages. No viral RNA was detected in liver (endodermal) whereas low but significant levels of 35S viral RNA and a 30S RNA were detected in brain (ectodermal). The precise nature of the 30S RNA has not been established. It may represent an aberrantly spliced viral mRNA.[13]

The expression of 35S and 21S viral transcripts in thymus and spleen but not liver has already been shown for mice containing genetically transmitted M-MuLV.[11,12] Surprisingly, however, mice containing ΔMo+Py M-MuLV showed expression of 35S and 30S viral RNA in brain[13] not observed in mice carrying a genetically transmitted M-MuLV.[11,12,51] Multiple infection of brain cells, however, can occur when the germ-line copy is transcriptionally activated during embryogenesis. This leads to reinfection of virtually all tissues in the developing embryo.[51] Proviruses that integrate at this stage of embryonic development do express viral RNA in every tissue in contrast to the genetically transmitted provirus.[11,12,51] Several findings argue against reinfection of tissues in mice carrying a ΔMo+Py M-MuLV. First, no amplification of the internal 5.6-kb *Sst*I fragment as compared to fragments that represent 5′ LTR-host DNA junctions was detected in the brain, the spleen, and other tissue DNAs tested (data not shown). Second, little or no infectious virus (100-1000-fold lower as compared to mice that express M-MuLV) was detected in the serum (TABLE 2). Third, no 21S-spliced envelope protein mRNA was detected in the brain,[13] making the presence of infectious virus within the brain unlikely. It thus appears that the presence of the PyF101 enhancer in the LTR imparts specificity for expression in thymus, spleen, and brain tissue. Polyoma viral DNA (PyF101) carries two distinct enhancer elements (A and B).[52] The virus ΔMo+Py M-MuLV harbors only element B in its LTR. Element A is the major enhancer in fibroblasts. In PyF101, only element B is changed as compared to wild-type polyoma. Enhancer B does not seem to be specific for EC cells, however, which may explain our findings of expression of ΔMo+Py M-MuLV RNA in lymphoid and brain tissue.

None of the F2 generation of mice carrying a ΔMo+Py M-MuLV provirus showed new proviral insertions.[13] Therefore, the enhancer element in the PyF101-*Pvu*II-4 DNA fragment does not seem to promote virus production during early embryogenesis resulting in reinfection of cells in the early embryo. Polyoma PyF101 grows on cells of the nullipotent EC cell line F9. The mutation in the enhancer element of PyF101 does not permit PyF101 to grow on more multipotential EC cell lines such as PCC4 and LT.[53] These cells, however, probably resemble early mouse embryo cells more than F9 cells, which seem to resemble a primitive endodermal stem cell. Therefore, experiments using enhancer sequences from polyoma mutants that grow in PCC4 and LT cells may result in gene expression in the early mouse embryo.

The experiments with ΔMo+Py M-MuLV allow us to conclude that LTRs of recombinant retroviruses introduced into preimplantation mouse embryos can be expressed. More importantly, this enlarges the scope of recombinant retroviral vectors containing a wide variety of tissue-specific enhancers for generating transgenic mice. To test this hypothesis in more detail we have now developed a series of retroviral vectors that carry an H-2 gene. Various LTR promoter-enhancer combinations are being tested for their potential to target the expression of the cell-surface H-2-marker molecule (F. Botteri, A. D. Miller, H. Fan, H. van der Putten, and R. M. Evans, unpublished results).

TABLE 2. Mouse Embryo Infections

Viruses and Vectors	Host Range	Embryo	Number Born/ Number Transferred	Number of Positive Tail DNAs and XC′ Tests	
M-MuLV (1×10^7 pfu/ml) LPHGL (1×10^6 cfu/ml) (Pronase/$-Ca^{2+}$)	Ecotropic	E1	56/292 (~19%)	U3 LTR GH or HPRT XC′	16/50 0/50 11/50
ΔMo+Py M-MuLV (1×10^5 pfu/ml) (Acidic Tyrode's medium/$+Ca^{2+}$)	Ecotropic	E2	62/228 (~30%)	*Pvu*II-4 MS-4 XC′	4/62 4/62 0/62
LDL1 (6×10^7 cfu/ml) M-MuLV (1×10^7 pfu/ml) (Acidic Tyrode's medium/$+Ca^{2+}$)	Ecotropic	E2	76/193 (~40%)	U3 LTR DHFR	32/74 15/74
LDL1 (5×10^5 cfu/ml) (Acidic Tyrode's medium/$+Ca^{2+}$)	Ecotropic	E1	58/197 (~30%)	U3 LTR DHFR	1/58 1/58
LPHGL (1×10^6 cfu/ml) (Pronase/$-Ca^{2+}$)	Amphotropic	E2	57/115 (~37%)	U3 LTR GH	0/46 0/46
LHMAL (1×10^6 cfu/ml) (Acidic Tyrode's medium/$+Ca^{2+}$)	Ecotropic	E3	64/200 (32%)	HPRT ADA	3/64 3/64
MLV(src) (1.5×10^5 cfu/ml) (Acidic Tyrode's medium/$+Ca^{2+}$)	Ecotropic	E3	86/254 (34%)	A U3 LTR src B U3 LTR src	0/54 0/54 5/30 5/30

NOTE: All infections were carried out for 16 hr in the presence of 4 μg/ml of Polybrene except for the MLV(src) series. Part of these infections (B series) were carried out by pretreating the embryos for 2 hr in Whitten's medium + 8 μg/ml of Polybrene. Subsequently, the infection was carried out in medium containing 0.4 μg/ml of Polybrene. Most cell lines have been described elsewhere.[13] The ψ-2 cells that produced the helper-virus-free MLV(src) virus were generated by transfection of ψ-2 cells and selecting for transformed colonies (H. Fan, unpublished results). The ψ-2 cells that produced LHMAL were provided by Scott McIvor (Genentech). Embryos were as follows: E1: C57BL/6 × SJL/J; E2: (C57BL/6 × SJL/J) × SJL/J; E3: (Balb/cJ × C57BL/6J) × SJL/J.

Insertion of Defective Recombinant Retroviruses

A major goal of our laboratory is to introduce foreign genes into the germ line by infection with replication-defective recombinant retroviruses. FIGURE 3 shows the proviral DNA structure of the recombinant retroviruses LDL1, LHMAL, and MLV-(src) containing, respectively, a mutant DHFR gene (LDL1), a human ADA gene expressed of a metallothionein promoter (LHMAL), and an src gene (MLV(src)) derived from an avian sarcoma virus. To address the issue of whether replication-defective recombinant retroviruses can be stably inserted into the mouse germ line, we initially performed mixed infections of a defective recombinant retrovirus (LDLI virus) and a helper virus (M-MuLV). To favor insertion of a defective recombinant virus, infections were performed using a cell line (TABLE 2) that produces a sixfold excess of a recombinant virus over M-MuLV helper. This cell line was generated after transfecting LDL1 and M-MuLV proviral DNA into NIH3T3TK$^-$ cells. Subsequent rounds of selection in medium containing increasing concentrations of methotrexate gave rise to the vector:helper ratio of 6:1.[54] The eventual cell line resistant to 10^{-5} M methotrexate produces an LDL1-derived virus that is 9 kb long in contrast to the original 3-kb LDL1 virus. This 9-kb LDL1 virus remains defective and appears to have acquired some M-MuLV-specific sequences. The *Sst*I restriction pattern of the 9-kb LDL1 provirus is similar but not identical to the pattern of the 3-kb LDL1 provirus when analyzed with the DHFR probe (F. M. Botteri, A. D. Miller, H. Fan, H. van der Putten, and R. M. Evans, unpublished results).

Embryo infections were performed using the high-virus-titer-producing cell line (6×10^7 cfu/ml 9-kb LDL1; 1×10^7 pfu/ml M-MuLV). Thirty-two out of a total of 74 mice born were found to harbor one or more proviruses as determined after screening with the U3 LTR probe.[13] Fifteen mice were found to harbor a DHFR-recombinant provirus.[13] Several mice harboring a DHFR-recombinant provirus were found to carry one or more M-MuLV proviruses. It is not clear if this implies a role for the helper virus in promoting infection of a defective recombinant virus. To pursue this question we examined the ability of several helper-virus-free recombinant viruses (LPHGL, the 9-kb LDL1, LHMAL, and MLV(src)) (FIG. 3) to infect preimplantation mouse embryos.

Recombinant retroviruses were prepared in cell lines that permit their helper-free propagation in either an amphotropic or an ecotropic viral coat. The ecotropic, helper-MuLV-free DHFR virus (the 9-kb LDL1) established its proviral DNA into preimplantation embryos indicating that a defective recombinant retrovirus can be inserted into the early mouse embryo in the absence of helper virus (TABLE 2).[13] Similarly, three mice were generated that harbor a defective provirus containing a human ADA cDNA sequence. Also, five mice were found to contain a defective recombinant MLV-(src) provirus (TABLE 2). Further, we tested a recombinant retrovirus (LPHGL) containing an amphotropic coat. Amphotropic viruses have a wider host range as compared to ecotropic viruses. None of 46 mice analyzed (TABLE 2) was found to be positive.

Therefore, we conclude that recombinant ecotropic viruses can be inserted into preimplantation mouse embryos. The presence of replicating helper virus is not required, though it could play a promoting role. It remains puzzling why relatively few embryos seem to be infected whereas there is a large excess of virus particles per embryo. This may be due to low levels of virus receptors on preimplantation embryos, partial occupancy of virus receptors by unknown factors in serum, or interference by expression of endogenous retroviral sequences in mouse embryos at this stage of development.

Expression of a Defective Recombinant DHFR Provirus

Retroviral LTR promoters are known to be inefficient in promoting gene activity in undifferentiated EC cells.[45,47] *Trans*-acting factors have been reported that reduce transcription in undifferentiated EC cells by interacting with viral enhancers.[50] Extensive methylation has also been shown to occur in undifferentiated EC cells and early mouse embryos, but demethylation alone is not sufficient to promote expression of the viral LTR enhancer-promoter.[49,55] Furthermore, the lack of LTR activity in EC cells is also partly due to positional effects at the site of integration.[56] Upon differentiation of undifferentiated EC cells, viral LTR promoters such as M-MuLV can become active.[49] Similarly, M-MuLV proviruses present as genetically transmitted copies in mice are repressed at the transcriptional level in early embryos. Activation can subsequently occur at developmental stages that are specific for each mouse strain, although some proviruses remain silent. Embryonic development and differentiation is accompanied by position-dependent and tissue-specific demethylation of M-MuLV-enhancer sequences in the LTR.[55] Whether activation of the M-MuLV LTR is a tissue-specific event or merely an event at random is not clear. Activation of an M-MuLV provirus results in a rapid spread of replication-competent virus. Therefore, the characterization of the cell type that initially activates the M-MuLV provirus was impossible. Replication-defective retroviruses such as the LDL1 9-kb cannot reinfect other cells when transcriptionally activated in the absence of replicating helper virus. Therefore, we investigated a possible tissue-specific activation of the M-MuLV LTR promoter in transgenic mice that harbor a single copy of a replication-defective DHFR provirus.

RNA analyses were performed on a variety of tissues of nine mice that harbor a 9-kb LDL1 provirus. One mouse was found to express DHFR viral RNA in thymus and spleen (F. Botteri, A. D. Miller, H. Fan, H. van der Putten, and R. M. Evans, unpublished results). Also, a low amount of viral RNA was detectable in liver, but not in brain. Therefore, it appears as if the expression of DHFR viral RNA is restricted to lymphoid cells. No expression of 35S and 21S M-MuLV RNA was found in mouse T473 as monitored by hybridization with the M-MuLV-specific U3LTR probe. The fact that no M-MuLV provirus was present in DNA of mouse T473 was further confirmed by hybridizations using a probe that detects M-MuLV but not the 9-kb LDL1 proviral DNA (MS-3) (FIG. 3). Therefore, our data suggest that activation of the M-MuLV LTR of the defective 9-kb LDL1 provirus in transgenic mice seems to be a tissue-specific event restricted to lymphoid cells. Because only one out of nine mice expressed DHFR viral RNA, it appears as if the chromosomal integration site is also a major determinant in whether activation can occur. Similar position effects have been reported to affect expression of genes inserted into the mouse germ line via DNA microinjection in the pronucleus of the fertilized egg.[15]

ACKNOWLEDGMENTS

We thank Dr. Ronald M. Evans, in whose laboratory these investigations were carried out, and Dr. Marguerite Vogt, for her help with some of the virus work. We would also like to thank Dr. Richard Mulligan for providing the ψ-2 cell line and Dr. Scott McIvor for providing the ψ-2 cells that produced the LHMAL virus.

REFERENCES

1. Brinster, R. L., H. Y. Chen., M. E. Trumbauer, M. K. Yagle & R. D. Palmiter. 1985. Proc. Natl. Acad. Sci. USA **82:** 4438-4442.
2. Mintz, B. & K. Illmensee. 1975. Proc. Natl. Acad. Sci. USA **72:** 3585-3589.
3. Illmensee, K. & B. Mintz. 1976. Proc. Natl. Acad. Sci. USA **73:** 549-553.
4. Hogan, B. L. M. 1977. Int. Rev. Biochem. **15:** 333-376.
5. Martin, G. R. 1980. Science **209:** 768-776.
6. Pellicer, A., E. F. Wagner, A. E. Karen, M. J. Dewey, A. J. Reuser, S. Silverstein, R. Axel & B. Mintz. 1980. Proc. Natl. Acad. Sci. USA **77:** 2098-2102.
7. Stewart, T. A. & B. Mintz. 1981. Proc. Natl. Acad. Sci. USA **78:** 6314-6318.
8. Bradley, A., M. Evans, M. H. Kaufman & E. Robertson. 1984. Nature **309:** 255-256.
9. Jaenisch, R., H. Fan & B. Croker. 1975. Proc. Natl. Acad. Sci. USA **72:** 4008-4012.
10. Jaenisch, R. 1976. Proc. Natl. Acad. Sci. USA **73:** 1260-1264.
11. Jaenisch, R., D. Jahner, P. Nobis, I. Simon, J. Lohler, K. Harbers & D. Grotkopp. 1981. Cell **24:** 519-529.
12. Jaenish, R. 1980. Cell **19:** 181-188.
13. van der Putten, H., F. M. Botteri, A. D. Miller, M. G. Rosenfeld, H. Fan, R. M. Evans & I. M. Verma. 1985. Proc. Natl. Acad. Sci. USA **82:** 6148-6152.
14. Jahner, D., K. Haase, R. Mulligan & R. Jaenisch. 1985. Proc. Natl. Acad. Sci. USA **82:** 6927-6931.
15. Palmiter, R. D. & R. L. Brinster. 1985. Cell **41:** 343-345.
16. Costantini, F. & E. Lacy. 1981. Nature **294:** 92-94.
17. Wagner, E. F., L. Covarrubias, T. A. Stewart & B. Mintz. 1983. Cell **35:** 647-655.
18. van der Putten, H., F. M. Botteri & K. Illmensee. 1984. Mol. Gen. Genet. **198:** 128-138.
19. Pinkert, C. A., G. Widera, C. Cowing, E. Heber-Katz, R. D. Palmiter, R. A. Flavell & R. L. Brinster. 1985. EMBO J. **4:** 2225-2230.
20. Palmiter, R. D., T. M. Wilkie, H. Y. Chen & R. L. Brinster. 1984. Cell **36:** 869-877.
21. Stewart, T. A., P. K. Pattengale & P. Leder. 1984. Cell **38:** 627-637.
22. Robertson, E. J., M. H. Kaufman, A. Bradley & M. H. Evans. 1983. *In* Teratocarcinoma Stem Cells. Cold Spring Harbor Conferences on Cell Proliferation. L. M. Silver, G. R. Martin & S. Strickland, Eds. Vol. **10:** 647-664. Cold Spring Harbor Laboratory. Cold Spring Harbor, NY.
23. Scholnick, S. B., B. A. Morgan & J. Hirsh. 1983. Cell **34:** 37-45.
24. Spradling, A. C. & G. M. Rubin. 1983. Cell **34:** 47-57.
25. Quint, W., H. van der Putten, F. Janssen & A. Berns. 1982. J. Virol. **41:** 901-908.
26. Teich, N. M., R. A. Weiss, G. R. Martin & D. R. Lowy. 1977. Cell **12:** 973-982.
27. Niwa, O., Y. Yokota, H. Ishida & T. Sughara. 1983. Cell **32:** 1105-1113.
28. Sorge, J., A. E. Cutting, V. D. Erdman & J. W. Gautsch. 1984. Proc. Natl. Acad. Sci. USA **81:** 6627-6631.
29. Gautsch, J. W. & M. C. Wilson. 1983. Nature **301:** 32-37.
30. Rubenstein, J. L. R., J. F. Nicolas & F. Jacob. 1984. Proc. Natl. Acad. Sci. USA **81:** 7137-7140.
31. White, J., M. Kielian & A. Helenius. 1983. Quant. Rev. Biophys. **16:** 151-195.
32. Andersen, K. B. 1985. Virology **142:** 112-120.
33. Andersen, K. B. & B. A. Nexo. 1983. Virology **125:** 85-98.
34. Gazdar, A. F., H. Oie, P. Lalley, W. W. Moss & J. D. Minna. 1977. Cell **11:** 949-956.
35. Kozak, C. A. 1983. J. Virol. **48:** 300-303.
36. Hartley, J. W. & W. P. Rowe. 1976. J. Virol. **19:** 19-25.
37. Bishayee, S., M. Strand & J. T. August. 1978. Arch. Biochem. Biophys. **189:** 161-171.
38. Johnson, M. H. 1981. Biol. Rev. **56:** 463-498.
39. Magnuson, T. & C. J. Epstein. 1981. Biol. Rev. **56:** 369-408.
40. Nicolson, G. L., R. Yanagimachi & H. Yanagimachi. 1975. J. Cell Biol. **66:** 263-274.
41. Ogou, S., T. S. Okada & M. Takeichi. 1982. Dev. Biol. **92:** 521-528.
42. Toyoshima, K. & P. K. Vogt. 1969. Virology **38:** 414-426.
43. Valerio, D., M. G. C. Duyvesteyn, B. M. M. Dekker, G. Weeda, M. Berkvens,

L. VAN DER VOORN, H. VAN ORMONDT & A. J. VAN DER EB. 1985. EMBO J. **4:** 437-443.
44. FEUERMAN, M. H., B. R. DAVIS, P. K. PATTENGALE & H. FAN. 1985. Mol. Cell. Biol. **54:** 804-816.
45. DAVIS, B., E. LINNEY & H. FAN. 1985. Nature **314:** 550-553.
46. FUJIMURA, F. K., P. L. DEININGER, T. FRIEDMANN & E. LINNEY. 1981. Cell **23:** 809-814.
47. LINNEY, E., B. DAVIS, J. OVERHAUSEN, E. CHAO & H. FAN. 1984. Nature **308:** 470-472.
48. JAENISCH, R. & D. JAHNER. 1984. Biochem. Biophys. Acta **782:** 1-9.
49. NIWA, O. 1985. Mol. Cell. Biol. **5:** 2325-2331.
50. GORMAN, C. M., P. W. J. RIGBY & D. P. LANE. 1985. Cell **42:** 519-526.
51. STUHLMANN, H., R. CONE, R. MULLIGAN & R. JAENISCH. 1984. Proc. Natl. Acad. Sci. USA **81:** 7151-7155.
52. HERBOMEL, P., B. BOURACHOT & M. YANIV. 1984. Cell **39:** 653-662.
53. TANAKA, K., K. CHOWDHURY, L. T. LIANG, K. S. S. CHANG, M. ISRAEL & Y. ITO. 1983. *In* Teratocarcinoma Stem Cells. Cold Spring Harbor Conferences on Cell Proliferation. L. M. Silver, G. R. Martin & S. Strickland, Eds. Vol. **10:** 295-305. Cold Spring Harbor Laboratory. Cold Spring Harbor, NY.
54. MILLER, A. D., M. F. LAW & I. M. VERMA. 1985. Mol. Cell. Biol. **5:** 431-437.
55. JAHNER, D. & R. JAENISCH. 1985. Mol. Cell. Biol. **5:** 2212-2220.
56. TAKETO, M., E. GILBOA & M. I. SHERMAN. 1985. Proc. Natl. Acad. Sci. USA **82:** 2422-2426.

Isolation of Clones Coding for the Catalytic Subunit of Phosphorylase Kinase[a]

EDGAR F. DA CRUZ E SILVA, GORDON C. BARR, AND PATRICIA T. W. COHEN[b]

Department of Biochemistry, Medical Sciences Institute University of Dundee Dundee DD1 4HN, Scotland

Two isoenzymes of phosphorylase kinase are present in approximately equimolar amounts in murine skeletal muscle. One isozyme has the structure $(\alpha\beta\gamma\delta)_4$, and the other has the structure $(\alpha'\beta\gamma\delta)_4$. The α, α', and β components are regulatory subunits; the γ subunit is catalytically active; and the δ subunit is identical to calmodulin. Deficiencies in phosphorylase kinase have been detected in murine muscle and in human liver, and both deficiencies show X linkage. We have previously demonstrated that in the murine muscle deficiency the α, α', β, and γ subunits are absent. In addition, we demonstrated that the β subunit gene is not located on the X chromosome, and were able to conclude that the defect either resides in the γ subunit, which consequently would have to be X linked, or in a control gene required for expression of the α, β, and γ subunits.[1] In order to differentiate between these two possibilities, cloning of the γ subunit coding sequence has been initiated.

A rabbit skeletal muscle complementary DNA library was constructed in dG-tailed pBR322,[2] and 50 000 clones from the library were screened with ^{32}P-labeled oligonucleotide γ_{252} (5′-TAA(G)TCA(G)TCCCAT(C)TC-3′), which was complementary to the coding sequence for amino acids 252-256 of the phosphorylase kinase catalytic subunit.[3] One positive clone was isolated. The plasmid DNA was purified,[4] and the insert was subcloned into phage M13. Sequencing of the DNA[5] indicated that the clone did not code for the phosphorylase kinase catalytic subunit, although it did have some similarities in sequence.

Therefore the 29-base oligonucleotide γ_{137} (5′-ACGATGTTCAGTTTGTGCA-GGGCGCAGAT-3′), which was complementary to the coding sequence for amino acids 137-146 of the phosphorylase kinase catalytic subunit, was used to screen[6] a rabbit genomic library in λ Charon 4A that had been previously amplified and kindly provided to us by R. C. Hardison.[7]

Six out of 160 000 clones screened were identified and replated to give positive plaques when washed in 6×SSC at 60° C (1 × SSC is 0.15 M NaCl, 0.015 M sodium citrate, pH 7.0). Two of these clones remained strongly positive after washing with 1

[a]This work was supported by Grant G 8126896 CA to Patricia T. W. Cohen from the Medical Research Council.

[b]To whom correspondence should be addressed.

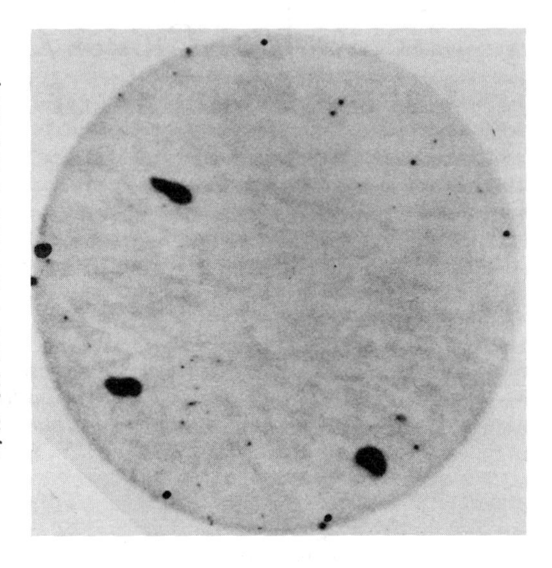

FIGURE 1. Nitrocellulose filter of replated positive clone 1/4 picked from a λ Charon 4A rabbit genomic library. Hybridization was carried out in 0.8 pmol/ml ^{32}P-labeled oligonucleotide γ_{137}, 0.09 M Tris-HC1 (pH 7.5), 0.9 M NaCl, 6 mM EDTA, 0.5% Nonidet P40, 2 × Denhardt's solution, 0.24% sodium dodecyl sulfate, 0.05% sodium pyrophosphate, 100 μg/ml *Escherichia coli* B DNA, and 70 μg/ml tRNA at 42° C for 24 hr. Washing in 6 × SSC was carried out at 60° C. Washing in 1 × SSC at 60° C did not change the appearance of the three positive plaques.

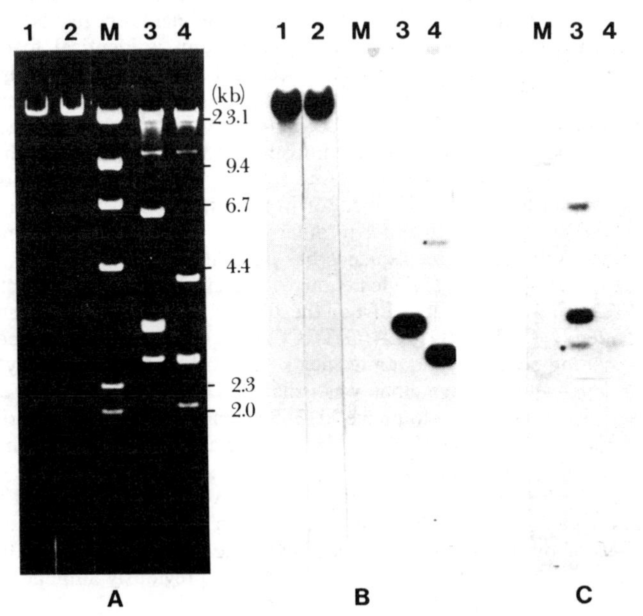

FIGURE 2. Restriction enzyme analysis of genomic clones 1/4 and 3/1. **(A)** An agarose gel stained with ethidium bromide. **(B)** A Southern blot hybridized with ^{32}P-labeled oligonucleotide γ_{137} and washed in 6 × SSC at 60° C. **(C)** A Southern blot hybridized with ^{32}P-labeled oligonucleotide γ_{252}. Hybridization conditions were as in FIGURE 1, except that hybridization was carried out at 20° C. The final wash was in 6 × SSC at 30° C. Lane 1: λ Charon 4A rabbit genomic clone 1/4; lane 2: λ Charon 4A rabbit genomic clone 3/1; lane 3: λ Charon 4A rabbit genomic clone 1/4 digested with *Eco*RI; lane 4: λ Charon 4A rabbit genomic clone 3/1 digested with *Eco*RI; lane M: λ DNA digested with *Hin*dIII.

× SSC at 60° C (FIG. 1). The DNA from the two latter clones, 3/1 and 1/4, was obtained by formamide extraction following cesium chloride density gradient purification of the phage.[4] It was digested with *Eco*RI and examined by agarose gel electrophoresis (FIG. 2). Five *Eco*RI fragments were obtained from the insert of clone 1/4 and six from the insert of clone 3/1, giving the total length of the inserts as 16 kb and 14 kb, respectively. The DNA was transferred from the gel to a nitrocellulose filter by Southern blotting and hybridized to the oligonucleotides γ_{137} and γ_{252}. A 2.7-kb fragment of the insert from clone 3/1 was positive with oligonucleotide γ_{137}; a 3.1-kb fragment of the insert from clone 1/4 was positive with both the oligonucleotides (FIG. 2).

Clone 1/4 is therefore likely to code for the γ subunit of phosphorylase kinase. Assuming this is the case, it can be used to elucidate whether or not the γ subunit gene is located on the X chromosome. This is not only a prerequisite for understanding the molecular basis of phosphorylase kinase deficiency, but may facilitate the identification of the gene for X-linked muscular dystrophy, which in mice is extremely closely linked to the gene for phosphorylase kinase deficiency (G. Bulfield, personal communication).

REFERENCES

1. COHEN, P. T. & P. COHEN. 1980. Carbohydrate Metabolism and Its Disorders. P. J. Randle, D. F. Whelan & W. J. Steiner, Eds. Vol. **3:** 119-138. Academic Press. New York, NY.
2. MANIATIS, T., E. F. FRITSCH & J. SAMBROOK. 1982. Molecular Cloning: A Laboratory Handbook. Cold Spring Harbor Laboratory. Cold Spring Harbor, NY.
3. REIMANN, E. M., K. TITANI, L. H. ERICSSON, R. D. WADE, E. H. FISCHER & K. A. WALSH. 1984. Biochemistry **23:** 4185-4192.
4. DAVIS, R. W., D. BOTSTEIN & J. R. ROTH. 1980. Advanced Bacterial Genetics: A Manual for Genetic Engineering. Cold Spring Harbor Laboratory. Cold Spring Harbor, NY.
5. SANGER, F., S. NICKLEN & A. R. COULSON. 1977. Proc. Natl. Acad. Sci. USA **74:** 5463-5467.
6. WOO, S. L. C. 1979. Methods Enzymol. **68:** 389-395.
7. MANIATIS, T., R. C. HARDISON, E. LACY, J. LAUER, C. O'CONNELL, D. QUON, G. K. SIM & A. EFSTRATIADIS. 1978. Cell **15:** 687-701.

Molecular Heterogeneity of McArdle's Disease

DOMINIQUE DAEGELEN, SOPHIE GAUTRON, FRANÇOIS MENNECIER, JEAN-CLAUDE DREYFUS, AND AXEL KAHN

*Institut National de la Santé
et de la Recherche Médicale Unité 214
75674 Paris Cedex 14, France*

McArdle's disease is a metabolic myopathy caused by the lack of glycogen phosphorylase activity in muscle.[1] Glycogen phosphorylase isoenzymes (muscle, liver, and brain types) show a specific tissue distribution such that the defect is limited to skeletal muscle. In most cases no altered inactive protein, that is, cross-reacting material (CRM), was found,[2] but in few other cases inactive protein could be detected.

To investigate the molecular defect responsible for the absence of CRM in most cases, we developed a specific human glycogen phosphorylase probe. Phosphorylase cDNA clones were isolated from a human cDNA library in *Escherichia coli* plasmid pBR 322[3] by hybridization with a rabbit muscle phosphorylase cDNA (gift from Putney). The most interesting clone covered 1300 bases: it overlapped half of the COOH-terminal coding region of the protein. After subcloning in the M13 vector this cDNA was partially sequenced using the method described by Sanger: the deduced amino acid sequence was compared with the known rabbit protein sequence and showed about a 94% homology. In Northern blot experiments this probe revealed one specific mRNA with a length of 3.4 kb, but only in tissues expressing muscle phosphorylase. Specific human aldolase A clones were isolated from the same library by using cross-hybridization with a human aldolase B cDNA.[4]

Detection and characterization of phosphorylase mRNA in muscle biopsies, which were taken from both normal and McArdle's disease patients, was realized by Northern blot analysis. The purification of mRNAs and the Northern blot analysis were scaled down to be compatible with the size of the clinical muscle biopsies. Subcloning of phosphorylase and aldolase cDNAs was performed in M13 bacteriophage in order to obtain highly labeled and improved monostrand probes.[5] Blots were hybridized with both phosphorylase and aldolase A probes.

Five McArdle's disease patients with no detectable CRM were investigated as follows: in three cases, no specific phosphorylase mRNA could be detected; in the two other cases, normal-sized mRNA was present, although in lower amounts (about 20% of normal as quantified with an internal constant mRNA, that is, an aldolase A mRNA). Typical results are shown in FIGURE 1. These results demonstrate that different molecular mechanisms can be responsible for the absence of CRM. Even in the case where mRNA is present no detectable protein is translated. Therefore, it must be recognized that these patients are compound heterozygotes carrying a mutation leading to no mRNA on one allele, and another type of lesion leading to nontranslatable RNA on the other allele.

Comparative studies of DNA from white blood cells from four patients and four controls by Southern blot analysis[6] showed no detectable deletion or rearrangement in the 20-kb region scanned by our probe. One case corresponded to a patient with no mRNA, and another case corresponded to a patient with a lowered amount of mRNA. Our observations seem to favor small or ponctual mutations leading to absent or unstable mRNA. For example, nonsense mutations are known to lead to lower amounts of RNA.

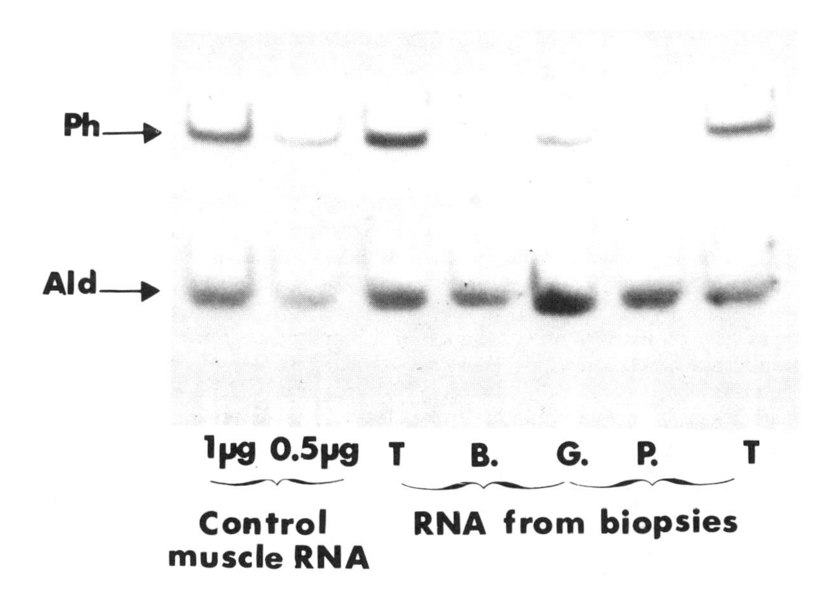

FIGURE 1. Northern blot analysis of RNAs from biopsies from a control (T) and three McArdle's disease patients (B, G, and P). Hybridization was carried out with both human muscle phosphorylase and aldolase A probes. About 1/4 of the total RNA purified from each biopsy was deposited in the corresponding slot. The intensities of the phosphorylase and aldolase A signals given by known amounts (0.5 and 1 μg) of human muscle RNA and the intensities corresponding to the control biopsy were used to determine the normal phosphorylase:aldolase A signal intensity ratio.

REFERENCES

1. McArdle, B. 1951. Myopathy due to a defect in muscle glycogen breakdown. Clin. Sci. **10:** 13-36.
2. Daegelen-Proux, D., A. Kahn & J. C. Dreyfus. 1981. Research on molecular mechanisms of McArdle's disease: Use of new protein mapping and immunological techniques. Ann. Hum. Genet. **45:** 113-120.
3. Hanauer, A., M. Levin, R. Heiling, D. Daegelen, A. Kahn & J. L. Mandel. 1983. Isolation and characterization of cDNA clones for human skeletal muscle actin. Nucleic Acids Res. **11:** 3503-3516.
4. Simon, M. P., C. Besmond, D. Cottreau, A. Weber, J. C. Dreyfus, J. Sala-Trepat, J. Marie & A. Kahn. 1983. J. Biol. Chem. **258:** 14576-14584.
5. Messing, J. 1983. Methods Enzymol. **101:** 20-78.
6. Southern, E. M. 1975. Deletion of specific sequences among DNA fragments separated by gel electrophoresis. J. Mol. Biol. **98:** 503-517.

The Isolation and Transfection of the Entire Rat β-Casein Gene

Y. DAVID-INOUYE, C. H. COUCH, AND J. M. ROSEN

Department of Cell Biology
Baylor College of Medicine
Houston, Texas 77030

The expression of the casein genes is developmentally and hormonally regulated in the mammary gland by insulin, prolactin, and hydrocortisone. DNA-mediated gene transfer experiments have been initiated to determine the sequences important for hormonal regulation of the β-casein gene. The β-casein gene was chosen because it is the best characterized of the casein genes, and expressed to the highest level during lactation when compared to other members of this multigene family.[1] The entire gene was transfected for two reasons: 1) Both 5′ flanking sequences and/or intragenic sequences may be important for correct gene expression and hormonal control. 2) The major effect of prolactin appears to be at a posttranscriptional level.[2] Thus the entire gene had to be transfected to ascertain whether hormonal effects are exerted on RNA processing and stability or on RNA transport.

The entire rat β-casein gene including 2.3 kb and 0.4 kb of 5′ and 3′ flanking DNA, respectively, plus 5.1 kb from the left arm of λ Charon 4A was cloned into pBMTH: an expression vector obtained from G. Pavlakis.[3] The vector pBMTH contains the human metallothionein gene making it possible to select for transfected cells on the basis of their resistance to cadmium poisoning.

Two recipient cell lines were used: T-47D cells, a human breast carcinoma cell line,[4] and COMMA-1D cells, a mouse mammary epithelial cell line.[5] Both cell lines respond to prolactin and glucocorticoids.[2,6] The β-casein BMTH gene was transfected either alone or with BPV-neo-delta, and, following selection with either cadmium or a combination of cadmium and G418, numerous transfectants were obtained.

The analysis of DNA from 25 T-47D and 6 COMMA-1D clones suggests that most copies of the transfected β-casein gene remained intact and unrearranged (FIG. 1). There were 2-40 copies of the transfected β-casein gene present.

Although the rat β-casein gene has been found to be expressed in the transfected cells analyzed thus far (FIG. 2), it does not appear to be hormonally regulated. Additionally, three major bands of approximately 1.15 (which may represent mature β-casein mRNA), 3.7, and 7.4 kb are evident. Control experiments designed to identify the origin of the three hybridization signals has demonstrated the following: 1) These bands do not represent hybridization to DNA. 2) They also do not represent transcription of lambda sequences upstream of the β-casein gene. Further analysis needs to be performed to determine whether these three bands represent transcripts initiated at the correct start site, "readthrough" transcripts that extend beyond the usual 3′ end, or unprocessed β-casein RNA. The observed lack of hormonal regulation may be due to either a requirement for more flanking sequences or a loss of hormone responsiveness of these cells as a result of transfection or cloning. Studies are underway to differentiate between these two possibilities.

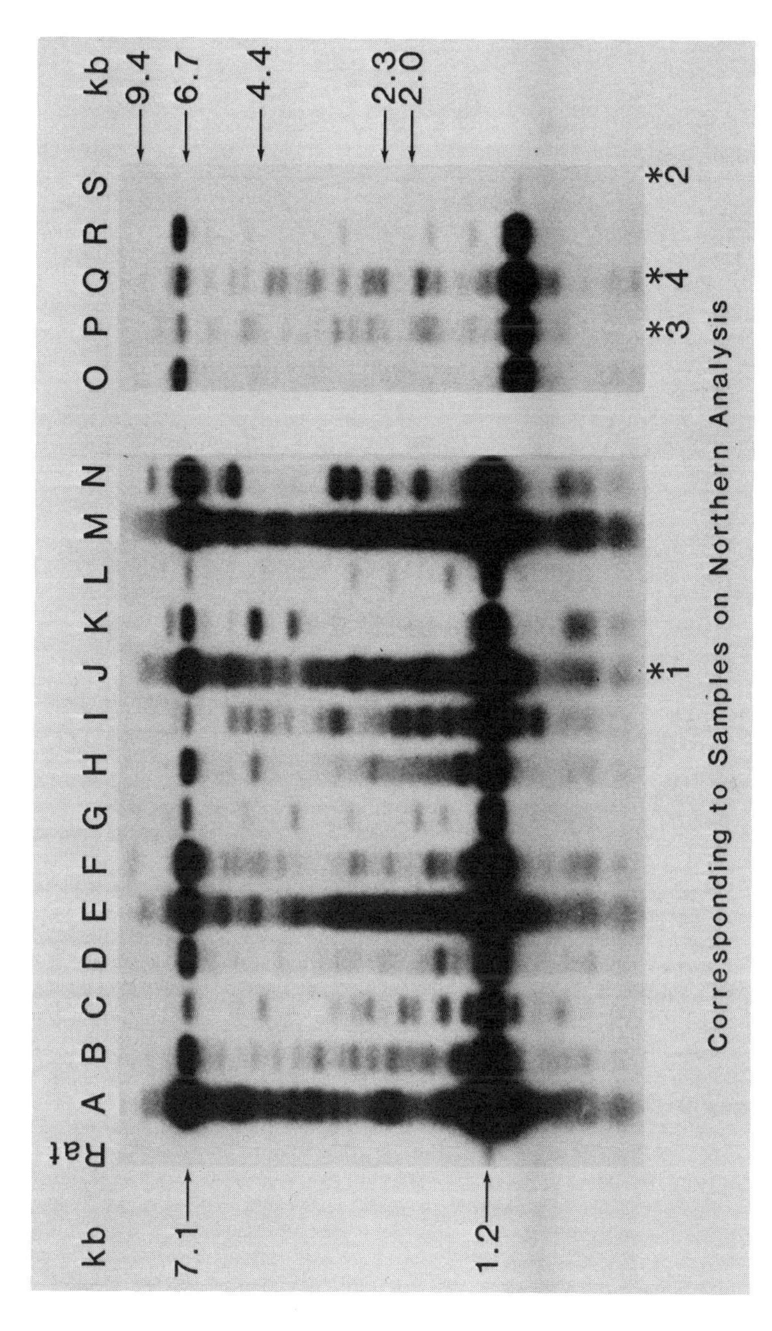

FIGURE 1. Transfection of the β-casein BMTH gene construction into T-47D cells. Total genomic DNA (20 μg, lanes A-N and the lane for rat DNA; 15 μg, lanes O-S) was digested with *MspI*, subjected to electrophoresis, and transferred to nitrocellulose. The DNA blot was probed with nick-translated β-1.9, a rat genomic probe that contains the largest exon VII. As expected, bands of 7.1 and 1.2 kb are evident in all 19 clones (lanes A-S). There are other bands, however, with molecular weights ranging between 1.2 and 7.1 kb, that also hybridized to β-1.9. These bands may represent partial digestion products or rearrangements of the β-casein gene.

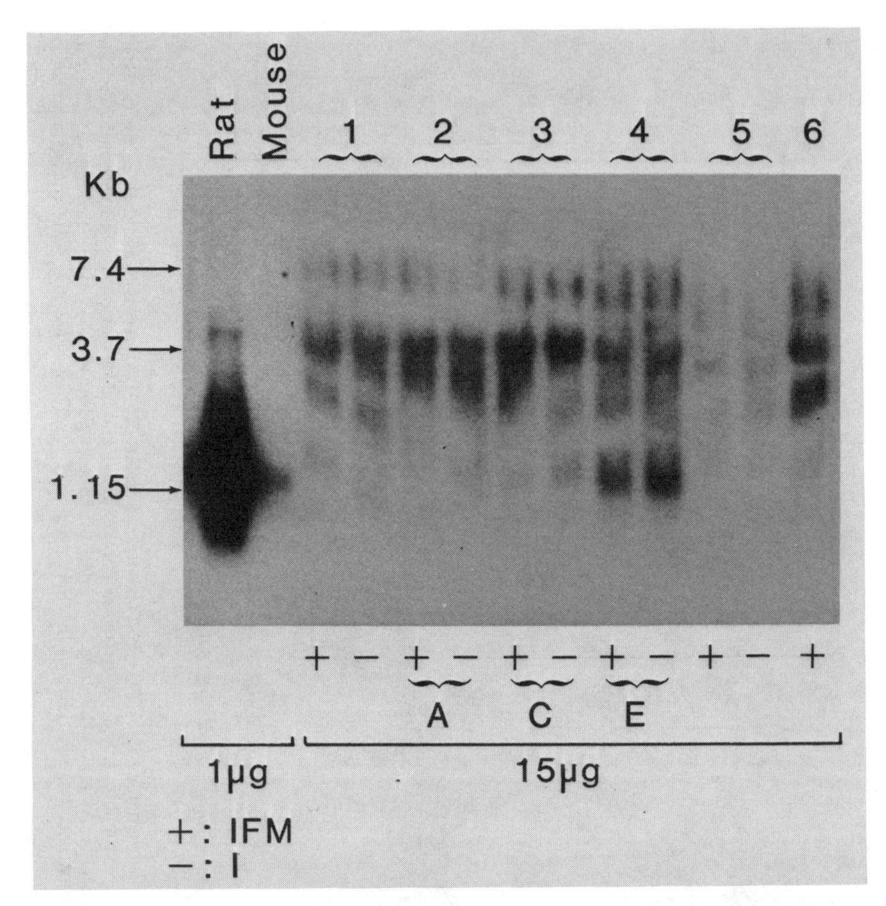

FIGURE 2. Rat β-casein gene expression in transfected T-47D and COMMA-1D cells. Cells were grown in the presence of I, that is, insulin (10 µg/ml for T-47D and 5 µg/ml for COMMA-1D cells), or IFM, that is, insulin, hydrocortisone (1 µg/ml), and prolactin (1 µg/ml), for 72 hr. COMMA-1D cells were cultured on floating collagen gels; 1 µg of total rat lactating and mouse lactating RNA and 15 µg of total RNA from transfectants (lanes 1-6) were subjected to electrophoresis on glyoxal agarose gels and transferred to nitrocellulose; the blot was hybridized to an Sp6 riboprobe. The Sp6 riboprobe was synthesized from a 500-bp *Pst*I fragment from the 3′ noncoding region of the rat β-casein cDNA, which is specific for rat casein transcripts. Lanes 1-4 and 6 represent RNA from transfected T-47D cells, and lane 5 represents RNA from a transfected COMMA-1D clone. A, C, and E refer to samples also analyzed on an RNA dot blot (data not shown).

REFERENCES

1. Jones, W. K., L.-Y. Yu-Lee, S. M. Clift, T. L. Brown & J. M. Rosen. 1985. J. Biol. Chem. **260:** 7042-7050.
2. Rosen, J. M., J. R. Rodgers, C. H. Couch, C. A. Bisbee, Y. David-Inouye, S. M. Campbell & L.-Y. Yu-Lee. 1986. Ann. N.Y. Acad. Sci. This volume.

3. ROSEN, J. M., W. K. JONES, J. R. RODGERS, J. R. COMPTON, C. A. BISBEE, Y. DAVID-INOUYE & L.-Y. YU-LEE. 1986. Ann. N.Y. Acad. Sci. **464:** 87-99.
4. HORWITZ, K. B., M. B. MOCKUS & B. A. LESSEY. 1982. Cell **28:** 633.
5. DANIELSON, K. G., C. J. OBORN, E. M. DURBAN, J. S. BUTEL & D. MEDINA. 1984. Proc. Natl. Acad. Sci. USA **81:** 3756.
6. SHIU, R. P. C. & B. M. IWASIOW. 1985. J. Biol. Chem. **260:** 11307.

Cloning and Expression of a Sucrose Operon in *Escherichia coli*

Mapping of Sucrose Metabolic Genes

J. M. DiRIENZO, J. ALEXANDER,
AND C. HARDESTY

Department of Microbiology
School of Dental Medicine
University of Pennsylvania
Philadelphia, Pennsylvania 19104

Strains of *Escherichia coli* lack the ability to ferment sucrose unless they contain a metabolic plasmid that encodes for the specific transport and enzymatic activities necessary for the uptake and catabolism of the disaccharide. A 78-kb conjugative plasmid, previously recovered from a clinical isolate of *Salmonella typhimurium,* was found to encode for two proteins.[1,2] The first protein, *scr*A, is a phosphoenolpyruvate-dependent carbohydrate: phosphotransferase system enzyme II^{scr}. The second enzyme, *scr*B, is a β-D-fructofuranoside fructohydrolase. The genes of this plasmid appear to be under the control of a promoter (*scr*P), an operator (*scr*O), and a regulator gene (*scr*R). *E coli* DS402 (pUR400) was obtained from Dr. R. Schmitt (Lehrstuhl für Genetik, Universität Regensburg, Regensburg, Federal Republic of Germany) to 1) study the regulation of sucrose metabolism in this operon and 2) construct cloning vectors and hybridization probes to facilitate the study of the expression and regulation of sucrose metabolic genes from oral bacteria.

The plasmid DNA was purified in cesium chloride gradients and digested to completion with *Bgl*II. Seven fragments ranging in size from 2 to 17 kb were obtained and cloned in the *Bam*HI site of the vector pBR328. The transformants were screened by plating on a MacConkey agar base supplemented with 1% sucrose and 50 μg/ml ampicillin. One transformant, pJD143, inducibly expressed both the hydrolase and permease (TABLE 1). Subcloning of the largest *Hin*dIII fragment resulted in a chimeric plasmid, pJD1086, that encoded for ampicillin resistance and sucrose fermentation. However, the hydrolase was now constitutively expressed. Thus, the *scr*R gene is thought to reside within a 6.5-kb region 3.8 kb upstream from the sucrose operon (FIG. 1). A single *Cla*I site in the insert DNA in pJD1086 was used to subclone each half of the plasmid to produce pJD1087 and pJD1088. Each of these plasmids lost the expression of both the hydrolase and permease. When the two *Cla*I fragments were re-ligated to produce pJD1089, both activities were restored (TABLE 1). It appears that the *Cla*I site may reside in the promoter/operator region and that the structural genes are located downstream from this site. A 5.7-kb *Sal*I fragment containing this region was subcloned into pBR328. The resulting plasmid, pJD2000, expressed high levels of both the hydrolase and permease (TABLE 1). When a *Bam*HI fragment was

TABLE 1. Phenotypes of Chimeric Sucrose Plasmids

Plasmid Designation[a]	Size (kb)		Sucrose 6-Phosphate Hydrolase Activity (nmol/min/ml culture)		[^{14}C]Sucrose Uptake (cpm × 10^3/min/ml culture)[d]	
	Plasmid	Insert	No inducer[b]	Inducer[c]	No inducer	Inducer
pUR400	78.0		0.7	12.3	49	440
pJD143	22.0	17.1	9.4	234.0	38	351
pJD1086	15.2	10.6	32.9	27.8		107
pJD1087	10.3	5.8	0	0.1		30
pJD1088	9.6	4.6	0	0.2		25
pJD1089	15.2	10.6	12.4	22.9		127
pJD2000	10.6	5.7		554.9		467
pJD2001	9.4	4.8	0.3	0.3		23
pBR328	4.9		0.2	0.1		19

[a] *E. coli* SA224, genotype Δ (*gal-chl*A), *rps*L, *rec*A, was the host strain for all of the plasmids except in the case of pUR400, where *E. coli* DS402 was used.

[b] Clone SA224 (pJD2000) did not grow in glycerol-supplemented medium.

[c] Inducible expression was determined by comparing cells grown in chemically defined medium supplemented with glycerol and sucrose.

[d] The specific activity of the [^{14}C]sucrose was 552 mCi/mmole, and the cultures were normalized to A_{490} = 2.0.

deleted from pJD2000, the new clone, pJD2001, no longer expressed the hydrolase and permease (TABLE 1). A detailed restriction map of pJD2000 is shown in FIGURE 1 and is compared to maps of pJD143 and pJD1086.

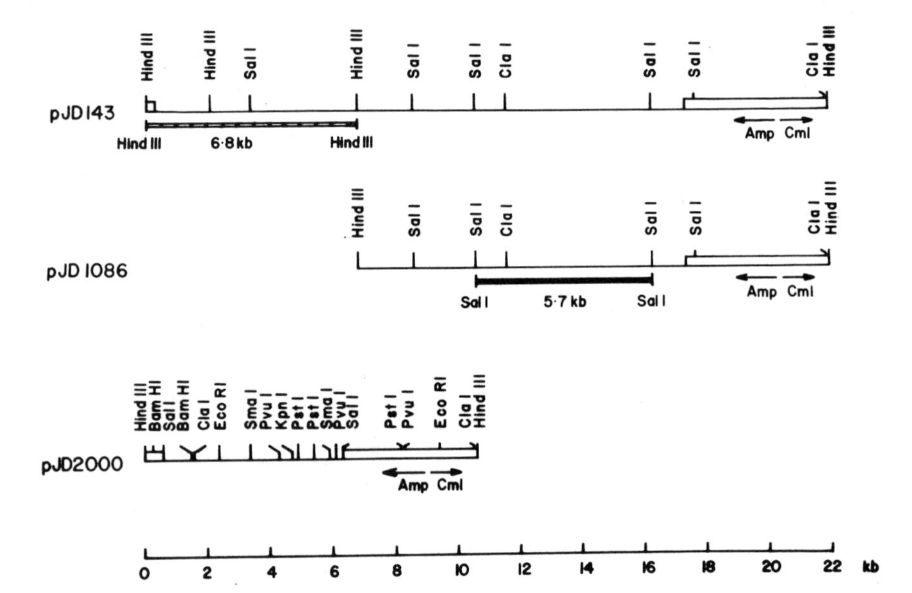

FIGURE 1. Cloning of the sucrose operon. A 17.1-kb *Bgl*II fragment containing the sucrose operon from pUR400 was cloned in *E. coli.* A restriction map of the chimeric plasmid is shown in the top of the figure. A 6.5-kb *Hind*III fragment was deleted to create pJD1086. A 5.7-kb *Sal*I fragment from pJD1086 was then subcloned as shown in the bottom of the figure. The sizes of the plasmids and inserts are listed in TABLE 1. The open areas on the maps are vector sequences, and the cloned fragments are indicated by the thin line.

REFERENCES

1. SCHMID, K., M. SCHUPFNER & R. SCHMITT. 1982. J. Bacteriol. **151:** 68-76.
2. LENGELER, J. W., R. J. MAYER & K. SCHMID. 1982. J. Bacteriol. **151:** 468-471.

A Sequence Analysis Program for the Macintosh Computer[a]

MANUEL J. GLYNIAS AND ALAN G. GOODRIDGE

Department of Pharmacology
Case Western Reserve University
Cleveland, Ohio 44106

We have developed a computer program for the analysis of DNA and protein sequences that is written in MacForth® (Creative Solutions, Silver Springs, MD) and runs on the Macintosh™ (Apple Computer, Cupertino, CA) computer. This program, which we call MacGene™, utilizes the standard Macintosh user-interface; it is both powerful and user-friendly. Commands are selected from menus or other screen images with the "mouse." The mouse also is used to select residues, groups of residues, areas of maps, etc., images of which can then be expanded or made available for further analysis.

MacGene consists of a core, which contains a sequence editor and utilities, and ten modules, each of which is selectable from the main menu.

Gel Entry simplifies the entry of data from DNA sequencing gels. The base sequence is entered either by typing into a full-screen text editor similar to that for MacWrite, or by selecting buttons on the screen labeled A, C, G, and T, using the mouse.

Alignment simultaneously aligns sequences from up to 50 gels. Both orientations of new sequences are tested for alignment to a consensus sequence. If a sequence aligns, it is added to the consensus. Spaces are inserted into each sequence to accommodate mismatched bases. A map is drawn on the screen showing the position of all aligning sequences relative to the consensus sequence.

Reading Frames finds all open reading frames in a DNA sequence. The amino acid composition and codon preference of each reading frame is calculated. DNA and protein sequences of any open reading frame can be "cut out" and "pasted" into one of the other modules for further analysis.

Restriction creates maps of restriction enzyme sites in a DNA sequence. Restriction enzymes are selected from menus. Linear or circular maps are drawn. Any region of a map can be selected and the base sequence of that area examined. All restriction enzyme sites in the sequence are listed.

Homology is a matrix program for DNA homologies. The user selects the window (number of residues) that will be used in the search for homologies. Homologies of greater than 50% are saved. For a given matrix, the user can select the level of homology to be displayed on the screen. Once the matrix is drawn, any part of it can be selected and expanded. Optimal alignment of selected regions also can be determined.

[a]Supported in part by Grant AM 21594 from the National Institutes of Health.

Communications allows the user to communicate by modem either with another Macintosh running MacGene or with Bionet™ (or other data bank). Sequences can be transferred in either direction.

Protein Structure uses the prediction rules of Garnier *et al.*[1] to calculate for each residue in a protein the probability that its structural conformation is in the form of an α helix, a β strand, a random coil, or a β turn. The results are displayed in tabular or graphic form, or as sequences with symbols for the predicted structures replacing the residue designations. This program also predicts which residues are in the interior of a protein and which are on the surface, using an algorithm that calculates the index of hydrophobicity.[2]

Peptides predicts the properties of peptides produced by specific enzymatic or chemical cleavage of a protein of known sequence. Peptides are listed by position in the protein sequence, length, molecular weight, isoelectric point, or elution properties during reverse-phase HPLC. In addition, the program displays the results as schematic diagrams of stained gels after electrophoresis or of elution profiles after chromatography.

Protein Homology is a matrix program for protein homologies. This program has the same features as the DNA homology module except that homologies are calculated using the accepted point mutation scores of Dayoff *et al.*[3]

Mutagenesis is designed to assist in the planning of site-directed mutagenesis. A DNA sequence is displayed in the "wild-type" window. The corresponding protein sequence and its predicted secondary structure are displayed beneath the DNA sequence. Any codon in the wild-type sequence can be selected and changed. The DNA sequence of the mutant, the predicted sequence of amino acids, and the predicted secondary structure of the new amino acid sequence are then displayed.

MacGene should be useful to laboratory researchers for analysis of data, and to students for learning the fundamentals of DNA and protein structure.

REFERENCES

1. GARNIER, F., D. J. OGSTHORPE & B. ROBSON. 1978. J. Mol. Biol. **120:** 97-120.
2. KYTE, J. & R. F. DOOLITTLE. 1982. J. Mol. Biol. **157:** 105-132.
3. DAYOFF, M. O., R. M. SCHWARTZ & B. C. ORCUTT. 1978. Atlas of Protein Sequence and Structure. Vol. **5**(Suppl. 3): 345-352. National Biomedical Research Foundation. Silver Springs, MD.

Sequence Analysis of the Genes for Yeast and Human Muscle Glycogen Phosphorylases

PETER K. HWANG, JOHN A. BURKE, AND ROBERT J. FLETTERICK

Department of Biochemistry and Biophysics
University of California
San Francisco, California 94143

Glycogen phosphorylase regulates intracellular glucose metabolism by catalyzing the phosphorolysis of glycogen into glucose 1-phosphate. Phosphorylases from different organisms and from different cell or tissue types display markedly contrasting regulatory features. Such contrasts reflect the evolution of the enzyme to suit the specific environment and energy demands of the cell in which it functions. In order to understand the structural and evolutionary basis for this functional divergence, we have cloned and analyzed gene sequences for yeast and human muscle phosphorylases.

The yeast phosphorylase gene was isolated from a genomic bank using a mixed oligonucleotide probe. The oligonucleotide sequence encoded a portion of a pyridoxal phosphate-bound peptide that was previously reported.[1] An amino acid sequence was deduced from the nucleotide sequence of a uninterrupted, 2.7-kb-long open reading frame. The protein sequence is 891 amino acid residues in length. Adenine-thymine base pairs accounted for 59.8% of the DNA sequence. TABLE 1 presents a summary of a preliminary comparison of this enzyme to the known three-dimensional structure of rabbit muscle phosphorylase. Yeast phosphorylase exhibits 50% overall amino acid homology to rabbit muscle enzyme over alignable sequences. There are, however, 14 positions of insertions or deletions in the yeast sequence relative to muscle. The NH_2-terminals of yeast and muscle enzymes are highly divergent; the yeast NH_2-terminal extends up to 25 residues beyond the NH_2-terminal of the muscle enzyme and contains the phosphorylation site. Major differences between muscle and yeast enzymes occur in the region of the AMP site, the subunit interface, and the serine 14 phosphorylation site. AMP regulation and intersubunit cooperativity, as we understand it in rabbit muscle, appears to be completely disrupted in the yeast enzyme. Highly conserved regions of the yeast structure include the active site and the β-barrel core of the COOH-terminal domain. Interestingly, the purine inhibition site also appears conserved: hence it may have a regulatory role in the yeast enzyme.

The gene for human muscle phosphorylase has been isolated using a human cDNA probe to the COOH-terminal portion[2] and is being analyzed by restriction mapping and DNA sequencing. Seven intron or exon boundaries have been determined to date; four of these are nearly coincident with positions of yeast insertion or deletions located

TABLE 1. Structural Comparison of Rabbit Muscle and Yeast Glycogen Phosphorylases

Site	Muscle	Yeast
Active	Binds the substrate glucose 1-phosphate or inhibitor glucose	Conserved
Storage	Binds glycogen; involved in allosteric activation	Mostly conserved; conservative replacements have occurred at several site residues
Phosphorylation	Involved in activation by phosphorylase kinase; specific target is serine 14	A homologous site no longer exists; principal activation is by cAMP-dependent protein kinase; new target is a threonine on the extended NH_2-terminal
AMP and subunit interface	Binds activator AMP; transmits allostery	Site is disrupted by a 14-residue insertion at position 114, also by the extended NH_2-terminal
Purine	Binds inhibitors such as caffeine, purines, and nucleosides; syngergistic inhibition with glucose	Principal residues, phenylalanine 285 and tyrosine 612, are conserved; may have a regulatory role

on the surface of COOH-terminal domain. This observation supports the notion that intron or exon junctions represent "hot spots" for evolutionary change.

REFERENCES

1. LERCH, K. & E. H. FISCHER. 1975. Biochemistry **14:** 2009-2014.
2. HWANG, P. K., Y. P. SEE, A. M. VINCENTINI, M. A. POWERS, R. J. FLETTERICK & M. M. CRERAR. 1985. Eur. J. Biochem. **152:** 267-274.

Tissue-specific Expression of Glucokinase

PATRICK B. IYNEDJIAN[a]

Institut de Biochimie Clinique
Centre Médical Universitaire
1211 Geneva 4, Switzerland

GISELA MÖBIUS AND HANS J. SEITZ

Physiologisch-Chemisches Institut
Universitäts Krankenhaus Eppendorf
2000 Hamburg 20, Federal Republic of Germany

The expression of the glucokinase gene appears to be a marker of terminal differentiation for the parenchymatous liver cell (see reference 1 for a review). The purpose of this work was to assess the tissue specificity of glucokinase gene expression by taking advantage of the low detection limit and high specificity of immunoblotting techniques.

FIGURE 1 illustrates a blot of total cytosol protein from rat liver and islets of Langerhans of the pancreas; the blot was revealed with a sheep antiserum produced against rat liver glucokinase. Both liver and pancreatic islets display an immunoreactive product with apparent M_r of 56 000; this product comigrates with purified hepatic glucokinase (FIG. 2). An excess of exogenous hepatic glucokinase, added to the antiserum before immunoblotting, competes effectively for antibody binding with this product, which was therefore tentatively identified as glucokinase (compare lanes 1 and 3 and lanes 2 and 4 in FIG. 1). Note that other stained bands remained unaffected in the presence of an excess of exogenous glucokinase and therefore represent unrelated proteins interacting nonspecifically with one of the immune reagents. The following evidence confirms that the 56 000-dalton immunoreactive islet protein is closely related or identical to glucokinase: 1) both proteins have similar behavior during batchwise diethylaminoethyl cellulose chromatography; 2) both proteins display similar isoelectric points in two-dimensional gel electrophoresis; 3) islet cytosol contains glucokinase enzyme activity that is suppressible by the sheep antiserum.[2] A transplantable rat insulinoma as well as the insulinoma-derived RINm5F cell line[3] were next examined for expression of glucokinase. As shown in FIGURE 2, the 56 000-dalton glucokinase protein was also identified by immunoblotting after sodium dodecyl sulfate-polyacrylamide gel electrophoresis of cytosol protein from these sources.

In contrast, all other tissues so far examined by the same protein blotting analysis were found to lack the glucokinase gene product, in spite of a detection limit on the

[a] Address for correspondence: Institut de Biochimie Clinique, Centre Médical Universitaire, Université de Genève, 1, rue Michel-Servet, 1211 Geneva 4, Switzerland.

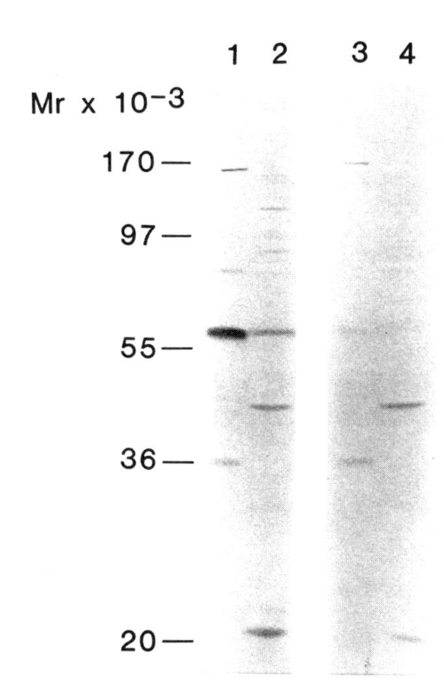

FIGURE 1. Sodium dodecyl sulfate-polyacrylamide gel electrophoresis of rat liver and pancreatic islet protein, and glucokinase immunoblot. Islets were isolated from glucose-fed rats by handpicking after collagenase digestion of the pancreas. Total cytosol protein was resolved in an 11% polyacrylamide gel and transferred electrophoretically to nitrocellulose filters. For detection of glucokinase, the filters were incubated with a sheep antiserum to rat liver glucokinase, followed by a rabbit antibody to sheep immunoglobulins and a goat antibody to rabbit immunoglobulins conjugated with horseradish peroxidase. The filter containing lanes 3 and 4 was incubated with sheep antiserum to glucokinase that had been pretitrated with an excess of added purified glucokinase. Lanes 1 and 3: 4 μg rat liver cytosol protein. Lanes 2 and 4: 42 μg islet cytosol protein.

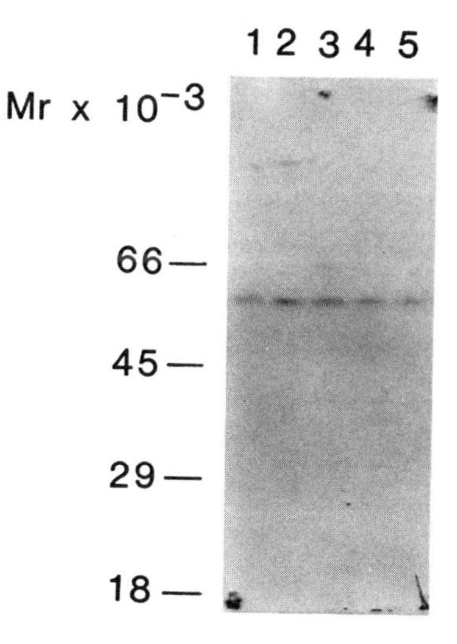

FIGURE 2. Immunoblotting of glucokinase in transplantable rat insulinoma and cultured RINm5F cells. Sodium dodecyl sulfate-polyacrylamide gel electrophoresis and immunoblotting were done as in FIGURE 1, except that detection was performed with ^{125}I-protein A instead of peroxidase-conjugated goat antibodies. Lane 1: 9 ng purified glucokinase from rat liver. Lane 2: 2.5 μg cytosol protein from liver. Lane 3: 25 μg cytosol protein from insulinoma. Lane 4: 25 μg cytosol protein from RINm5F cells. Lane 5: 13 μg cytosol protein from pancreatic islets.

order of 10 ppm of total soluble protein. These tissues include exocrine pancreas, intestinal mucosa, adipose tissue, spleen, kidney, and brain.[2]

Three main conclusions emerge from this work. First, the data strengthen the view that the expression of the glucokinase gene is restricted to a very few cell types in the body, perhaps only to liver and pancreatic islets. Second, the identification of the gene product in pancreatic islets extends the work of Meglasson and Matschinsky[4] and is consistent with a role for glucokinase in islet glucose metabolism. Third, the expression of glucokinase is maintained in tumor cells derived from the β cell of the islet of Langerhans.

REFERENCES

1. WEINHOUSE, S. 1976. Curr. Top. Cell. Regul. **110:** 1-50.
2. IYNEDJIAN, P. B., G. MÖBIUS, H. J. SEITZ, C. B. WOLLHEIM & A. E. RENOLD. 1986. Proc. Natl. Acad. Sci. USA **83:** 1998-2001.
3. GAZDAR, A. F., W. L. CHICK, H. K. OIE, H. L. SIMS, D. L. KING, G. C. WEIR & V. LAURIS. 1980. Proc. Natl. Acad. Sci. USA **77:** 3519-3523.
4. MEGLASSON, M. D. & F. M. MATSCHINSKY. 1984. Am. J. Physiol. **246:** E1-E13.

Molecular Cloning of a Hormone-regulated Isoform of the Regulatory Subunit of Type II cAMP-dependent Protein Kinase from Rat Ovaries

T. JAHNSEN,[a,b] L. HEDIN,[a] S. M. LOHMANN,[c] U. WALTER,[c] V. J. KIDD,[a] J. LOCKYER,[a] S. L. RATOOSH,[a] J. DURICA,[a] T. Z. SCHULZ,[a] AND J. S. RICHARDS[a]

[a]*Department of Cell Biology*
Baylor College of Medicine
Houston, Texas 77030

[b]*Institute of Pathology*
Rikshospitalet
Oslo 1, Norway

[c]*Department of Physiological Chemistry and Medicine*
University of Würzburg
8700 Würzburg, West Germany

It has been reported that the regulatory subunit of cAMP-dependent protein kinase type II (RII) is present in high concentrations in rat granulosa cells of preovulatory follicles and is induced by the synergistic actions of estradiol and of follicle-stimulating hormone (FSH)[1] or cAMP.[2,3] The 10-fold increase in RII in granulosa cells or rat preovulatory follicles is associated with the induction of receptors for luteinizing hormone (LH) and increased steroidogenesis, and therefore may be important for ovulation and luteinization.

To gain further insight into the molecular events regulating the action of FSH (and cAMP) on granulosa cell function, and upon RII content specifically, we have recently purified and characterized the ovarian RII subunits.[4,5] Rat ovarian RII_{51-52} (M_r: 51-52 000) is structurally and immunologically similar to rat brain (neural) RII_{52} (M_r: 52 000), but is distinct from RII_{54} (M_r: 54 000) present in rat heart, brain, and ovary. RII_{51-52} but not RII_{54} is hormonally regulated in the rat ovarian follicle. To characterize the gene for the ovarian hormone-regulated RII_{51-52}, we have constructed a λgt11 cDNA expression library using poly$(A)^+$ RNA obtained from granulosa cells of estradiol-FSH-primed rats. A polyclonal antibody against bovine heart RII was affinity purified against ovarian R subunits, and allowed us to detect 1 ng RII_{51-52} by a dot blot assay. Using this antibody we screened the λgt11 library and identified a positive recombinant clone. A 1.6-kb cDNA insert was isolated and subcloned into pBR322 and M13 mp18 vectors, and these vectors were used for hybrid-selected translation and DNA sequence analysis, respectively. Results of the hybrid-selected *in vitro* translation are shown in FIGURE 1. Note that the pBR322 vectors containing the 1.6-kb cDNA insert (lanes 5-7) but not the pBR322 vectors themselves (lanes 8-10) are shown to selectively bind the mRNA that, when translated *in vitro,* synthesized authentic RII_{51}. This was demonstrated by 1) immunoprecipitation with affinity-purified rat brain RII antibody (lane 6); 2) cAMP-Sepharose affinity chromatography (lane 7); and 3) comigration on one-dimensional sodium dodecyl sulfate-

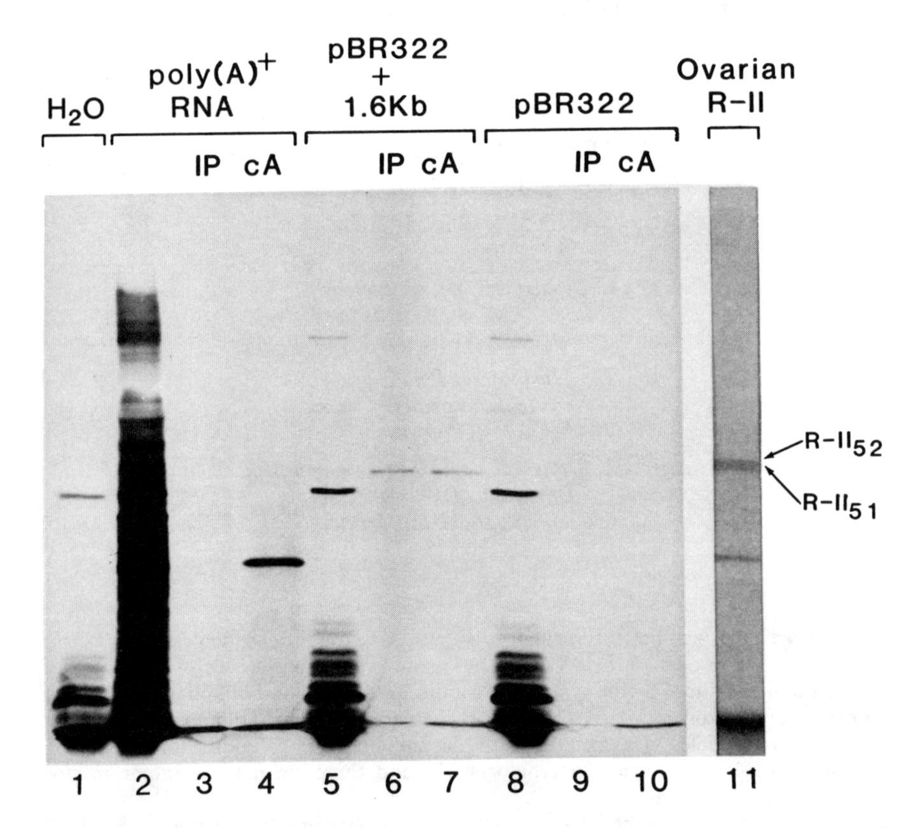

FIGURE 1. Identification of RII cDNA by hybrid-selected translation. Rat granulosa cell poly(A)$^+$ RNA from hypophysectomized rats treated with estradiol and FSH was translated *in vitro* before (lanes 1-4) or after hybridization and elution from pBR322 containing the 1.6-kb cDNA (lanes 5-7) or pBR322 alone (lanes 8-10). Authentic RII was identified by immunoprecipitation with an affinity-purified antibody directed against rat brain RII or by cAMP-Sepharose affinity chromatography followed by one-dimensional sodium dodecyl sulfate-polyacrylamide gel electrophoresis and fluorography. Total translation products were obtained from each of the following: lane 1: rabbit reticulocyte lysate alone; lane 2: 2 μg of granulosa cell poly(A)$^+$ RNA; lane 3: same as lane 2, followed by immunoprecipitation; and lane 4: same as lane 2, followed by cAMP-Sepharose affinity chromatography. Translational products from mRNA that hybridized to pBR322 containing the 1.6-kb cDNA (lanes 5-7) or pBR322 alone (lanes 8-10): lanes 5 and 8: total hybrid-selected translation products; lanes 6 and 9: same as lanes 5 and 8, respectively, followed by immunoprecipitation; lanes 7 and 10: same as lanes 5 and 8, respectively, followed by cAMP-Sepharose affinity chromatography; lane 11: purified ovarian RII_{51-52} stained with Coomassie blue.

P

1. Arg Arg Val Ser Val Cys Ala Glu Thr Tyr Asn Pro Asp Glu Glu Glu Glu Asp Thr Asp
2. Arg Arg Ala Ser Val Cys Ala Glu Ala Tyr Asn Pro Asp Glu Glu Glu Asp Asp Ala Glu
3. Arg Arg Ala Ser Val Cys Ala Glu Ala Tyr Asn Pro Asp Glu Glu Glu Asp Asp Ala Glu

FIGURE 2. Comparison of amino acid sequences around the autophosphorylation site of RII. Lines 1 and 2: Amino acid sequences for bovine heart RII[6] and bovine brain RII,[7,8] respectively. Line 3: Amino acid sequence deduced from a partial nucleotide sequence of the rat granulosa cell 1.6-kb RII cDNA. Differences between the rat ovary RII$_{51}$, bovine brain RII, and bovine heart RII are underlined. The autophosphorylation site is marked with a P.

polyacrylamide gel electrophoresis with authentic purified ovarian RII$_{51}$, which was stained with Coomassie blue (lane 11).

DNA sequence analysis of the 1.6-kb fragment revealed an open reading frame of nucleotide sequences coding for amino acids homologous to bovine brain RII (FIG. 2).

Based on these observations, we conclude that a cDNA for the RII$_{51-52}$ isoform of RII present in rat ovary and brain has been cloned. This clone will be used to analyze the mechanisms by which estradiol and FSH regulate RII$_{51-52}$ mRNA expression in rat granulosa cells.

REFERENCES

1. RICHARDS, J. S. & A. I. ROLFES. 1980. J. Biol. Chem. **225:** 5481-5489.
2. DARBON, F.-M., M. KNECHT, T. RANTA, M. L. DUFAU & K. J. CATT. 1984. J. Biol. Chem. **259:** 14778-14782.
3. RATOOSH, S. L. & J. S. RICHARDS. 1985. Endocrinology **117:** 917-927.
4. JAHNSEN, T., S. M. LOHMANN, U. WALTER, L. HEDIN & J. S. RICHARDS. 1985. J. Biol. Chem. **260:** 15980-15987.
5. JAHNSEN, T., L. HEDIN, S. M. LOHMANN, U. WALTER & J. S. RICHARDS. 1985. J. Biol. Chem. **261:** 6637-6639.
6. TAKIO, K., S. B. SMITH, E. G. KREBS, K. A. WALSH & K. TITANI. 1982. Proc. Natl. Acad. Sci. USA **79:** 2544-2548.
7. WELDON, S. L., M. C. MUMBY & S. S. TAYLOR. 1985. J. Biol. Chem. **260:** 6440-6448.
8. STEIN, J. C. & C. S. RUBIN. 1985. J. Biol. Chem. **260:** 10991-10995.

Modulation of mRNA Levels for Urea Cycle Enzymes in Rat Liver by Diet and Hormones[a]

SIDNEY M. MORRIS, JR. AND
CAROLE L. MONCMAN

Department of Microbiology, Biochemistry, and Molecular Biology
University of Pittsburgh
Pittsburgh, Pennsylvania 15261

WILLIAM E. O'BRIEN

Department of Pediatrics
Baylor College of Medicine
Houston, Texas 77030

The enzymes of the urea cycle have been shown to undergo adaptive responses to altered dietary protein intake[1,2] and to certain hormones, such as glucocorticoids[3,4] and glucagon.[4,5] Moreover, the reported changes in activity of the various urea cycle enzymes appear to occur in a roughly coordinate fashion. The levels at which these changes are implemented in the liver of the intact animal have not yet been fully investigated.

Diet-induced changes in levels of translatable mRNA have been reported for the first two enzymes of the pathway.[6] Relative levels of mRNA for carbamyl phosphate synthetase I have been reported to increase in response to dietary protein[7] and to glucocorticoid and cAMP.[8] However, no comprehensive study of dietary or hormonal regulation of mRNA levels for all five urea cycle enzymes has been reported. Using cloned cDNA probes, we have begun such a study as a first approach to elucidating possible regulatory mechanisms. The initial results presented here are addressed to several general questions: 1) Do diet-induced changes in level of urea cycle enzyme reflect changes in relative abundance of the corresponding mRNAs? 2) Are the relative levels of mRNAs for the urea cycle enzymes regulated by glucocorticoids and cAMP, either singly or in combination? 3) Are the steady state levels of mRNAs for the different urea cycle enzymes regulated in a coordinate or parallel fashion?

Our initial results are presented in TABLES 1 & 2 and may be briefly summarized as follows: 1) Changes in relative abundance of mRNAs for urea cycle enzymes are sufficient to account for reported increases in enzyme levels due to changes in dietary protein intake. However, relative abundances after prolonged fasting or a switch to a

[a] This work was supported by grants from the National Institutes of Health to Sidney M. Morris, Jr. and William E. O'Brien and by Basil O'Connor Starter Research Grant 5-466 from the March of Dimes Birth Defects Foundation to Sidney M. Morris, Jr.

TABLE 1. Effect of Bt_2cAMP and Dexamethasone on Relative Abundance of mRNAs for Urea Cycle Enzymes in Rat Liver

mRNA	Relative Abundance of mRNA[a]: Bt_2cAMP (hr) 1	Bt_2cAMP (hr) 5	Dexamethasone[b]	Dexamethasone and Bt_2cAMP[b]
Carbamyl phosphate synthetase I	2.1	3.8	1.6	4.3
Argininosuccinate synthetase	3.4	4.6	3.3	8.1
Argininosuccinate lyase	1.5	4.7	2.0	5.2

NOTE: Male Sprague-Dawley rats (150-225 g) were maintained on a 0% casein diet (ICN Nutritional Biochemicals, Cleveland, OH) for 5-6 days before hormone treatment. Rats received multiple intraperitoneal injections of Bt_2cAMP (25 mg/kg, at 0 hr, 2 hr, and 4 hr) or a single injection of dexamethasone (5 mg/kg) or both. Total liver RNA was extracted with guanidine thiocyanate.[9] Relative levels of specific mRNAs were determined by hybridization with radiolabeled cloned cDNA probes in Northern blot or dot blot analyses, followed by quantitation by liquid scintillation spectrometry.

[a]Average of measurements for three rats.

[b]Relative abundance at 5 hr.

TABLE 2. Effect of Diet on Relative Abundance of mRNA for Urea Cycle Enzymes in Rat Liver[a]

mRNA	0% Casein (6 days)	Starve (days) 1	Starve (days) 5	60% Casein (days) 1	60% Casein (days) 6
Carbamyl phosphate synthetase I	3	10	100	16	73
Argininosuccinate synthetase	3.4	14	100	13	77
	27% Casein (6 days)	**Starve (days) 1**	**Starve (days) 5**	**60% Casein (days) 1**	**60% Casein (days) 6**
Carbamyl phosphate synthetase I	6.8	11	70	19	55
Argininosuccinate synthetase	8.5	8	65	19	67

NOTE: Male Sprague-Dawley rats (150-175 g) were maintained on a 0% casein (ICN) or 27% casein diet (ICN) for 6 days before switching to a 60% casein diet (ICN) or starvation for varying periods of time. Extraction of total RNA and analysis of mRNA levels were as described in the footnote to TABLE 1.

[a]Average of measurements for three rats at each point.

high-protein diet are much greater than reported increases in enzyme levels. 2) Relative abundance of mRNAs for urea cycle enzymes increases severalfold in response to dexamethasone and Bt_2cAMP. The data indicate a small additional increase in relative mRNA levels when the two stimuli are present in combination rather than singly. 3) The rapidity of increases in the mRNA levels indicate that the half-lives of these mRNAs may be less than 1 hr. 4) In general, the relative mRNA levels for the different urea cycle enzymes change in an approximately parallel fashion in both time and extent, although there are differences in the degree of change in some instances.

REFERENCES

1. SCHIMKE, R. T. 1962. J. Biol. Chem. **237:** 459-468.
2. SCHIMKE, R. T. 1962. J. Biol. Chem. **237:** 1921-1924.
3. MCLEAN, P. & M. W. GURNEY. 1963. Biochem. J. **87:** 96-104.
4. MCLEAN, P. & F. NOVELLO. 1965. Biochem. J. **94:** 410-422.
5. SNODGRASS, P., R. C. LIN, W. A. MULLER & T. T. AOKI. 1978. J. Biol. Chem. **253:** 2748-2753.
6. MORI, M., S. MIURA, M. TATIBANA & P. P. COHEN. 1981. J. Biol. Chem. **256:** 4127-4132.
7. RYALL, J., R. A. RACHUBINSKI, M. NGUYEN, R. ROZEN, K. E. BROGLIE & G. C. SHORE. 1984. J. Biol. Chem. **259:** 9172-9176.
8. DE GROOT, C. J., A. J. VAN ZONNEVELD, P. G. MOOREN, D. ZONNEVELD, A. VAN DEN DOOL, A. J. W. VAN DEN BOGAERT, W. H. LAMERS, A. F. M. MOORMAN & R. CHARLES. 1984. Biochem. Biophys. Res. Commun. **124:** 882-888.
9. CHIRGWIN, J. M., A. E. PRZYBLA, R. J. MACDONALD & W. J. RUTTER. 1979. Biochemistry **18:** 5294-5299.

Transcriptional and Posttranscriptional Regulation of Glycolytic Enzyme Gene Expression in the Liver

ARNOLD MUNNICH, SOPHIE VAULONT, JOËLLE MARIE, MARIE-PIERRE SIMON, ANNE-LISE PICHARD, CLAUDE BESMOND, AND AXEL KAHN

Institut National de la Santé
et de la Recherche Médicale Unité 129
75674 Paris Cedex 14, France

L-Type pyruvate kinase (L-PK) and aldolase B are two enzymes in the glycolytic pathway whose activities rapidly decline during fasting and increase upon refeeding with a carbohydrate-rich diet in the liver. We have previously shown that L-PK mRNAs are barely detectable in the liver of fasted animals. Feeding fasted rats the carbohydrate-rich diet causes a 40-100-fold increase of L-PK mRNA abundance in the liver, with a maximum occurring after 18 hr of refeeding.[1,2] In order to elucidate the mechanisms that regulate the induction of glycolytic enzyme mRNAs in response to dietary stimuli in the liver, we have studied the respective roles of hormones and glycolytic substrates in L-PK and aldolase B gene expression.

METHODS

Using recombined M13 phages as probes, we first quantitated the amount of cytoplasmic RNAs in the liver of animals refed the carbohydrate-rich diet for 18 hr by dot blot and Northern blot hybridization. In order to assess whether the regulation of mRNA accumulation occurs at the transcriptional or at the posttranscriptional level, we measured the rate of glycolytic enzyme gene transcription on isolated nuclei from carbohydrate-fed rats under different dietary and hormonal conditions.[3]

RESULTS

Glycolytic enzyme mRNAs were barely detectable in the liver of both fasted and protein-fed rats. Both the maltose-rich and the fructose-rich diets induced maximum expression of the mRNAs, results similar to those obtained with the sucrose-rich diet.[2] This dietary induction by carbohydrates resulted from a dramatic activation of the gene transcription (relative rate of L-PK gene transcription: 0 and 138 ± 53 ppm for fasted and carbohydrate-fed rats, respectively), which peaked 7-12 hr into the refeeding period (FIG. 1). Glucose, in preliminary experiments, seemed to be a much more powerful transcriptional activator (2-4-fold) than fructose (data not shown).

When fasted adrenalectomized (ADX), thyroidectomized (TX), or diabetic animals were refed the carbohydrate-rich diet, the amount of glycolytic enzyme mRNAs present was about 10-25% of the amount present in the controls, despite adequate food consumption.[4,5] When animals were given the missing hormone, the level of hybridizable material returned to normal. The level of gene transcription in both ADX and TX animals matched that of controls, suggesting that their defective mRNA induction results from posttranscriptional events (FIG. 2).

Finally, fasted normal rats were refed the carbohydrate-rich diet under glucagon or cAMP administration. Glucagon resulted in a dramatic decrease of glycolytic enzyme mRNA levels, which fell to fasting values in the liver of refed animals.[4,5] This was due to a rapid and sustained blockade of gene transcription.[6]

DISCUSSION

The regulation of glycolytic enzyme gene expression in the liver occurs at both transcriptional and posttranscriptional levels. The transcription of glycolytic enzyme genes is under strong control by two major effector systems: one negative, namely glucagon and cAMP,[6] and one positive, namely carbohydrates and, probably, insulin.

The glucocorticoids and thyroid hormones play a permissive role on gene expression: they are devoid of any effect by themselves in fasted rats,[4,5] but they are required for the full expression of the genes in fed rats. Both hormones act at the posttran-

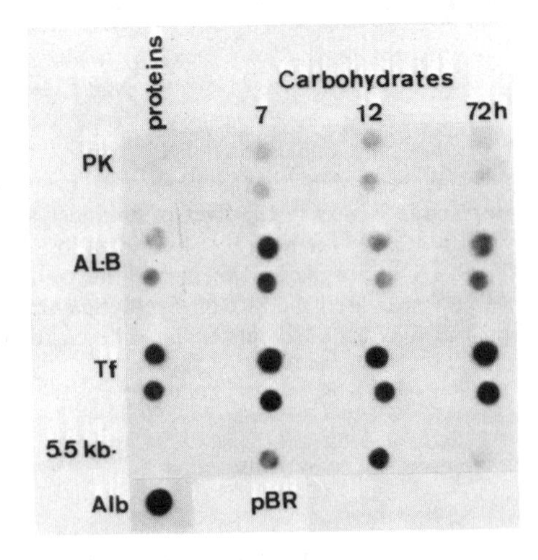

FIGURE 1. Dietary regulation of glycolytic enzyme gene transcription. Liver nuclei were isolated from rats that were either fed on a high-protein diet (proteins) or refed with the sucrose-rich diet (carbohydrates) for 7, 12, or 72 hr. Labeled transcripts were hybridized to immobilized probes in excess (6 μg of recombined plasmids). PK: L-type pyruvate kinase; AL-B: aldolase B; Tf: transferrin; 5.5 kb: liver-specific unassigned mRNA that is induced by the carbohydrate-rich diet; Alb: albumin; pBR: nonrecombined plasmid.

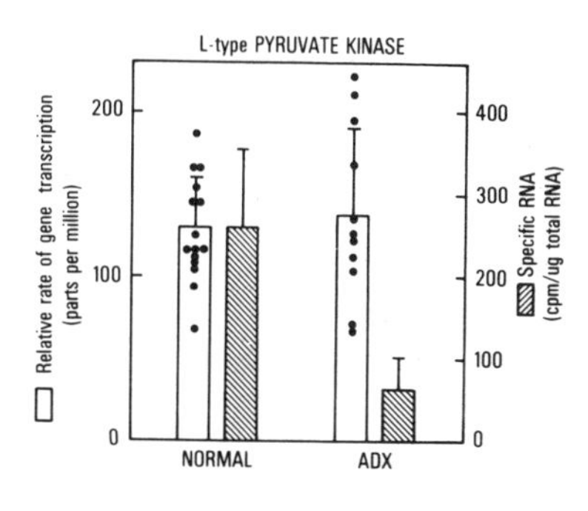

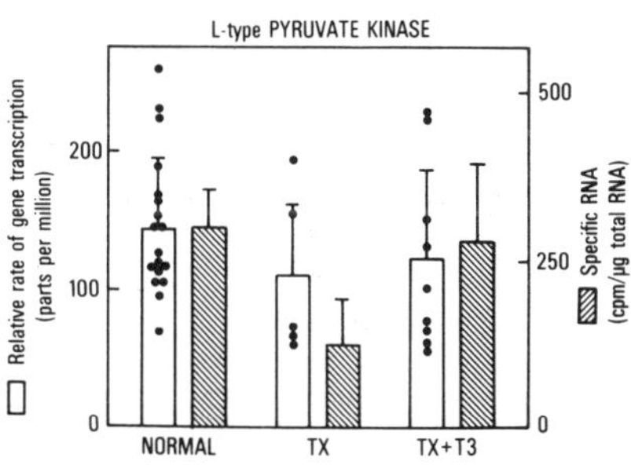

FIGURE 2. Relative rate of gene transcription as compared to the level of specific RNA in ADX and TX animals. ADX or TX animals were refed the carbohydrate-rich diet for 12 hr after a 24-hr starvation period. The level of gene transcription and the amount of cytoplasmic RNAs were assayed in the same livers and compared to controls. TX animals were given triiodothyronine (50 μg intraperitoneally) after a 12-hr refeeding period and were sacrificed 3 hr later (TX + T3).

scriptional level because their absence does not alter the level of the gene transcription. Thus the permissive role of hormones in our system appears to be a posttranscriptional phenomenon.

REFERENCES

1. SIMON, M. P., C. BESMOND, D. COTTREAU, A. WEBER, P. CHAUMET-RIFFAUD, J. C. DREYFUS, J. SALA TREPAT, J. MARIE & A. KAHN. 1983. Molecular cloning of cDNA for rat L-type pyruvate kinase and aldolase B. J. Biol. Chem. **258:** 14576-14584.
2. WEBER, A., J. MARIE, D. COTTREAU, M. P. SIMON, C. BESMOND, J. C. DREYFUS & A. KAHN. 1984. Dietary control of aldolase B and L-type pyruvate kinase mRNAs in rat. J. Biol. Chem. **259:** 1798-1802.
3. VAULONT, S., A. MUNNICH, J. MARIE, G. REACH, A. L. PICHARD, M. P. SIMON, C. BESMOND, P. BARBRY & A. KAHN. 1984. Cyclic AMP as a transcriptional inhibitor of glycolytic enzyme gene expression. Biochem. Biophys. Res. Commun. **125:** 135-141.
4. MUNNICH, A., J. MARIE, G. REACH, S. VAULONT, M. P. SIMON & A. KAHN. 1984. *In vivo* hormonal control of L-type pyruvate kinase gene expression in the liver. J. Biol. Chem. **259:** 10228-10231.
5. MUNNICH, A., C. BESMOND, S. DARQUY, G. REACH, S. VAULONT, J. C. DREYFUS & A. KAHN. 1985. Dietary and hormonal regulation of aldolase B gene expression. J. Clin. Invest. **75:** 1045-1052.
6. VAULONT, S., A. MUNNICH, A.-L. PICHARD, C. BESMOND, D. TUIL & A. KAHN. 1986. Cyclic AMP controls hepatic gene expression at both transcriptional and posttranscriptional levels. Ann. N.Y. Acad. Sci. This volume.

Structure of Mammalian Isozymes of Glycogen Phosphorylase

Implications for Function

CHRISTOPHER B. NEWGARD, KENICHI NAKANO,
PETER K. HWANG, AND ROBERT J. FLETTERICK

Department of Biochemistry and Biophysics
University of California
San Francisco, California 94143

Glycogen phosphorylase (GP) is the primary enzyme of glycogenolysis and plays a major role in regulating metabolic processes in a variety of mammalian tissues. Tissue-specific forms of GP exist in mammalian liver, muscle, and brain. Both liver and muscle isozymes of GP are regulated in part by a classical cAMP-mediated phosphorylation-dephosphorylation mechanism; inactive muscle phosphorylase *b,* however, responds to a greater range of allosteric effectors than does its liver counterpart. For example, AMP causes a nearly complete activation of muscle phosphorylase *b* but has little or no effect on the liver form.[1] Also, no evidence exists that the glycogen storage (activation) site, which has been well characterized in muscle GP,[2] is functional in liver GP. The resolution of the amino acid sequence and crystallographic structure of rabbit muscle GP, in this and other laboratories,[3] has allowed regions of the protein responsible for activity and allosteric regulation to be characterized. Before this study was begun, however, essentially no structural information of any kind was available for liver GP. Thus, the purpose of this study was to use molecular cloning techniques to elucidate the primary sequence of liver and muscle GP in an effort to 1) understand the evolutionary relationship of mammalian isozymes of GP and 2) identify the structural features responsible for functional differences between the liver and muscle forms of GP.

A cDNA fragment encoding a COOH-terminal portion of rabbit muscle GP, prepared as previously described,[4] was used to isolate a full-length muscle GP clone from a cDNA library (generously provided by D. MacLennan, Charles H. Best Institute, Toronto, Canada). The complete nucleotide sequence of muscle GP was determined by dideoxy sequencing. When the amino acid sequence derived from the nucleotide sequence was compared with the previously published amino acid sequence of rabbit muscle GP,[5] differences were found at eight positions. The accuracy of the present data was confirmed by analysis of a human muscle GP gene fragment and from X-ray crystallographic data in our laboratory.

The same rabbit muscle GP probe used to isolate the full-length muscle cDNA was used to screen a λ gtll human liver cDNA library (courtesy of S. Woo and A. DiLella, Baylor College of Medicine, Houston, Texas). A partial clone was isolated and shown to encode amino acids 660 through 845 (the COOH-terminal) of human liver GP. The partial liver clone was in turn used to isolate four longer clones, which together encoded 92% of the liver GP sequence, stretching from amino acid 72 to the COOH-terminal.

Comparison of human liver and rabbit muscle GP sequences reveals 1) an overall sequence homology of 78% at the amino acid level and 70% at the nucleotide level and 2) an overall GC content of 60% in the rabbit muscle cDNA sequence compared with 47% for that of human liver.

The changes identified in the liver GP sequence relative to muscle correspond primarily to the solvent-exposed surface of the protein, where such changes would be expected to have little functional effect. Exceptions to this trend include significant alteration of the glycogen storage site in liver GP and divergence in residues that may be involved in the allosteric transmission of AMP activation, but not the binding of AMP.

REFERENCES

1. Graves, D. J. & J. H. Wang. 1972. *In* The Enzymes. Vol. **7:** 435-482.
2. Kasvinsky, P. J., N. B. Madsen, R. J. Fletterick & J. Sygusch. 1978. J. Biol. Chem. **253:** 1290-1296.
3. Fletterick, R. J. & N. B. Madsen. 1980. Annu. Rev. Biochem. **49:** 31-61.
4. Hwang, P. K., Y. P. See, A. M. Vincentini, M. Powers, R. J. Fletterick & M. M. Crear. 1985. Eur. J. Biochem. In press.
5. Titani, K., A. Koide, J. Hermann, L. H. Ericcson, S. Kumar, R. D. Wade, K. A. Walsh, H. Neurath & E. H. Fischer. 1977. Proc. Natl. Acad. Sci. USA **74:** 4762-4766.

Regulation of Chick Embryo/ Neonate Hepatic Tyrosine Aminotransferase in Development

I. O. ONOAGBE AND A. J. DICKSON

Department of Biochemistry
University of Manchester
Manchester M13 9PT, England

In chicks, within 2-3 days of hatching, several key hepatic enzymes undergo drastic changes in activity.[1] An effect at the level of gene expression has been identified in detailed studies of these changes. In rodents, hepatic tyrosine aminotransferase (TAT) (E.C. 2.6.1.5) is subject to control by multiple signals acting at different sites.[3] The distinct nutrient environments of chick embryos and neonatal chicks may impose differential regulatory properties upon hepatic TAT. Thus the activity of TAT before and after hatching has been determined, and the role of glucagon in regulation of TAT activity may be highlighted.

Throughout embryonic life hepatic TAT activity remained constant but hatching produced a (transient) threefold increase in activity (FIG. 1). A second increase, 7-8 days after hatching, was also observed. Immunotitration, to remove "true" TAT protein, was performed on extracts at different stages of development (FIG. 1). (We have found pure TAT from rat and chicken livers to be immunochemically indistinguishable.) Thus increases observed in TAT activity reflect alterations in true TAT.

Although the chick embryo exists within a gluconeogenic "starvation" situation,[1] TAT activity is low, unlike the situation expected for starvation in rodents.[3] It is feasible that, as a consequence of the enclosed environment of the embryo, tyrosine catabolism is minimized to ensure an adequate supply of tyrosine for protein (at a time of rapid growth) and for tyrosine derivatives. The hepatic activity of phenylalanine hydroxylase in embryos alone may not be sufficient to satisfy tyrosine demand.[6]

To investigate the mechanisms responsible for TAT activity changes after hatching, hepatocytes were isolated from 17-day-old embryos[7] and incubated in a complex medium.[8] Hepatocyte TAT activity (and cell viability) remained constant throughout incubations of up to 6 hr. Glucagon addition stimulated TAT activity with a doubling apparent within 4 hr. (FIG. 2). A lag period of 1.5-2.5 hr was noted before glucagon effects were expressed. This contrasts with the rapidity of glucagon effects on TAT in suspensions of adult rat hepatocytes[8] and the action of glucagon on other processes in chick embryo hepatocytes (cAMP levels increased 20-fold within 5 min; glycogenolysis product levels increased 50% within 15 min). Incorporation of radiolabeled leucine into TAT and subsequent immunoprecipitation showed TAT synthesis to be low under control conditions (0.04% of total protein synthesis) but to be stimulated by glucagon (0.18% of total protein synthesis) (FIG. 2). A lag period of 1.5-2 hr was observed before the expression of an effect of glucagon on TAT synthesis.

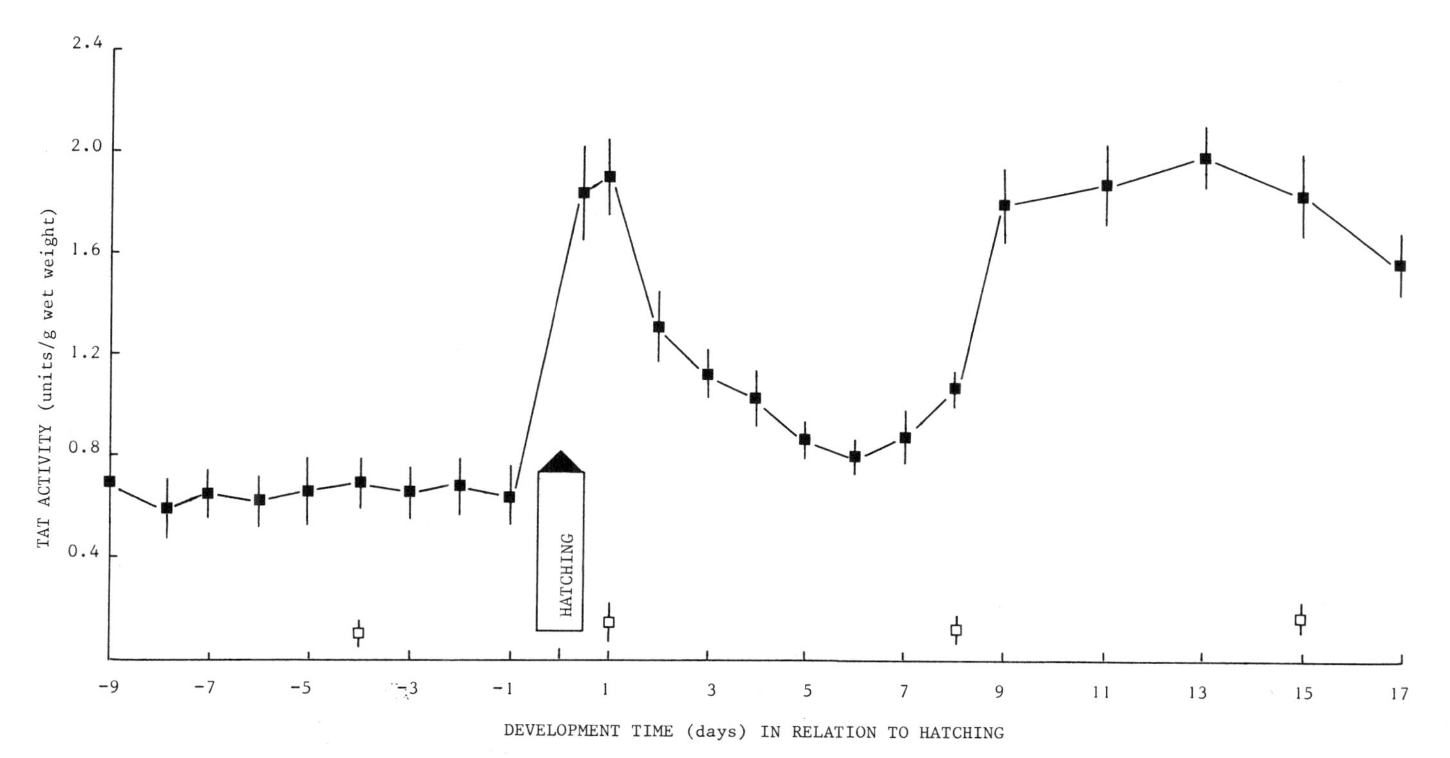

FIGURE 1. TAT activity in the liver of embryonic and neonatal chicks. Livers were homogenized in four volumes of buffer (50 mM potassium dihydrogen orthophosphate, pH 7.6, containing 3 mM 2-oxoglutarate, 1 mM EDTA, 1 mM dithiothreitol, and 0.2 mM pyridoxal 5′-phosphate); the homogenate was centrifuged at 40 000 g for 45 min at 2° C. TAT activity in the supernatant fraction was assayed by a radiometric technique.[4] The activities (■) are means ± SEM for at least six individual livers. At certain stages of development, samples of supernatants were subjected to immunotitration with antirat TAT antiserum.[5] The residual activity that remained after complete immunotitration (□) represents the activities of other aminotransferases towards tyrosine.

Glucagon can increase hepatic TAT activity after injection into neonatal chicks *in vivo*,[10] and plasma concentrations of glucagon increase at hatching.[11] It appears likely that glucagon has a physiological role to play in the expression of TAT activity in chick development. The major effect of glucagon is expressed at the level of TAT synthesis, but the reason for the lag before action on synthesis remains to be determined.

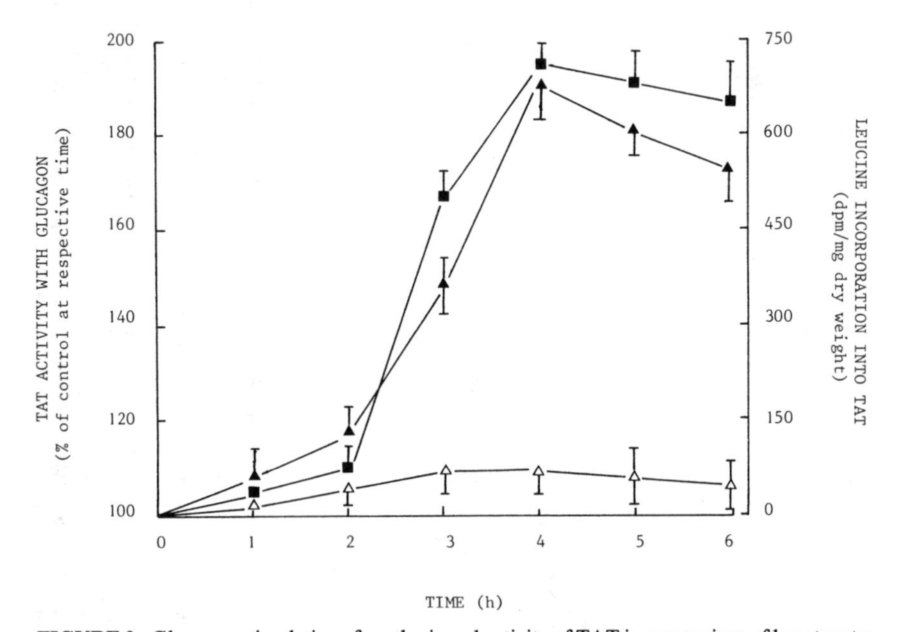

FIGURE 2. Glucagon stimulation of synthesis and activity of TAT in suspensions of hepatocytes isolated from chick embryos on the 17th day of incubation. TAT activity (■) was measured by a radiometric assay after lysis, during which three freezing-thawing cycles in 0.2 ml homogenization buffer were carried out (as in FIG. 1). Extracts were centrifuged at 12 000 g for 10 min to provide a supernatant for analysis of TAT activity. Results show the effect of glucagon (final concentration: 2.87×10^{-6} M) as a percentage of the respective control value at each time point. The values are the means ± SEM for four individual cell preparations. The TAT activity of hepatocytes at the start of incubations was 1.72 ± 0.11 mU/mg dry weight (approximately 0.466 U/g wet weight). Synthesis of TAT was measured by incorporation of L-[4,5-^{3}H]leucine (final specific activity: 25 Ci/mol) into immunoprecipitable TAT. TAT was extracted from cells as described above and mixed with 300 mU of carrier TAT in the presence of sufficient antirat TAT antiserum to precipitate 450 mU of TAT. Immunoprecipitates were washed extensively and, after solubilization with NCS tissue solubilizer,[9] counted for radioactivity. Values are expressed as means ± SEM for four individual hepatocyte preparations and represent incorporation either under control conditions (△) or in the presence of glucagon (▲) (final concentration: 2.87×10^{-6} M).

REFERENCES

1. PEARCE, J. 1977. Int. J. Biochem. **8:** 269-275.
2. MORRIS, S. M., L. K. WINBERRY, J. E. FISCH, D. W. BACK & A. G. GOODRIDGE. 1984. Mol. Cell. Biochem. **64:** 63-68.
3. GROENEWALD, J. V., S. E. TERBLANCHE & W. OELOFSEN. 1984. Int. J. Biochem. **16:** 1-18.

4. MARSTON, F. A. O. & C. I. POGSON. 1977. FEBS Lett. **83:** 277-280.
5. STANLEY, J. C., A. R. NICHOLAS, A. J. DICKSON, I. M. THOMPSON & C. I. POGSON. 1984. Biochem. J. **220:** 341-344.
6. STRITTMATTER, C. F. & G. OAKLEY. 1966. Proc. Soc. Exp. Biol. Med. **123:** 427-432.
7. PICARDO, M. & A. J. DICKSON. 1982. Comp. Biochem. Physiol. **71B:** 689-693.
8. DICKSON, A. J., F. A. O. MARSTON & C. I. POGSON. 1981. FEBS Lett. **186:** 28-32.
9. DICKSON, A. J. & C. I. POGSON. 1980. Biochem. J. **186:** 35-45.
10. HORVATH, I., P. ARANYI, A. NARAY, I. FOLDES & A. GYURIS. 1975. Int. J. Cancer **16:** 897-904.
11. BENZO, C. A. & S. B. STEARNS. 1976. Am. J. Anat. **142:** 515-518.

Regulation of ATP-Citrate Lyase in Rat Adipose Tissue

SEETHALA RAMAKRISHNA AND
WILLIAM B. BENJAMIN

Department of Physiology
Diabetes Research Laboratory
School of Medicine
State University of New York
Stony Brook, New York 11794

ATP-citrate lyase (EC 4.1.3.8), a cytosolic enzyme-producing acetyl CoA from citrate, is affected by insulin and adrenaline actions antithetically, even though both induce an increase in structural site phosphorylation. This lipogenic enzyme is induced in liver and adipose tissue of starved rats by feeding a fat-free high-carbohydrate diet. To determine whether varying levels of ATP-citrate lyase enzyme activity observed under different dietary conditions are partly due to some change in enzyme activity, we purified the enzyme from rat adipose tissue from rats that were either starved, starved and refed with a fat-free high-carbohydrate diet (diet-fed rats), or fed with a normal diet (control rats).

To improve the recovery of ATP-citrate lyase during purification, the enzyme was bound to an affinity column of coenzyme A hexane-agarose and eluted with 0.1 mM coenzyme A in buffer. This purification method afforded 65-fold purification from adipose tissue of diet-fed rats with a total recovery of 30-35%. Sodium dodecyl sulfate gel electrophoresis of the enzyme gave a single band corresponding to a polypeptide with an M_r of 115 000. Sucrose density gradient centrifugation of the native enzyme yields a single peak of ATP-citrate lyase and autophosphorylation activities at the sedimentation coefficient of 14.8S.

A comparative purification of ATP-citrate lyase from adipose tissue of starved, control, and diet-fed rats is given in TABLE 1. The specific activity of the enzyme increased sixfold in crude extracts of diet-fed compared to control rats and was reduced to about one-third in starved compared to normal rats. The total adipose tissue enzyme activity was increased about 56-fold in diet-fed compared to starved rats and decreased to one-fourth in starved compared to control rats. ATP-citrate lyase was about 0.2%, 0.5%, and 2.8% of total cytosolic protein in adipose tissue of starved, control, and diet-fed rats, respectively. The simultaneous purification of ATP-citrate lyase yielded a homogenous enzyme that had a specific activity of 3.2, 4.4, and 9.0 U/mg protein when obtained from starved, control, and diet-fed rats, respectively. To determine whether, during dietary manipulation, changes in ATP-citrate lyase activity were in part due to activation or deactivation of the enzyme, kinetic constants for the purified enzyme from starved, control, and diet-fed rats were determined. The adipose tissue enzyme like liver lyase was specific for ATP whereas other nucleotides were poor substrates. ATP-citrate lyase showed normal Michaelis-Menton kinetics with respect to ATP concentrations. The Lineweaver-Burk plot gave K_m values for ATP of 179

TABLE 1. Purification of ATP-Citrate Lyase under Different Dietary Conditions

Purification Step	Specific Activity of ATP-Citrate Lyase (units/mg protein)		
	Starved	Control	Diet-fed
Crude extract	0.0066	0.023	0.139
$(NH_4)_2SO_4$			
0-60%	0.024	0.070	0.470
0-40%	0.076	0.180	0.860
DE-52	0.569	1.080	4.440
Coenzyme A hexane-agarose	3.200	4.380	8.980
Yield (%)	30.0	33.4	30.4
Purification (fold)	485	190	65
Purified protein (mg)	0.135	0.435	2.570
Total protein (%)	0.21	0.52	2.84
Total activity (units)	1.360	5.69	75.99
Activity/g (units)	0.138	0.210	4.780

NOTE: ATP-citrate lyase was purified from adipose tissue from seven rats that were either starved three days (starved); fed with normal diet (control); or starved for three days and fed with a fat-free high-carbohydrate diet for three days (diet-fed). Each value is an average of the values from two independent experiments.

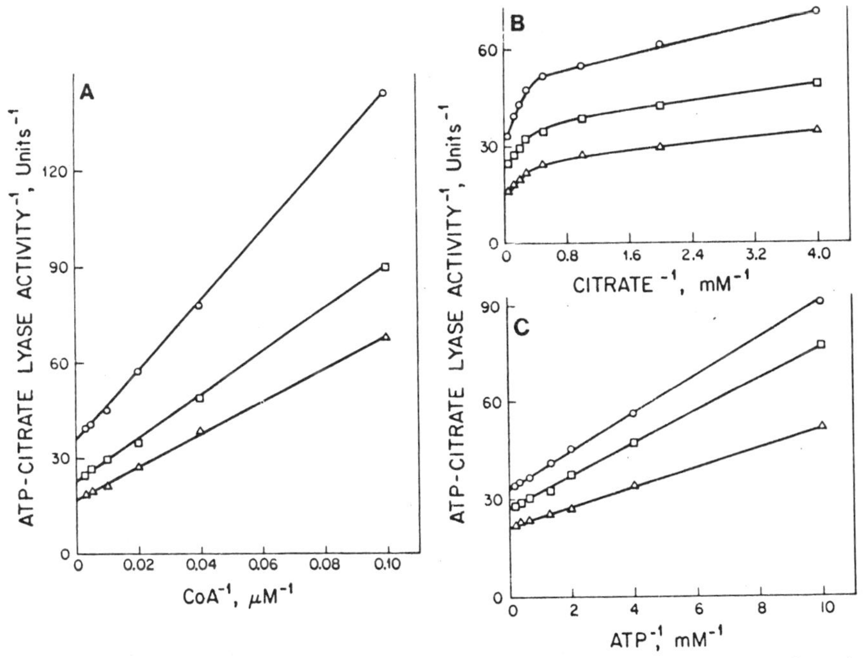

FIGURE 1. Effect of substrate concentration on ATP-citrate lyase activity. Double reciprocal plots of velocity versus (**A**) ATP concentration, (**B**) citrate concentration, and (**C**) coenzyme A concentration. ATP-citrate lyase was purified enzyme from adipose tissue of starved (○——○), control (□——□), or diet-fed rats (△——△).

μM for control, 181 μM for starved, and 145 μM for diet-fed rats (FIG. 1A). At low Cl^- concentrations, when citrate was used as the varying substrate, fat ATP-citrate lyase did not follow normal Michaelis-Menton kinetics (FIG. 1B), as has been found with the liver enzyme. Two apparent K_m values for citrate were obtained for the adipose tissue enzyme at low Cl^- concentrations: 105 μM and 1.5 mM from control, 109 μM and 1.8 mM from starved, and 107 μM and 1.4 mM from diet-fed rats. ATP-citrate lyase demonstrated normal kinetics as coenzyme A concentrations were varied and yielded K_m values of 27 μM for control, 29.5 μM for fasted, and 25 μM for diet-fed rats (FIG. 1C).

There were no significant differences in the K_m values for ATP, citrate, and coenzyme A for the purified enzymes from starved, control, and diet-fed rats. Differences in the specific activities of the purified enzymes were due to differences in V_{max} values. Thus, it appears, during induction of ATP-citrate lyase by a fat-free high-carbohydrate diet, the observed increase in activity of the enzyme was mostly due to increased synthesis of the enzyme and to a lesser extent due to increased V_{max} values.

Effect of Insulin on the Expression of Genes Encoding Tyrosine Aminotransferase, Tryptophan Oxygenase, and Phosphoenolpyruvate Carboxykinase in Cultured Rat Hepatocytes[a]

ULRICH K. SCHUBART

Department of Medicine
Diabetes Research and Training Center
Albert Einstein College of Medicine
New York, New York 10461

Tyrosine aminotransferase (TAT), the rate-limiting enzyme for tyrosine catabolism has been shown, in several studies, to be induced by insulin both *in vivo* and *in vitro.* The present study was undertaken to examine this effect at the level of mRNA using a cloned cDNA probe.[1] Adult rat hepatocytes maintained in primary culture were incubated in serum-free medium containing various concentrations of insulin. The effect of insulin was also studied in the presence of other known inducers of TAT, that is, dexamethasone (Dex), a synthetic glucocorticoid, and 8pClPhS-cAMP (CPT-cAMP), a potent analogue of cAMP.[2] Following the incubations, total cellular RNA was extracted, and the abundance of TAT mRNA was determined by Northern blot analysis. The same RNA samples were also examined for the abundance of mRNAs encoding tryptophan oxygenase (TO), a hepatocyte-specific enzyme that has been reported to be suppressed by insulin,[3] and phosphoenolpyruvate carboxykinase (PEPCK), a key enzyme in hepatic gluconeogenesis. Granner *et al.* have recently demonstrated in H4IIE hepatoma cells that insulin decreases PEPCK mRNA by inhibiting transcription of the gene.[4]

Dex (10^{-7} M) increased the levels of TAT mRNA and TO mRNA 10-20-fold and 5-15-fold, respectively. The level of TAT mRNA was also increased by CPT-cAMP (up to 36-fold at 5 μM) whereas TO mRNA was not affected. Insulin (10^{-11}-10^{-7} M) did not increase the levels of TAT mRNA or TO mRNA, regardless of the presence of other inducers using incubation periods ranging from 2.5 to 24 hr. In some experiments insulin moderately reduced TAT mRNA in the presence of Dex (FIGS. 1A & 2). The responsiveness of the hepatocytes to insulin was demonstrated by measuring its effect on PEPCK mRNA. In contrast to TAT mRNA and TO mRNA, the level of PEPCK mRNA was dramatically decreased by insulin. This

[a]This work was performed during a sabbatical stay at the Institute of Cell and Tumor Biology, German Cancer Research Center, Heidelberg, Federal Republic of Germany.

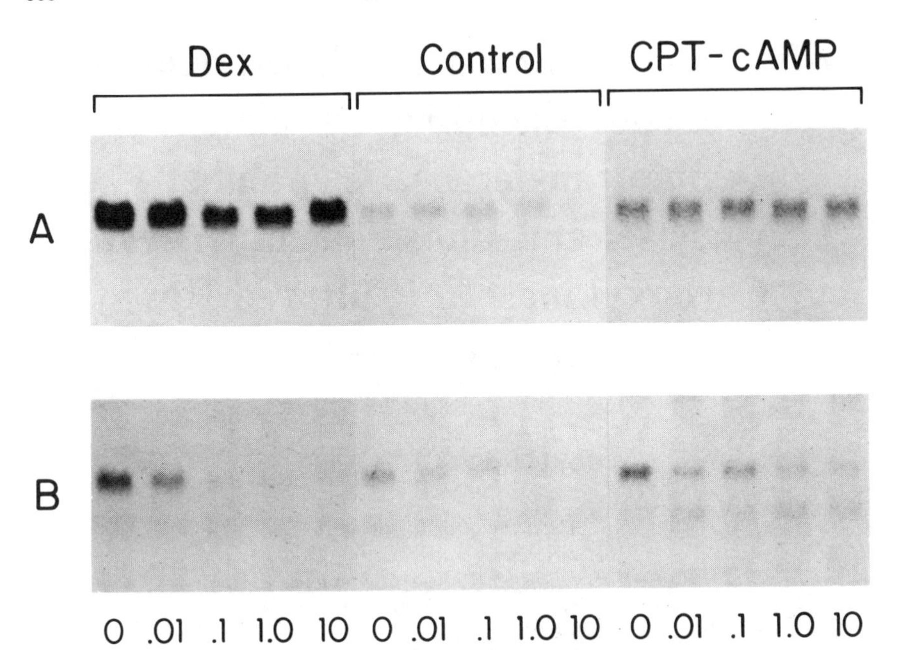

FIGURE 1. Northern blot analysis of the effect of insulin on levels of **(A)** TAT mRNA and **(B)** PEPCK mRNA in adult rat hepatocytes in the absence or presence of 10^{-7} M Dex or 0.5 μM CPT-cAMP. Hepatocytes were isolated from adult male rats by *in situ* perfusion with collagenase and allowed to attach to tissue culture dishes in Ham's F-12 medium containing fetal bovine serum (10%) and porcine insulin (1 μg/ml). After 25 hr in culture, the cells were rinsed and incubated in serum- and insulin-free Ham's F-12 overnight, followed by changing the medium to Ham's F-12 containing the additions indicated and reincubating the cells for 12 hr. Total cellular RNA was then extracted,[5] resolved by electrophoresis on 1% agarose formaldehyde gels, and transferred to nitrocellulose sheets. The blots were hybridized with [^{32}P]RNA transcripts of TAT cDNA[1] and PEPCK cDNA[6] subcloned into pSP64[7] in an antisense orientation. The predominant bands visible on the autoradiographs shown in **A** and **B** correspond to the hybridization signals of TAT mRNA (2.4 kb) and PEPCK mRNA (3.0 kb), respectively.

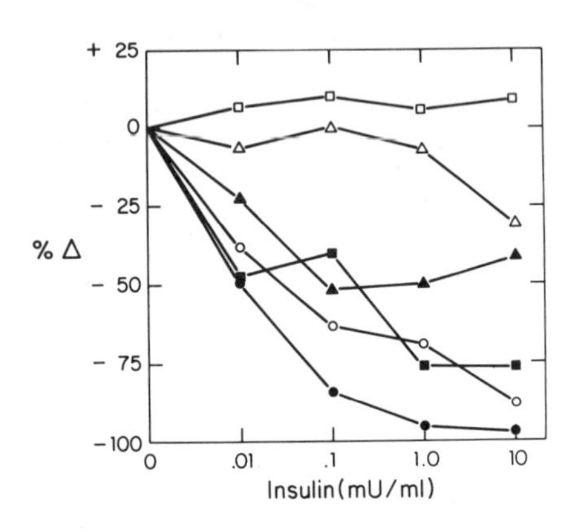

FIGURE 2. Quantitation of the results shown in FIGURE 1. The bands containing TAT mRNA and PEPCK mRNA were excised from the nitrocellulose filters and their ^{32}P radioactivity was determined by liquid scintillation counting. The results are expressed as percent change (%Δ) from the values obtained from cells not exposed to insulin. △——△: TAT mRNA, control; ▲——▲: TAT mRNA, Dex; □——□: TAT mRNA, CPT-cAMP; ○——○: PEPCK mRNA, control; ●——●: PEPCK mRNA, Dex; ■——■: PEPCK mRNA, CPT-cAMP.

effect was enhanced by Dex (FIGS. 1B & 2) and, in most experiments, reduced by CPT-cAMP.

FTO-2B rat hepatoma cells,[8] which do not express TO mRNA, showed responses similar to those observed in hepatocytes. Insulin caused a slight reduction in TAT mRNA, but, in these cells, the effect was only observed in the presence of CPT-cAMP. The level of PEPCK mRNA in FTO-2B cells was suppressed in a manner closely resembling the effects observed in hepatocytes in the present study and in H4IIE hepatoma cells as previously reported.[4]

The data demonstrate that, in hepatocytes that are highly responsive to insulin and other hormones, insulin does not increase the concentration of TAT mRNA and TO mRNA. Instead, under some experimental conditions insulin causes a moderate decrease in the level of TAT mRNA. The data strongly suggest that the induction of TAT observed in previous studies occurs independent of the concentration of TAT mRNA.

ACKNOWLEDGMENTS

I thank Günther Schütz for his generous support and Daryl Granner for providing the PEPCK cDNA.

REFERENCES

1. SCHERER, G. *et al.* 1982. Proc. Natl. Acad. Sci. USA **79:** 7205.
2. MILLER, J. P. *et al.* 1975. J. Biol. Chem. **250:** 426.
3. NAKAMURA, T. *et al.* 1980. J. Biol. Chem. **255:** 7533.
4. GRANNER, D. *et al.* 1983. Nature (London) **305:** 549.
5. AUFFRAY, C. & F. ROUGEON. 1980. Eur. J. Biochem. **107:** 303.
6. BEALE, E. G. *et al.* 1985. J. Biol. Chem. **260:** 10748.
7. MELTON, D. A. *et al.* 1984. Nucleic Acids Res. **12:** 7035.
8. KILLARY, A. M. & R. E. K. FOURNIER. 1984. Cell **38:** 523.

Activation of cAMP-dependent Protein Kinases with Defective cAMP-binding Sites[a]

ROBERT A. STEINBERG, JOANNE L. RUSSELL, CAROLINE S. MURPHY, AND DAVID A. YPHANTIS

Department of Molecular and Cell Biology
University of Connecticut
Storrs, Connecticut 06268

An S49 mouse lymphoma cell subline hemizygous for expression of the regulatory (R) subunit of type I cAMP-dependent protein kinase[1] was used to isolate cAMP-resistant mutants with structural lesions in the R subunit. Mutations that altered the R subunit charge were localized by two-dimensional gel electrophoretic analysis of partial proteolysis peptides.[2,3] On the basis of these mapping studies, two mutants—each with a lesion at one of the sites thought to be important for cAMP binding—were chosen for further analysis. The apparent affinities of wild-type and mutant R subunits for [^{3}H]cAMP were not appreciably different. Nevertheless, abnormalities in cAMP-binding properties of the mutant proteins were revealed by competition experiments with site-selective analogues of cAMP and studies of dissociation of bound [^{3}H]cAMP. These studies suggested that the R subunit lesions prevent binding of cAMP to the mutated sites, but, because of complex interactions of cAMP-binding sites in the wild-type R subunit, interpretation of these results was not unequivocal.

Cyclic AMP and six site-selective analogues of cAMP were used in kinase activation studies to compare properties of wild-type and mutant enzymes. Kinases from both mutant sublines were fully activable by cyclic nucleotides, but they required higher concentrations of activators than did wild-type kinase for half-maximal stimulation. Activation of mutant kinases was highly resistant to analogues selective for mutated sites, but relatively sensitive to analogues selective for nonmutated sites; sensitivity to cAMP was intermediate. Consistent with their harboring defects that prevented binding of cyclic nucleotides to their mutated sites, the mutant enzymes exhibited no apparent interaction between cAMP-binding sites I and II: activation curves for the mutant enzymes were less steep than those for wild-type kinase, and the mutant enzymes showed negligible synergism when activated by mixtures of site I- and site II-selective analogues. From these results we conclude that occupation of either site I or site II is sufficient for activation of cAMP-dependent protein kinase: the presence of four functional cAMP-binding sites in wild-type kinase enhances the cooperativity and sensitivity of cAMP-mediated activation. Since the mutant enzymes were relatively more resistant to cAMP than to analogues selective for their nonmutated sites, it

[a] Supported by Grant AM33977 from the National Institute of Arthritis, Diabetes, and Digestive and Kidney Diseases and by a grant from the University of Connecticut Research Foundation.

appears likely that, in wild-type kinase, activation by site-selective analogues is not fully cooperative.

ACKNOWLEDGMENTS

We thank Drs. Theodoor van Daalen Wetters and Jon P. Miller for their generosity in providing us, respectively, with "hemizygous" sublines of S49 cells and otherwise unavailable analogues of cAMP.

REFERENCES

1. VAN DAALEN WETTERS, T. & P. COFFINO. 1983. Mol. Cell. Biol. **3:** 250-256.
2. STEINBERG, R. A. 1984. Anal. Biochem. **141:** 220-231.
3. STEINBERG, R. A. 1984. Mol. Cell. Biol. **4:** 1086-1095.

Concert Effect of Thyroid and Glucocorticoid Hormones on Hepatic Phosphoenolpyruvate Carboxykinase Gene Expression[a]

WERNER SÜSSMUTH, WOLFGANG HÖPPNER,
CHRISTINE O'BRIEN, AND HANS J. SEITZ[b]

Institut für Physiologische Chemie
Universitäts-Krankenhaus Eppendorf
D-2000 Hamburg 20, Federal Republic of Germany

Recent data indicate that the induction of the rat liver regulatory glucogenic enzyme phosphoenolpyruvate carboxykinase (EC 4.1.1.32) (PEPCK) is primarily regulated at the transcriptional level. Although glucocorticoids and cAMP stimulate PEPCK gene transcription,[1] their effect is rapidly counteracted by insulin.[2] Recently we[3,4] and others[5] could demonstrate that in addition thyroid hormones stimulate the synthesis and the mRNA activity of this important enzyme. In the present study we investigate the effect and interaction of triiodothyronine (T3), glucocorticoids, and cAMP on PEPCK mRNA, under *in vivo* conditions and in hepatocytes from hypothyroid rats and from serum- and hormone-free cultures.[3]

In hypothyroid and hypothyroid-adrenalectomized rats, T3 significantly enhanced PEPCK mRNA activity within 3 to 6 hr. This effect was further enhanced by the presence of glucocorticoids, yielding a maximum response to cAMP at 6 hr (TABLE 1). The decay rate of translatable PEPCK mRNA was identical in starved hypo-, eu-, and hyperthyroid rats (T/2 $\sim$ 40 min), as measured in cordycepin-injected rats.

In cultured hepatocytes, T3 (10^{-7} M), dexamethasone (10^{-7}M), T3 plus dexamethasone, or cAMP by itself had only a minor effect on the increase in translatable PEPCK mRNA. Both T3 and dexamethasone stimulated the cAMP-mediated induction of PEPCK mRNA by a factor of 2-3; T3 plus dexamethasone showed an additive effect on both parameters (a factor of 4-5) (FIG. 1).

Thus our results *in vivo* and *in vitro* demonstrate that both glucocorticoids and T3 1) predominantly act at a pretranslational level, 2) exert a predominantly enhancing effect on the cAMP-provoked increase in PEPCK mRNA, and 3) have an additive effect (TABLE 1 & FIG. 1). The additive effect of thyroid and glucocorticoid hormones indicates their independent and, therefore, possibly different modes of action. Their effect is most pronounced during the early induction period, that is, 2 hr after the

[a] This work was supported by Deutsche Forschungsgemeinschaft, Sonderforschungsbereich 232.

[b] Address for correspondence: Institut für Physiologische Chemie, Abteilung Biochemische Endokrinologie, Martinistrasse 52, D-2000 Hamburg 20, Federal Republic of Germany.

TABLE 1. Effect of T3 on Hepatic PEPCk mRNA *in Vivo*

Experimental Condition[a]	mRNA Activity (Percentage of Total)
Hypothyroid state	
0	0.21 ± 0.02 (6)
3	0.25 ± 0.03* (4)
6	0.35 ± 0.04** (4)
12	0.37 ± 0.04** (4)
Hypothyroid state plus Bt_2cAMP (90 min)	
0	0.19 ± 0.01 (7)
3	0.23 ± 0.01* (3)
6	0.51 ± 0.04** (4)
12	0.35 ± 0.04** (4)
24	0.33 ± 0.01** (4)
Hypothyroid-adrenalectomized state plus Bt_2cAMP (90 min)	
0	0.16 ± 0.04 (4)
3	0.30 ± 0.03** (4)
6	0.32 ± 0.02** (4)
12	0.31 ± 0.04** (4)

NOTE: T3 (50 μg/100 g body weight) was injected intraperitoneally at zero time into rats that had been starved for 48 hr. Bt_2cAMP (Bt_2cAMP plus theophylline, 2 μg/100 g body weight, each) was injected 90 min before sacrifice. The mRNA measurement was performed as described previously.[4] Data are given as means ± SEM, and the number of animals is shown in parenthesis. Statistical significance: * $p < .05$, ** $p < .01$.

[a] Number of hours after the addition of T3.

addition of cAMP (FIG. 1), and occurs at physiological circulating levels of glucocorticoid[5] and thyroid hormones.[3] Although it is generally accepted that glucocorticoids stimulate the PEPCK gene transcription within about 1 hr, T3 exerts its action within 3-6 hr (TABLE 1). Thus the PEPCK enzyme is the earliest protein of known biological function to be stimulated by T3.

A similar "permissive" or enhancing effect of T3 has recently been described for the set of enzymes responsible for the conversion of carbohydrates into lipids, that

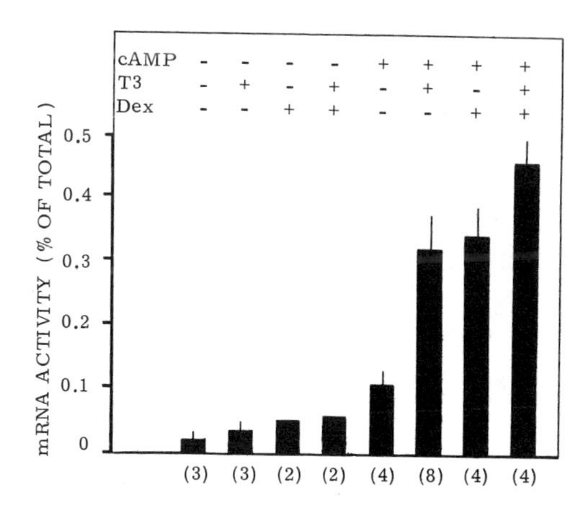

FIGURE 1. Enhancement of the cAMP-mediated PEPCK mRNA by T3 and dexamethasone. Hepatocytes were kept in serum- and hormone-free culture for 40 hr.[cf3] The addition of Bt_2cAMP (0.2 mM) 2 hr and of T3 (10^{-7} M) or dexamethasone (10^{-7} M) 6 hr before harvesting was carried out and translatable mRNA was measured as previously described.[4] Values represent the mean or mean ± SE from two to eight independent measurements. Statistical significance: Bt_2cAMP plus T3 plus dexamethasone versus Bt_2cAMP plus dexamethasone: $p < .05$; versus Bt_2cAMP plus T3: $p < .025$.

is, glucokinase,[6] acetyl-CoA carboxylase,[7] fatty acid synthetase,[7,8] and malic enzyme.[7,8] The lag period of thyroid hormones here, however, is considerably extended (> 10 hr).

To summarize, T3 and glucocorticoids rapidly, significantly, and additively enhance the cAMP-mediated increase in PEPCK mRNA at the nuclear level, an increase that is followed by a corresponding increase in enzyme synthesis. Thus these hormones contribute considerably towards enabling the liver cell to switch rapidly from glycolysis to gluconeogenesis and, probably, vice versa.

ACKNOWLEDGMENTS

We thank Mrs. A. Harneit and Mrs. D. Luda for expert technical assistance and Mrs. A. Smigilski for expert typing of the manuscript.

REFERENCES

1. Lamers, W. H., R. W. Hanson & H. M. Meister. 1982. cAMP stimulates transcription of the gene for cytosolic phosphoenolpyruvate carboxykinase in rat liver nuclei. Proc. Natl. Acad. Sci. USA **79:** 5137-5141.
2. Sasaki, K., T. P. Cripe, S. R. Koch, T. L. Andreone, D. D. Petersen, E. G. Beale & D. K. Granner. 1984. Multihormonal regulation of phosphoenolpyruvate carboxykinase gene transcription: The dominant role of insulin. J. Biol. Chem. **259:** 15242-15251.
3. Süssmuth, W., W. Höppner & H. J. Seitz. 1984. Permissive action of thyroid hormones in the cAMP-mediated induction of phosphoenolpyruvate carboxykinase in hepatocytes in culture. Eur. J. Biochem. **143:** 607-611.
4. Höppner, W., W. Süssmuth & H. J. Seitz. 1985. Effect of thyroid state on cyclic AMP-mediated induction of hepatic phosphoenolpyruvate carboxykinase. Biochem. J. **226:** 67-73.
5. Iynedjian, P. & A. Salavert. 1984. Effects of glucagon, dexamethasone and triiodothyronine on phosphoenolpyruvate carboxykinase (GTP) synthesis and mRNA levels in rat liver cells. Eur. J. Biochem. **145:** 489-497.
6. Sibrowski, W. & H. J. Seitz. 1984. Rapid action of insulin and cyclic AMP in the regulation of functional messenger RNA coding for glucokinase in rat liver. J. Biol. Chem. **259:** 343-346.
7. Goodridge, A. G. 1983. Regulation of malic enzyme in hepatocytes in culture: A model system for analyzing the mechanism of action of thyroid hormone. *In* Molecular Basis of Thyroid Hormone Action. J. H. Oppenheimer & H. H. Samuels, Eds.: 245-263. Academic Press. New York, NY.
8. Mariash, C. N. & J. H. Oppenheimer. 1983. Thyroid hormone-carbohydrate interaction. *In* Molecular Basis of Thyroid Hormone Action. J. H. Oppenheimer & H. H. Samuels, Eds.: 265-292. Academic Press. New York, NY.

Stimulation of *Trypanosoma brucei* Pyruvate Kinase by Fructose 2,6-Bisphosphate

E. VAN SCHAFTINGEN,[a] F. R. OPPERDOES,[b]
AND H. G. HERS[a]

[a]*Laboratory of Physiological Chemistry*
Catholic University of Louvain
Louvain, Belgium

[b]*Research Unit for Tropical Diseases*
International Institute of Cellular and Molecular Pathology
Brussels, Belgium

Fructose 2,6-bisphosphate is a potent stimulator of 6-phosphofructo-1-kinase (PFK-1) in higher animals and in fungi, and of pyrophosphate:fructose 6-phosphate 1-phosphotransferase in plants. It is also an inhibitor of fructose 1,6-bisphosphatase from various sources.[1]

We recently observed that fructose 2,6-bisphosphate was present in the bloodstream form of the hemoflagellate *Trypanosoma brucei,* the causative agent of sleeping sickness in man and of nagana in cattle. In this protozoan, in which the first seven steps of glycolysis occur in a special organelle called the glycosome,[2] PFK-1 is known to be unaffected by fructose 2,6-bisphosphate.[3]

In contrast, the activity of *T. brucei* pyruvate kinase was found to be greatly increased by fructose 2,6-bisphosphate, which converted the saturation curve for phosphoenolpyruvate from a sigmoid into a hyperbola with no change in V. Phosphate and arsenate had an effect opposite to that of fructose 2,6-bisphosphate, and the approximate K_a for fructose 2,6-bisphosphate was shifted from 75 nM to 1.5 μM by the presence of 5 mM phosphate. Fructose 1,6-bisphosphate had effects similar to those of fructose 2,6-bisphosphate but at approximately 4000-fold higher concentrations.

The pyruvate kinases of *Crithidia luciliae* and of *Leishmania major,* two trypanosomatids that are like *T. brucei* in containing glycosomes,[5] were also stimulated by fructose 2,6-bisphosphate and inhibited by phosphate. In contrast, the activities of pyruvate kinases from rat liver, spinach leaves, *Saccharomyces cerevisiae,* and *Euglena gracilis* were unaffected by micromolar concentrations of fructose 2,6-bisphosphate.

REFERENCES

1. HERS, H. G. & E. VAN SCHAFTINGEN. 1982. Biochem. J. **206:** 1-12.
2. OPPERDOES, F. R. & P. BORST. 1977. FEBS Lett. **80:** 360-364.
3. CRONIN, C. N. & K. F. TIPTON. 1985. Biochem. J. **227:** 113-124.
4. VAN SCHAFTINGEN, E., F. R. OPPERDOES & H. G. HERS. 1985. Eur. J. Biochem. **153:** 403-406.
5. OPPERDOES, F. R. 1985. Br. Med. Bull. **41:** 130-136.

Control of Hepatic Gene Expression at Both Transcriptional and Posttranscriptional Levels by cAMP

SOPHIE VAULONT, ARNOLD MUNNICH, ANNE-LISE PICHARD, CLAUDE BESMOND, DAVID TUIL, AND AXEL KAHN

Institut National de la Santé et de la Recherche Médicale Unité 129
75674 Paris Cedex 14, France

The goal of this work was to investigate the influence of glucagon and cAMP on the expression of different genes specifically active in liver.

The cDNA probes for three carbohydrate-induced mRNAs (L-type pyruvate kinase (L-PK), aldolase B, and an unidentified 5.4-kb mRNA species) and for transferrin were isolated[1-3] and used to measure gene transcription rates (by elongation of nascent RNAs in isolated nuclei) and mRNA concentrations.

We found that low doses of glucagon (0.5 μg/rat) or Bt_2cAMP quickly blocked gene transcription in all four cases.[4,5] The blockade persisted as long as glucagon was administered for the carbohydrate-induced genes; in contrast, it was profound but transient for the transferrin gene, transcription starting again after the second hour of infusion.[5]

To evaluate whether the transcriptional control by cAMP accounted for all of the changes in mRNA concentrations, we compared the level of gene transcription in isolated nuclei derived from glucagon-treated animals to the level of its specific mRNA.

FIGURE 1 shows this comparison for the L-PK gene. Although the L-PK gene transcription is totally blocked after 15 min of glucagon administration, the L-PK mRNA remains stable up to the third hour of treatment with a half-life higher than 5 hr. After the third hour, L-PK mRNA becomes very unstable and is degraded with a half-life lower than 1 hr (as shown on the semilog representation in the insert). This result indicates that cAMP first blocks L-PK gene transcription, then, after a 3-hr lag period, leads to mRNA destabilization. This lag-time period suggests an indirect effect of cAMP, and that the immediate transcriptional blockade of the L-PK gene favors a primary effect of cAMP. The same result showing a biphasic pattern of mRNA degradation after gene transcriptional inhibition is obtained for the 5.4-kb mRNA species.

Thus, it seems that glucagon and cAMP could mediate alterations in message stability as a means to regulate cytoplasmic mRNA concentration. In agreement with our data, Jungman *et al.*[6] have demonstrated that cAMP regulates not only the rate of transcription of lactate dehydrogenase mRNA in the rat C6 glioma cells but that the stability of mRNA is increased during the induction phase.

In conclusion, cAMP, which was already known as a transcriptional activator of genes encoding neoglucogenic enzymes,[7,8] is also able to inhibit transcription of various other genes, especially genes encoding lipogenic and glycolytic enzymes.

In addition, the regulation of carbohydrate metabolism by cAMP appears to be integrated at the different level of gene expression: transcriptional, posttranscriptional, and, possibly, posttranslational regulation of enzyme activities.

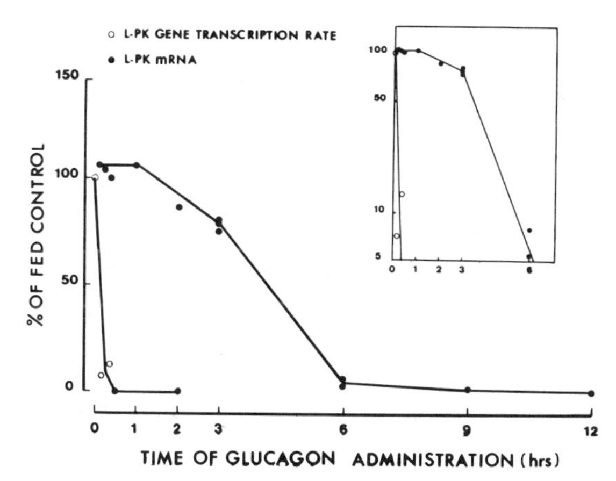

FIGURE 1. Time course of the effect of glucagon on L-PK gene transcription and mRNA accumulation in carbohydrate-fed rats. Rats refed the carbohydrate diet for 12 hr were injected intraperitoneally with both rapid glucagon (500 μg/kg) and zinc glucagon (7.5 mg/kg) and subcutaneously with long-acting glucagon (7.5 mg/kg) for the times indicated. The transcription rate was measured in isolated nuclei as previously described.[4] Incorporation, corrected for gene and cDNA length and for hybridization efficiency, is given in ppm. Quantification of mRNA was performed by Northern blot analysis.[1] The intensity of the mRNA bands detected was measured by scanning the autoradiograms using a Shimadzu densitometer. The results are expressed in scanning values.

REFERENCES

1. SIMON, M. P., C. BESMOND, D. COTTREAU, A. WEBER, P. CHAUMET-RIFFAUD, J. C. DREYFUS, J. SALA-TREPAT, J. MARIE & A. KAHN. 1983. Molecular cloning of cDNA for rat L-type pyruvate kinase and aldolase B. J. Biol. Chem. **258:** 14576-14584.
2. PICHARD, A. L., A. MUNNICH, M. C. MEIENHOFER, S. VAULONT, M. P. SIMON, J. MARIE, J. C. DREYFUS & A. KAHN. 1985. Characterization and metabolic regulation of a liver-specific 5.4-kb mRNA whose synthesis is transcriptionally induced by carbohydrates and repressed by glucagon and cyclic AMP. Biochem. J. **226:** 637-644.
3. UZAN, G., M. FRAIN, I. PARK, C. BESMOND, G. MAESSEN, J. SALA-TREPAT, M. M. ZAKIN & A. KAHN. 1984. Molecular cloning and sequence analysis of cDNA for human transferrin. Biochem. Biophys. Res. Commun. **119:** 273-281.
4. VAULONT, S., A. MUNNICH, J. MARIE, G. REACH, A. L. PICHARD, M. P. SIMON, C. BESMOND, P. BARBRY & A. KAHN. 1984. Cyclic AMP as a transcriptional inhibitor of glycolytic enzyme gene expression. Biochem. Biophys. Res. Commun. **125:** 135-141.
5. TUIL, D., S. VAULONT, M. J. LEVIN, A. MUNNICH, M. MOGUILEWSKY, M. M. BOUTON, P. BRISSOT, J. C. DREYFUS & A. KAHN. 1985. *In vivo* control of rat transferrin gene expression. FEBS Lett. **189:** 310-314.

6. JUNGMANN, R. A., D. C. KELLEY, M. F. MILES & D. M. MILKOWSKI. 1983. Cyclic AMP regulation of lactate dehydrogenase. J. Biol. Chem. **258:** 5312-5318.
7. LAMERS, W. H., R. W. HANSON & H. M. MEISNER. 1982. cAMP stimulates transcription of the gene for cytosolic phosphoenolpyruvate carboxykinase in rat liver nuclei. Proc. Natl. Acad. Sci. USA **79:** 5137-5141.
8. HASHIMOTO, S., W. SCHMID & G. SCHÜTZ. 1984. Transcriptional activation of the rat liver tyrosine aminotransferase gene by cAMP. Proc. Natl. Acad. Sci. USA **81:** 6637-6641.

Index of Contributors